AF389414

Innovative Technologies and Materials for the Production of Mechanical, Thermal and Corrosion Wear-Resistant Surface Layers and Coatings

Innovative Technologies and Materials for the Production of Mechanical, Thermal and Corrosion Wear-Resistant Surface Layers and Coatings

Editors

Artur Czupryński
Claudio Mele

Basel • Beijing • Wuhan • Barcelona • Belgrade • Novi Sad • Cluj • Manchester

Editors
Artur Czupryński
Department of Welding
Engineering
Silesian University
of Technology
Gliwice
Poland

Claudio Mele
Department of Engineering
for Innovation
University of Salento
Lecce
Italy

Editorial Office
MDPI
St. Alban-Anlage 66
4052 Basel, Switzerland

This is a reprint of articles from the Special Issue published online in the open access journal *Materials* (ISSN 1996-1944) (available at: www.mdpi.com/journal/materials/special_issues/layers_coatings).

For citation purposes, cite each article independently as indicated on the article page online and as indicated below:

Lastname, A.A.; Lastname, B.B. Article Title. *Journal Name* **Year**, *Volume Number*, Page Range.

ISBN 978-3-0365-9093-6 (Hbk)
ISBN 978-3-0365-9092-9 (PDF)
doi.org/10.3390/books978-3-0365-9092-9

Contents

About the Editors

Artur Czupryński

Dr. Artur Czupryński is an Associate Professor at the Silesian University of Technology in Gliwice. He has been working in the Department of Welding since 1997 and, in 2001, he completed the course of the European Welding Engineer (EWE). He has completed scientific internships at the Welding Institute in Gliwice (Poland), University of Warwick (UK), VSB Technical University of Ostrava (Czech Republic), Óbuda University (Hungary), University of Ruse (Bulgaria), and University of West Bohemia (Czech Republic). He is a member of the Polish Welding Society and a member of the Programme Council of the Persons Certification Body TÜV Rheinland Polska. In his scientific activity, he deals with technologies of welding, brazing, thermal cutting, surfacing, and thermal spraying, as well as quality control in welding. Dr. Artur Czupryski is the author or coauthor of 3 books and academic textbooks, as well as more than 200 articles and 2 patents. In 2018, he received the Stanisław Olszewski Medal awarded by the SIMP Welding Section to the most outstanding Polish and foreign welding engineers "in recognition of their contribution to Polish welding". For his scientific, didactic, and organisational achievements, he has been repeatedly awarded individual and team awards by the Rector of the Silesian University of Technology and decorated by the President of the Republic of Poland.

Claudio Mele

Claudio Mele is an Associate Professor of Applied Physical Chemistry at the Department of Engineering for Innovation of the University of Salento. He is a teacher of the courses of "Electrochemical technologies" and of "Laboratory of applied physical chemistry". He is the Head of the Applied Electrochemistry laboratory at the Department of Engineering for Innovation of the University of Salento. His research activity is mainly focused on electrochemical preparation and on the kinetic, structural, compositional, optical, mechanical, and corrosion characterization of metals, alloys, and oxides. The following topics have been especially studied: (a) electrodeposition and corrosion of metals and alloys; (b) fabrication and functional characterization of materials for energy: fuel cells (PEMFC and SOFC), supercapacitors, and metal–air batteries; and (c) fabrication and functional characterization of metallic biomaterials.

Preface

The cost of the regeneration of an element is much lower than the cost of its production. In addition, regeneration restores its functional properties and can also increase durability several times. The coatings used from innovative materials on elements of machines and devices operating in conditions of abrasive, erosive, and corrosive wear are an effective technique for increasing the element's durability. The coating allows you to combine the beneficial properties of the core with the wear resistance, hardness, and heat resistance coating and separate the function of load transfer from the protection against the influence of the working environment of the element. This reprint characterizes the methods of thermal coating application and presents the research results on coating properties and application areas. There are two main coating method groups, cladding methods and spraying methods. During cladding, the substrate melts. An increase in the temperature of the welded element causes deterioration of the properties shaped by the previous treatment. They are mixed substrate materials and the alloy, as well as an intermediate composition between the chemical composition of the substrate and the alloy formed. The expected performance of the properties is obtained in the case of two- or three-layer thick welds. The main advantage of welded coatings is good adhesion to the substrate. Thermal spray technology enables the coating production from a wide group of materials. The technology advantage is the slight heating of the substrate during the coating application, which practically excludes microstructural transformations and deformation of the substrate. The disadvantage of heat-sprayed coatings is their lower adhesion and specific microstructure, which may reduce corrosion resistance. Undercoatings are used in very diverse operating conditions. One study focuses on coatings resistant to abrasive wear, erosion wear, and corrosion. In individual chapters, the phenomenon of abrasive wear, theoretical models of a phenomenon, and applied experimental methods are characterized. In addition, the general characteristics of the erosion process are given, and wear mechanisms, experimental methods, the influence of test parameters on erosion intensity, and erosion theoretical models are discussed. The characteristics of thermal spray processes are also included, as well as the materials used for spraying, the microstructural structure of coatings, and technology development trends.

Artur Czupryński and Claudio Mele
Editors

Article

Microstructure and Abrasive Wear Resistance of Metal Matrix Composite Coatings Deposited on Steel Grade AISI 4715 by Powder Plasma Transferred Arc Welding Part 1. Mechanical and Structural Properties of a Cobalt-Based Alloy Surface Layer Reinforced with Particles of Titanium Carbide and Synthetic Metal–Diamond Composite

Artur Czupryński

Department of Welding Engineering, Faculty of Mechanical Engineering, Silesian University of Technology, Konarskiego 18A, 44-100 Gliwice, Poland; artur.czuprynski@polsl.pl

Citation: Czupryński, A. Microstructure and Abrasive Wear Resistance of Metal Matrix Composite Coatings Deposited on Steel Grade AISI 4715 by Powder Plasma Transferred Arc Welding Part 1. Mechanical and Structural Properties of a Cobalt-Based Alloy Surface Layer Reinforced with Particles of Titanium Carbide and Synthetic Metal–Diamond Composite. *Materials* **2021**, *14*, 2382. https://doi.org/10.3390/ma14092382

Academic Editor: Hendra Hermawan

Received: 18 April 2021
Accepted: 1 May 2021
Published: 3 May 2021

Publisher's Note: MDPI stays neutral with regard to jurisdictional claims in published maps and institutional affiliations.

Abstract: The article discusses test results concerning an innovative surface layer obtained using the cladding with powder plasma transferred arc welding (PPTAW) method. The above-named layer, being a metal matrix composite (MCM), is characterised by high abrasive wear resistance, resistance to pressure and impact loads, and the possibility of operation at elevated temperatures. The layer was made using powder in the form of a cobalt alloy-based composite reinforced with monocarbide TiC particles and superhard spherical particles of synthetic metal–diamond composite provided with tungsten coating. The surface layer was deposited on a sheet made of low-alloy structural steel grade AISI 4715. The layer is intended for surfaces of inserts of drilling tools used in the extraction industry. The results showed the lack of the thermal and structural decomposition of the hard layer reinforcing the matrix during the cladding process, its very high resistance to metal-mineral abrasive wear and its resistance to moderate impact loads. The abrasive wear resistance of the deposited layer with particles of TiC and synthetic metal–diamond composite was about than 140 times higher than the abrasive wear resistance of abrasion resistant heat-treated steel having a nominal hardness of 400 HBW. The use of diamond as a metal matrix reinforcement in order to increase the abrasive resistance of the PPTAW overlay layer is a new and innovative area of inquiry. There is no information related to tests concerning metal matrix surface layers reinforced with synthetic metal–diamond composite and obtained using PPTAW method.

Keywords: PPTAW; cladding; deposition; abrasion; impact load; titanium carbide; synthetic metal–diamond composite

1. Introduction

Processes taking place during the boring of oil and gas wells and the mining of rock in underground workings are extremely complex and difficult. Among other things, the aforesaid situation results from the mechanical properties of mined ground or rock layers and their inhomogenous geological structure (responsible for the fast wear of drilling tools used in the extractive industry). The necessity of the frequent replacement of worn-out mining blades (drills, boring crowns, cone cutters, etc.) significantly increases the costs of excavated raw materials. The properties of structural or tool materials depend both on the microstructure of the core of a given element and the condition of its surface layer. In cases of elements that do not transfer significant loads or are not exposed to intense abrasive wear during operation, the condition of the surface layer is of lesser importance. However, tools and machinery elements made of steel are exposed to abrasion combined with high unit pressure, impact loads, a corrosive environment, and high operating temperature [1–4].

In particular, the above-presented operating conditions affect tools being in direct contact with abrasives such as rock, sand, clay or other hard components present, among other things, in the ground. Globally, the aforesaid problems are present in the fossil fuels excavation industry. Power engineering and machine-building sectors compete intensively to develop modern technologies, making it possible to obtain a longer service life of tools and machinery elements used in coal mines, quarries, oil rigs and climate engineering, and during the construction of motorways. There is a high demand for spare parts of mining machinery and in particular for drilling and geological tools, which wear quickly and, consequently, lose their operational properties. Such tools have to be replaced very often, generating additional and, frequently, high costs, connected not only with the purchase or the refurbishment of new tools but, primarily, with the time needed to replace them. In addition, the dismantling and the reassembly of tools are responsible for costly downtimes. Presently, it is possible to observe a tendency of extending the service life of drilling and geological tools, even at the expense of significantly higher prices. Mining concerns find it more beneficial to buy more expensive tools characterised by higher quality than to stop production (several times) in order to retool machinery. Being a specific abrasive, the ground is not easy to define explicitly. This fact results mainly from the geological and engineering conditions of a given excavation area as well as from weather conditions present during the operation of the tool (affecting friction conditions in the ground-tool system). Abrasive wear, to which drilling and geological tools are exposed during operation in the ground, translates directly into the service life and the reliability of mining machinery. The inspection and the forecasting of tool wear in the ground prove very difficult. As of today, related engineering knowledge is limited to experimentation and the development of the so-called neural networks. The more detailed identification of the destruction of materials being in motion during extraction requires the combination of many elementary wear-related phenomena.

This issue was addressed by, among others, Kenny et al., 1976; Gharahbagh et al., 2013; Dewangan et al., 2014, 2015, Amoun et al., 2017, and Nahak et al., 2018 [5–10]. The problems encountered by the extractive industry necessitate the search for methods making it possible to reduce the wear of tools and machinery parts used in the extractive industry. Researchers and engineers constantly try to develop new ranges of tools, changing both design-related solutions and materials. However, the improvement of operational properties remains primarily connected with the improvement of the properties of the surface layer. An increase in hardness and abrasive wear resistance as well as surface processing involving the use of chemical elements improving corrosion resistance enable the extension of tool service life. It should also be noted that surface processing belongs to the most economically effective and useful methods applied in widely defined materials engineering. The making of layers characterised by new and unique properties may entirely change the operational parameters of every base material. Related publications concerning the subject discuss the obtainment of the increased abrasive wear resistance of tools primarily through the application of ceramic materials [11–13], diffusive carbide [14–16], boride coatings [17–19] and thermally sprayed layers [20–22]. Some of the above-named methods fail to produce desirable results and, in addition, are both energy-consuming and laborious. The aforesaid coatings are usually deposited on the entire surface of a given product, which is not always economically justified. Many researchers believe that the most promising technologies enabling the fabrication of abrasive wear resistant coatings should be based on high-energy density methods, including plasma or laser cladding.

Presently, in developed countries such methods are used to extend the service life of mining and drilling tools as well as to make corrosion resistant layers. The powder plasma transferred arc welding (PPTAW) or the laser metal deposition (LMD) are used by, among others, General Electric Oil&Gas and Honeywell International, i.e., leading oil and gas producers [23,24], as well as by many manufacturers of plasma arc welding systems. Various plasma arc welding methods can be applied to make surface layers using nearly any metallic material. In such cases, a deposited material becomes the primary component of the surface layer and, because of high process temperature, melts along with the substrate. Issues concerning the powder plasma transferred arc cladding process were discussed, among others, by Khaskin et al. (2016), Brunner-Schwer et al. (2018, 2019), Xia et al. (2010) and Gao et al. (2020) [25–29]. Recently, it has been possible to observe very high interest in the application of plasma and laser cladding processes to make composite surface layers on steels and alloys of non-ferrous metals. The above-named layers are composed of the metallic matrix reinforced with hard particles of interstitial compounds. The matrix is usually made of iron [30,31], nickel [32,33] or alloys containing the aforesaid elements. The Ni-Cr-B-Si alloy is usually applied through thermal spraying [34]. There are also numerous publications concerning the matrix containing cobalt and its alloys [31,35] (e.g., stellites [36,37]). However, stellites are less frequently referred to as matrix materials in composite layers and more often as homogenous layers. Particles reinforcing composite surface layers are various interstitial compounds, usually carbides but also nitrides and borides [15,17]. The most commonly used carbides include tungsten carbide (WC) [30,38], silicon carbide (SiC), boron carbide (B_4C) [39] and titanium carbide (TiC) [40].

Titanium carbide (TiC) particles are widely used to reinforce structural materials (both as volume and surface reinforcement). The plasma arc melting of metallic powder and TiC particles (the granularity of which was restricted within the range of 10 μm to 14 μm) on the surface of elements made of steel AISI 304 was investigated by Bober and Grześ (2015) [41]. The use of the powder plasma transferred arc made it possible to obtain the proper joint of the deposited material components and provided appropriate adhesion to the substrate. Kindrachuk et al. (2016) [42] made a TiC-Co composite layer deposited on the surface of steel 2Cr13. The layer, composed of several variedly structured sub-layers, was applied to extend the service life of machinery elements and power generation equipment. The substrate, affected by the plasma arc, underwent self-hardening, whereas the remaining part of the layer contained the zone of molten material and the dilution zone. The structure of the layer was composed of supersaturated cobalt dendrites with dispersed TiC particles. Depending on a steel grade, it is possible to obtain its reinforcement through self-hardening. However, in cases of superalloys, the surface layer changes its chemical composition and structure (as a result of diffusion and dilution), which is an undesired phenomenon. The chromium-nickel matrix is a very popular material of the MMC composite reinforced with TiC particles. According to Onuoha (2016) [43], in the Cr-Ni alloy reinforced with TiC particles having a granularity restricted within the range of 4 μm to10 μm and 70–90 vol%, the larger grain size of the hard phase was responsible for increased abrasive wear. The mechanism of the aforesaid wear consisted primarily of micro-cutting. Sakamoto et al. (2015) [44] used 2 wt% of TiC particles to significantly improve the mechanical properties of the Cr-Ni alloy. The TiC-Co-type composite layers were also examined by Jung et al. (2015) [45]. The specimens were prepared through high-energy ball milling and liquid phase sintering. The size of the TiC particles was restricted within the range of 7 μm to 10 μm. The researchers emphasized the significance of the size of the particles of the hard phase and the type of the matrix material. It was demonstrated that fine-grained powders based on cobalt alloys inhibited the growth of TiC grains during sintering. The work does not contain any results of tribological tests.

Karantzalis et al. (2013) [46] made a cobalt alloy-based composite reinforced with TiC particles [46]. The components of the composite were melted using the vacuum arc melting method. It was found that a greater amount of the hard TiC phase favoured the grain growth in the metallic matrix. The results of the abrasive wear test proved promising. Another type of a sintered alloy, i.e., Co-TiC, was investigated by Jung et al. (2015) [45]. It was revealed that the modification of the initial powder material through the addition of cobalt nanoparticles decreased the powder sintering temperature. The obtained alloy was characterised by favourable thermal stability and advantageous mechanical properties. In their research, Anasori et al. (2016) [47] demonstrated that a magnesium alloy containing 5, 20 and 50 vol% of TiC and Ti_2AlC sintered carbides was characterised by excellent energy absorbability. The favourable size of the particles reinforcing the matrix was restricted within the range of 5 μm to 15 μm. According to the Authors, the damping effect was obtained owing to the natural ability of materials to absorb energy, the large matrix-carbide contact area and a different thermal expansion coefficient. However, the aforesaid factors also contributed to an increased number of dislocations.

The above-presented research results concerning the making of metal matrix composite layers reinforced with TiC particles justified the formulation of the following conclusions:

- TiC is a very promising material reinforcing the matrix of metallic materials (MMC) in applications requiring high abrasive wear resistance under conditions of dry sliding friction. Matrix materials include many metals and alloys, e.g., cobalt, nickel, magnesium and aluminum;
- All tests were based on the application of various combinations of the hard reinforcing phase and the metal matrix;
- Abrasive wear resistance tests of Me-TiC-type composite layers were primarily performed at room temperature and under conditions of dry sliding friction;
- In most tests, the composite material was obtained using sintering methods, where the volume fraction of carbides was restricted within the range of 40% to 60%. The microstructural tests revealed the proper dilution and the uniform distribution of alloying elements and carbide particles in the matrix material;
- Depending on the type of the metal matrix and test conditions, the effect of the high content of TiC in the composite can be both favourable [42] and disadvantageous [48]. However, it should be noted that reinforcing the metal matrix with TiC particles favourably reduces abrasive wear regardless of the type of the matrix and the reinforcement-matrix ratio in the composite.

The analysis of related reference publications and the results of individual research led to the conclusion that it was possible to obtain a composite surface layer in the ceramic reinforcement–metal matrix system of phases, the microstructure and abrasive wear resistance of which would be similar to those of sintered carbides [49].

2. Materials and Methods

2.1. Materials

The surface layer was deposited using the powder plasma transferred arc welding (PPTAW) method on specimens having dimensions of 75 mm × 25 mm × 10 mm, made of low-alloy structural steel grade AISI 4715 (Table 1). The cladding process was performed using metal-matrix composite (MMC) powder belonging to the group of Co3 alloys (in accordance with EN 147000:2014) [50]. The powder contained superhard phases in the form of ceramic particles made of crushed sharp-edged titanium carbide (TiC) (see Table 2, Figure 1a) and spherical particles made of synthetic metal–diamond composite sinter in the tungsten lagging (PD-W) (Harmony Industry Diamond, Zhengzhou, China) (see Table 2, Figure 1b). The components of the powder were mixed in a Turbula T2F laboratory powder mixer-shaker (Glen Mills Inc., Clifton, NY, USA) using ceramic balls.

Table 1. Chemical composition of low-alloy structural steel AISI 4715 according to the manufacturer data (TimkenSteel Ltd., Canton, OH, USA).

Chemical Composition, wt.%								
C	Mn	S	P	Si	Cr	Mo	Ni	Fe
0.12–0.18	0.65–0.95	≤0.015	≤0.015	0.15–0.35	0.40–0.70	0.45–0.60	0.65–1.00	Bal.

Table 2. Chemical composition of Co3+TiC+PD-W powder.

Chemical Composition of Co 3 Alloy Matrix, wt.%									Ceramic Reinforcement of the Matrix, wt.%	
C	Si	Mn	Cr	Ni	Mo	W	Fe	Co	TiC	PD-W
2.5–3	≤1	≤2	24–28	≤3	≤1	12–14	<5	Bal.	90	10
Carbide-to-matrix ratio: 60/40 (wt.%)										

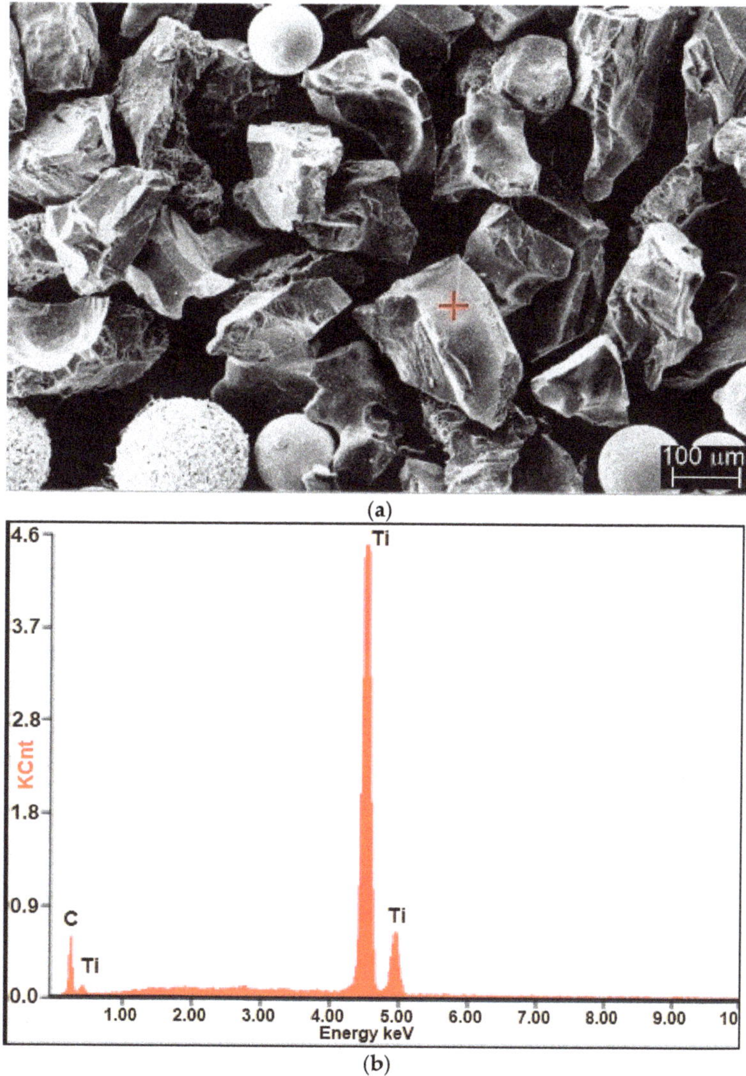

Figure 1. *Cont.*

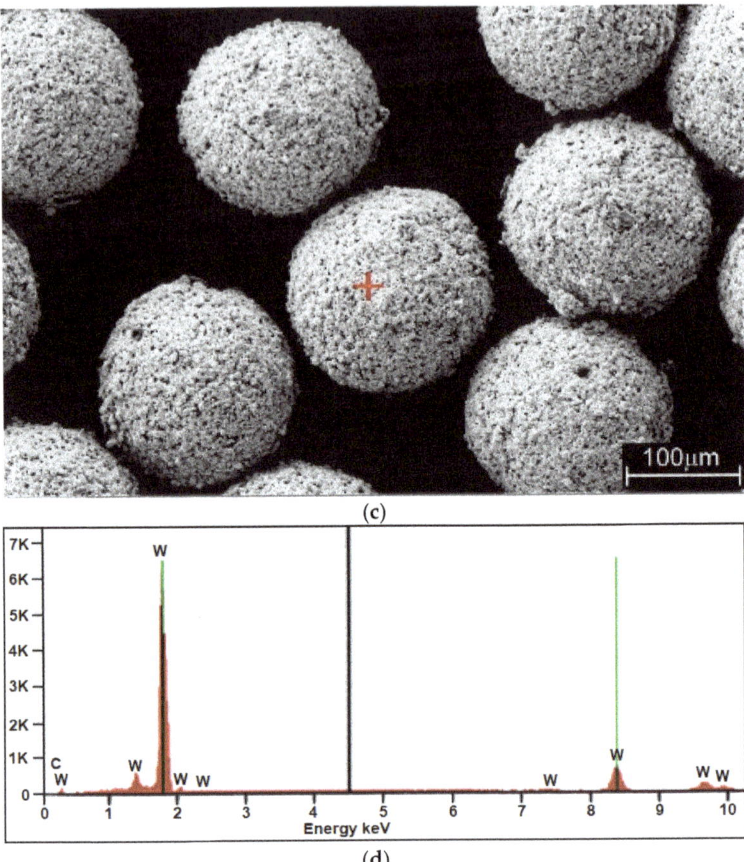

Figure 1. Components of the hard phase in the Co3+TiC+PD-W powder: (**a**) sharp-edged titanium carbide, TiC (mag. ×300); and (**c**) synthetic metal–diamond composite in the tungsten lagging, PD-W (mag. ×500), and diagram of the energy of scattered X-radiation with energy lines present in the area of components: (chemical elements (ceramic particle of TiC and metal matrix)) subjected to analysis (**b**) ceramic particle of TiC; (**d**) tungsten-coated synthetic metal–diamond composite (PD-W).

2.2. Plasma Processing

The plasma deposition process was carried out with surfacing machine Durweld 300T PTA with a maximal current of about 300A. For experiment, the machine powder plasma surfacing torch PT 300AUT (Durum Verschleiss-Schutz GmbH, Willich, Germany) with a thoriated tungsten cathode 4 mm in diameter (Figure 2) was used, mounted on industrial robot Fanuc R-2000iB (FANUC Ltd., Oshino-mura, Japan) arm.

PTA welding system Durweld 300T PTA was PLC-controlled and equipped with a HMI-interface, gas mass flow meter and powerful water cooling unit. PLC provides reliable operation and allows for easy integration in robot cells. The powder cladding system consisted of a computer-controlled powder feeding system PFU 4 (4th generation Powder Feeding Unit design) and a PTA torch integrated with a six-axis robot. Powder feeder was intended for applications that require feeding of different powders in the weld pool, i.e., matrix and carbides. Feeding rate step was controlled via feeding wheel speed directly from PLC. The coaxial injection of the powder was performed using the plasma, carrier and shielding gas. The cladding process was performed using the following gas flow rates: plasma gas (Ar) = 1.6 L/min, carrier gas (Ar + 5% H$_2$) = 4 L/min and shielding gas (Ar + 5% H$_2$) = 12 L/min.

Figure 2. View of the powder plasma transferred arc welding (PPTAW) processes.

The determination of the optimum range of cladding parameters required the making of a series of weave-bead claddings using a main arc current of 40, 60, 80, 100 and 120 A; a cladding speed restricted within the range of 1 mm/s to 4 mm/s; and a powder feed rate restricted within the range of 10 g/min to 30 g/min. The optimal processing parameters for robotic plasma cladding were established based on NDT and metallographic tests. The cladding parameters identified as optimum (Table 3) were those ensuring the uniform distribution of the powder over the entire liquid metal area in the melt pool, the proper depth of penetration g = 1.2 mm, layer height h = 3 mm and the dilution of the base material in the cladding amounting to D = 4.5%.

Table 3. Optimum processing parameters of robotic plasma powder transferred arc cladding of Co3+TiC+PD-W composite powder deposition on steel AISI 4715.

Process Parameters	Value of Parameter
Main arc current, I_a (A)	80
Pilot arc current, I_p (A)	15
Arc voltage, U (V)	25
Cladding speed, S (mm/s)	2.5
Powder feed rate, q (g/min)	15
Plasma gas flow rate, Q_p [1] (L/min)	1.6
Shielding gas flow rate, Q_o [2] (L/min)	12
Carrier gas flow rate, Q_s [2] (L/min)	4
Nozzle-workpiece distance, l (mm)	5
Overlap ratio, O (%)	33
Heat input, E_u [3] (J/mm)	480

Notes: [1] Argon 5.0 (99.999%) acc. ISO 14175—I1: 2009 was used as plasma, [2] argon/hydrogen 5% H2, Ar (welding mixture ISO 14175-R1-ArH-5) was used as shielding and carrier gas, [3] calculated acc. to the formula: $E_u = k \cdot (U \times I)/v$ The thermal efficiency coefficient for plasma transferred arc k = 0.6 was used.

2.3. Testing Methodology

The analysis of the morphology and the size of the composite powder (MMC) was based on images obtained using a scanning electron microscope. The assessment of the quality of the layer and the detection of cladding imperfections (if any) such as cracks, porosity, spikes, undercuts, and shape and dimension-related imperfections required the performance of non-destructive tests, including visual tests (VT), penetrant tests (PT) and tests concerning the roughness (Ra) of the deposited layer. The assessment of surface properties was based on the analysis of macro and microscopic metallographic test results, chemical composition analysis, X-ray diffraction results, hardness and roughness measurement results, and results of tests concerning metal-mineral abrasive wear resistance and impact resistance.

2.3.1. Composite Powder Morphology, and the Structure and Chemical Composition of the Deposited Layer

The assessment of the surface and the size of the particles of the composite powder as well as of the structure of the deposited layer were performed using a Zeiss Supra 25 scanning electron microscope (Carl Zeiss AG, Oberkochen, Germany). The tests were performed using a detector of secondary electrons (SE), an accelerating voltage of 20 kV and a probe current of 5 nA. The chemical composition of the powder components and of the deposited layer was identified on the basis of tests performed using a Zeiss Supra 25 scanning electron microscope featuring an EDS and UltraDry EDS detector (for X-ray microanalysis) (ThermoFisher Scientific, Waltham, MA, USA). The tests were performed on the surface of the specimens using a point or an area-based analysis.

Results of composite powder (Co3+TiC+PD-W) morphology tests are presented (in the form of the SEM images and diagrams of scattered X-radiation) in Figure 3. The tests revealed that the size of the powder particles was restricted within the range of 60 μm and 250 μm (mediana Q_{50} = 152 μm) and constituted the mixture of irregular and spherical components. The tests were performed on the surface of the powder particles using point or micro-area-based analysis.

(a)

Figure 3. *Cont.*

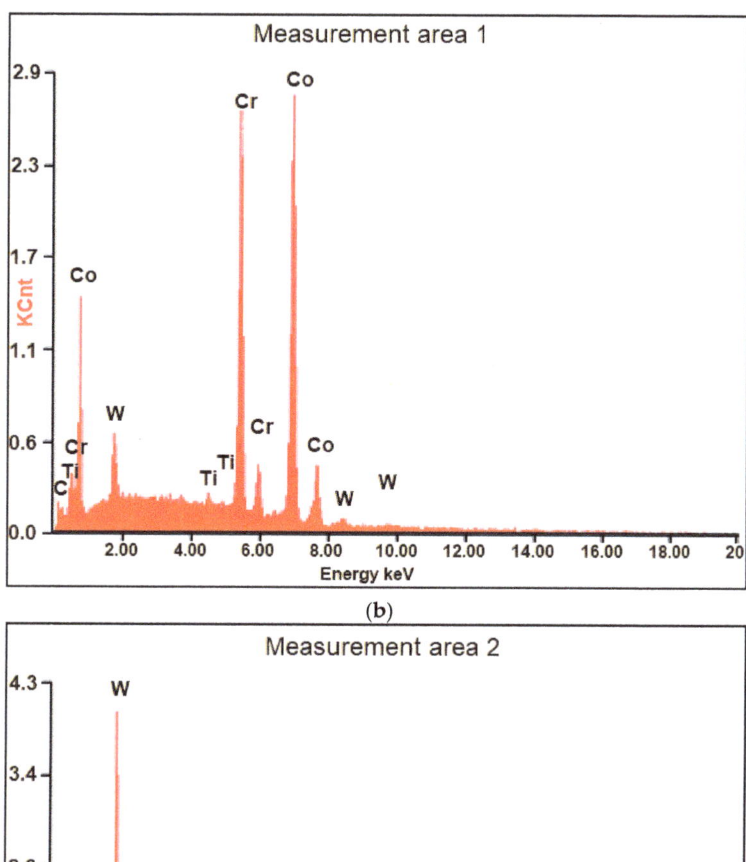

(b)

(c)

Figure 3. *Cont.*

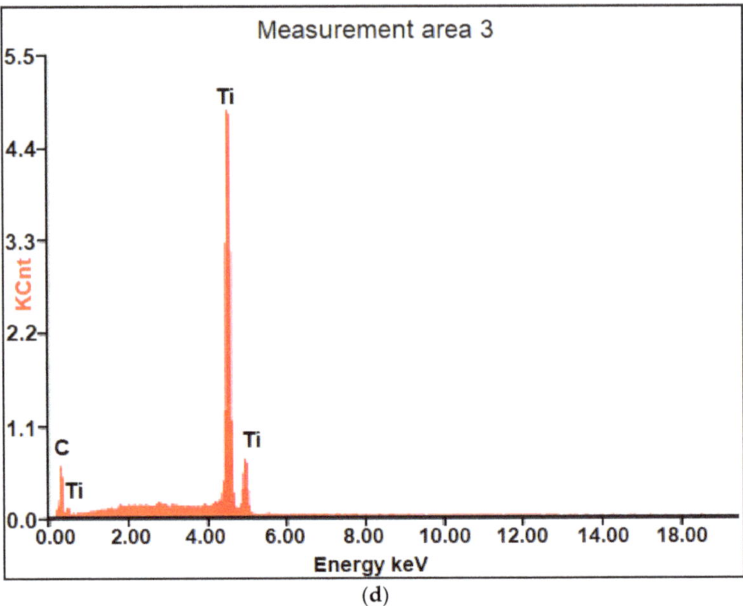

(**d**)

Figure 3. Results of the microanalysis of the chemical composition of the Co3+TiC+PD-W composite powder: (**a**) SEM image of the morphology of the powder particles with the area subjected to analysis and the diagrams of scattered X-radiation with energy lines present in the area of components (chemical elements) subjected to analysis: (**b**) matrix, (**c**) ceramic particle (TiC) and (**d**) tungsten-coated synthetic metal–diamond composite (PD-W).

The mixer-shaker used for the mixing of the components of the Co3+TiC+PD-W composite powder enabled the making of the homogenous composition of the composite material. The use of an additional element facilitating the stirring process (i.e., ceramic balls) precluded the segregation of the components and the formation of larger agglomerates of the plastic phase. Methods used previously to mix the powder components, e.g., in a conical mixer, failed to produce desirable results.

2.3.2. Non-Destructive Tests

The visual tests (external visual inspection) and the penetrant tests were performed in accordance with the requirements specified in related standards, i.e., ISO 17637 [51] and ISO 3452-2, respectively [52]. The visual tests involved the verification and the assessment of the condition of the deposited layer by the unaided eye (direct visual test). Before the test, the surface to be inspected was subjected to thorough cleaning and drying. The penetrant tests were performed using a system of dye penetrants (System Designation Type II, Sensitivity 2) Cd-2 PT ISO 3452-2 II Cd-2 and EN 571-1 (Figure 4). The surface roughness measurements were performed in five areas of the deposited layer, using a Surtronic 3+ surface roughness tester (Ametek Taylor Hobson, Berwyn, PA, USA).

2.3.3. Metallographic Examination and X-ray Diffraction Analysis

The microscopic tests were performed using metallographic specimens subjected to standard preparation. The etchant was the so-called "aqua regia", i.e., the 3:1 mixture of concentrated hydrochloric acid and nitric acid; the time of etching was determined experimentally. The observation and the recording of macro and microstructural images were performed using an Olympus SZX9 stereoscopic microscope (Olympus Corporation, Tokyo, Japan) equipped with a Moticam 5.0+ digital camera and a Motic Images plus 3.0 software programme as well as an Olympus GX 71 inverted metallographic microscope

(Olympus Corporation, Tokyo, Japan). The phase composition of the deposited layer was determined on the basis of X-ray diffraction analysis performed using a Panalytical X'Pert Pro MPD diffractometer (Malvern Panalytical Ltd., Malvern, UK) and the filtered radiation (filter Kβ Fe) of a cobalt anode lamp ($\lambda K\alpha$ = 0.179 nm). The diffraction patterns were recorded in the Bragg–Brentano geometry, using a PIXcell 3D detector and the axis of the beam deflected within the angle range of 20 to 110 (2θ) (increment = 0.05°, counting time per increment = 100 s). The diffraction patterns were subjected to analysis involving the use of a dedicated Panalytical High Score Plus software programme and a PAN-ICSD structural database. The X-ray quantitative phase analysis was performed using the Rietveld refinement method.

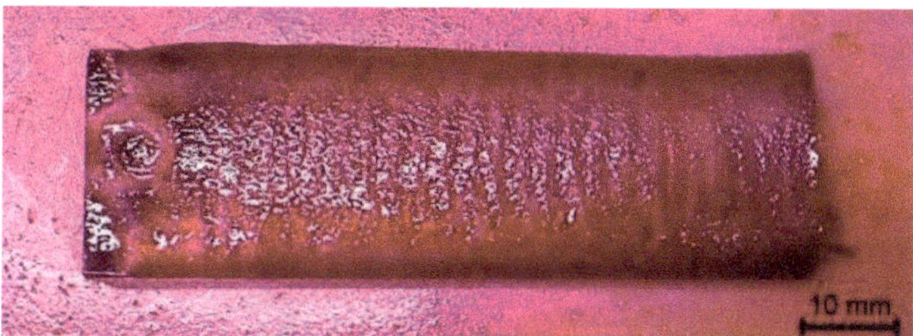

Figure 4. View of the sample during the penetrant testing (PT).

2.3.4. Density and Porosity of the Deposited Layer

The density of the surface layer was measured using the Archimedes method in accordance with the ISO ASTM D792 standard [53]. The measurement involved the use of a Radwag AS 220.R2 analytical laboratory balance (Radwag, Warsaw, Poland) along with a set for Archimedes-method-based density measurements (Figure 5).

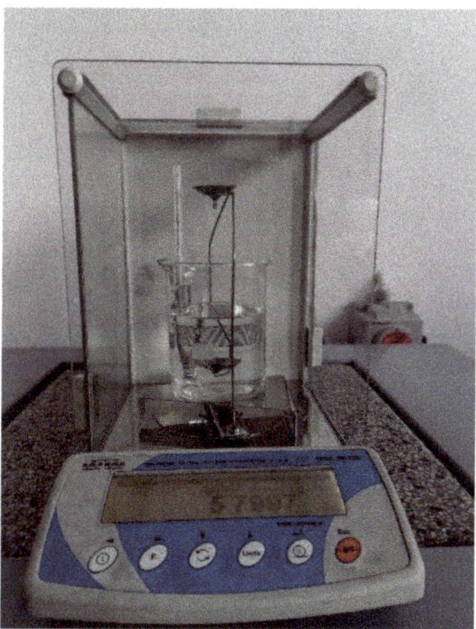

Figure 5. View of an analytical balance for density of the deposited layer measurement.

The analysis of the surface layer roughness degree was performed using a μCT Nanotom 180N micro-tomograph (Ge Sensing & Inspection Technologies GmbH, Wunstorf, Germany) equipped with an X-ray tube having a maximum voltage of 180 kV. Tomographic images were recorded using a Hamamastu 2300 × 2300 pixel decoder. The virtual reconstruction of tested objects was performed using a GE datosX ver.2.1.00 software programme. All of the tomographic images were made using a source voltage of 140 kV and 200 μA; the element was rotated by 360°, in 2400 steps. The time of exposure amounted to 500 ms, exposure averaging amounted to 3, the image refresh rate amounted to 1 and the time of scanning a single element was $t = 80$ min. The analysis of roughness was performed using an MyVGL programme software.

2.3.5. Hardness Measurements

The hardness measurements concerning the external surface and the cross-section of the surface layer and of the reference material (abrasion-resistant heat-treated steel having a nominal hardness of 400 HBW) were performed using the Vickers hardness test (in accordance with the procedure referred to in the ISO 6507-1 standard) [54] and a Future-Tech FM-ARS 9000 hardness tester with an automatic measurement line and an image analysis system (Future-Tech Corp., Kawasaki, Japan). The specimen surface hardness was determined within the HV10 scale using a test load of 10 kgf (total test force 98 N) and time $t = 30$ s. The hardness tests were performed at five test points on the surface layer subjected to grinding. Exemplary locations of measurement points on the surface of the abrasive wear resistant layer are presented in Figure 6.

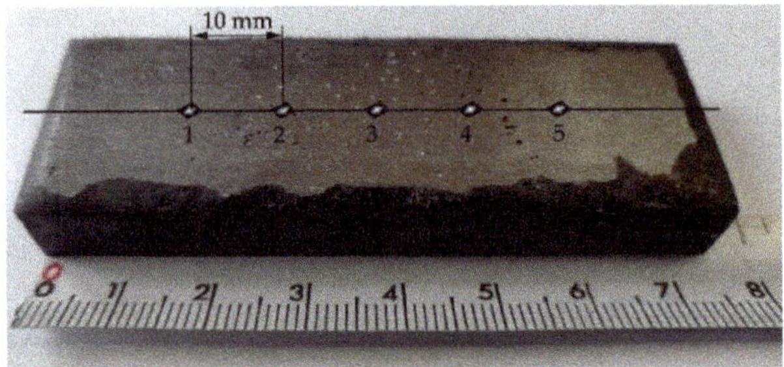

Figure 6. Exemplary locations of measurement points on the surface of the wear resistant layer.

The microhardness testing was done using Vickers method HV0.5. The hardness measurements concerning the cross-section of the surface layer were performed on metallographic specimens, at 10 measurement points, separately for the particles of the hard phase and the matrix.

2.3.6. Abrasive Wear Resistance Tests

The metal-mineral abrasive wear resistance test of the surface layer and of the reference material (abrasion-resistant steel grade AR400) was performed in accordance with ASTM G 65-00, Procedure A [55], using the, "rubber wheel" machine (Figure 7).

The "rubber wheel" abrasive wear test governed by the ASTM G65 standard is the most popular test used in materials engineering to assess metal-mineral abrasive wear resistance. The abrasive used in the test was quartz sand, the grain size of which was restricted within the range of 50 mesh to 70 mesh (0.297–0.210 mm); the sand was fed gravitationally to the friction zone. The experimental tests concerning the deposited layer and the reference material involved the preparation of two specimens having dimensions of 75 mm × 25 mm × 10 mm. During an approximately 30-min-long test, the rubber wheel made 6000 revolutions. The test material was subjected to a pressure force of 130 N, whereas the feed rate of the abrasive (A. F. S. Testing Stand 50–70 mesh) amounted to 335 g/min.

Before and after the abrasive wear test, the specimens were weighed on the laboratory balance with an accuracy of up to 0.0001 g. The average density of the deposited layer and that of the reference material was determined on the basis of three measurements of the specimen density, sampled and weighed at room temperature in air and liquid. The volume loss was calculated on the basis of the measured average density of the deposited surface layer and the average specimen mass loss after abrasion, using the following formula (1):

$$\text{volume loss } \left[\text{mm}^3\right] = \frac{\text{mass loss } [\text{g}]}{\text{density } \left[\frac{\text{g}}{\text{cm}^3}\right]} \times 1000 \tag{1}$$

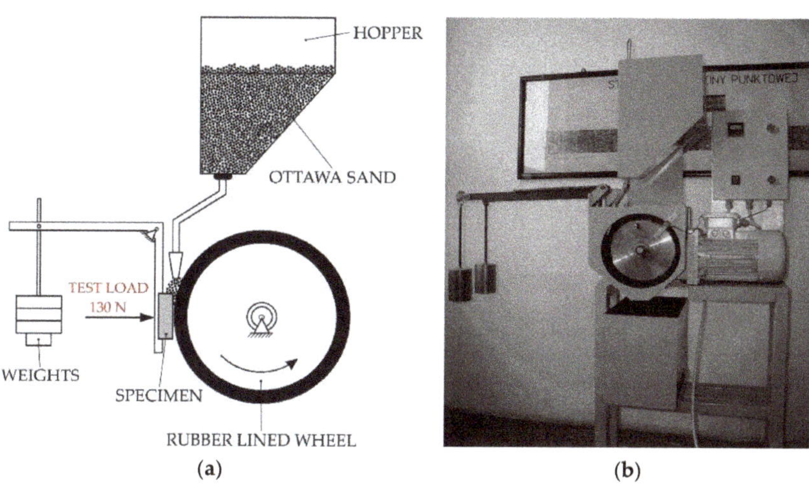

Figure 7. Testing station for metal-mineral abrasive wear resistance tests in performed in accordance with ASTM G65: (**a**) schematic diagram and (**b**) main view [30].

The specimen surface abrasion area was subjected to microscopic examination performed using a Zeiss Smartproof 5 confocal microscope (Carl Zeiss AG, Oberkochen, Germany). The type of abrasive wear was identified using a criterion based on the quotient of the cross-sectional areas of the sum of the two-sided upsetting of the material near micro-scratch F1 and the depth of micro-scratch F2.

The material loss in the surface layer during abrasive wear was classified in relation to micro-ridging, i.e., the plastic deformation of contact areas and the upsetting of the material on both sides of the micro-ridge, where F1/F2 = 1; micro-cutting, where F1/F2 = 0; and micro-scratching, if the material was partly deformed plastically and partly cut in the form of chips as wear products, where $0 \leq$ F1/F2 ≤ 1 [56].

2.3.7. Impact Resistance Tests

The impact resistance tests of the deposited layer were performed in laboratory conditions using a dedicated testing station (Figure 8). The tests involved the use of a specimen previously subjected to penetrant testing (aimed to identify already existing cracks and surface imperfections). A criterion adopted in the impact resistance test was the number of cracks and chips as well as the general comparative assessment of damage caused by the multiple strokes of the deposited layer with a 20 kg tool (ram) released freely from a height of 1.02 m (impact energy of 200 J). The condition of the deposited layer was identified on the basis of visual tests after 5, 10 and 20 strokes.

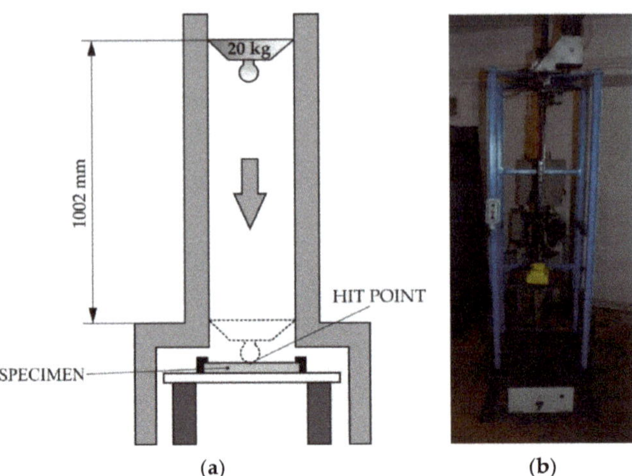

Figure 8. Test rig used in the tests of the impact resistance of the surface layer: (**a**) schematic diagram and (**b**) main view [21].

3. Results and Discussion

3.1. Non-Destructive Tests–Visual Test Results

The visual and penetrant tests of the surface layer only revealed the presence of welding imperfections in the form of shallow crater at the end of weld pass (Figure 9b). The above-named tests did not reveal such imperfections as cracks, porosity, spikes, undercuts, spatter, or shape and dimension-related imperfections. The deposited layer was characterised by relatively low roughness (average value Ra = 14 μm), uniform surface and the symmetric overlapping of successive cladding beads (Figure 9a). According to Przestacki et al. (2014) [57], surface roughness parameter (Ra) of deposits of tungsten carbide surface layer obtained using the cladding with Direct Laser Deposition (DLD) method is up to 32 μm. Additionally, cobalt matrix composite coatings obtained by thermal spraying can present a high surface roughness Ra, often significantly exceeding 30 μm [58]. PPTAW cladding with a weaving bead trajectory, at a relatively high value of the welding current (80A), resulted in a good thermal activation of the hard reinforcing phase wetting process by the liquid matrix metal. The composite layer formed correctly with a relatively low surface roughness, despite the rather unfavourable morphology of the TiC grains, which had an irregular shape and a very expanded surface.

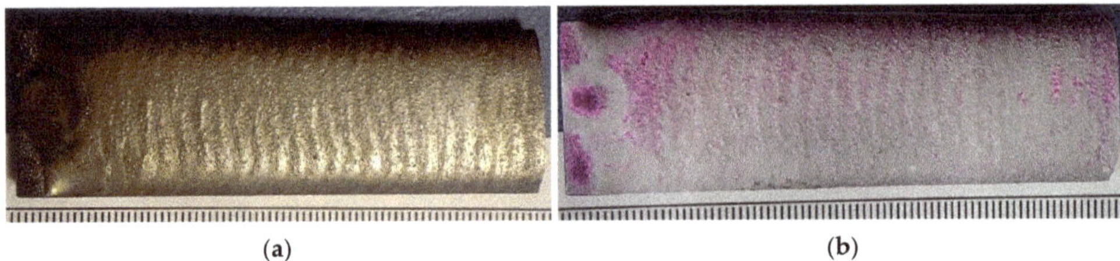

Figure 9. Surface layer made using the Co3+TiC+PD-W composite powder: (**a**) layer after the visual tests (VT) and (**b**) layer after the penetrant tests (PT).

Because of the fact that the cladding process is allied to welding and qualified as special, it is necessary to subject a given cladding technology to verification based on adopted standards, e.g., ISO 15614-7 [59]. However, the application of the above-named standard in relation to deposited composite layers may prove problematic. The aforesaid composite layers could pose a problem in terms of inspection in accordance with quality standards and standard-related quality levels as the requirements specified in the standards preclude the acceptance of surface elements containing imperfections. On the other hand, it should be noted that welding imperfections, for example, in the form of cracks contribute to the reduction of stresses in elements subjected to cladding and, consequently, improve the greasing conditions of the interacting surfaces of friction association. With respect to the special application of a given product, the transverse crack in the composite layer can be considered as acceptable. Powder plasma surfacing enabled the formation of surfaced layer with quality level B. According to ISO 5817 [60] norm, level B corresponds to the highest quality of manufactured layers.

3.2. Metallographic Test Results and the Results of XRD Analysis

The results of the microscopic metallographic observations made it possible to identify the structure of the matrix as well as the type, distribution and dimensions of the surface layer reinforcement. The observations were performed using magnification restricted within the range of 50 times to 500 times. The observations were concerned with the subsurface as well as the middle zone and the dilution zone of the cladding and of the HAZ of the substrate. The results of the microstructural observations are presented in Figure 10. The SEM photographs are presented in Figure 11. The SEM observations were performed using a magnification of 80, 500 and 1500 times. The results concerning the microanalysis of the chemical composition of the deposited layer are presented in Figure 12.

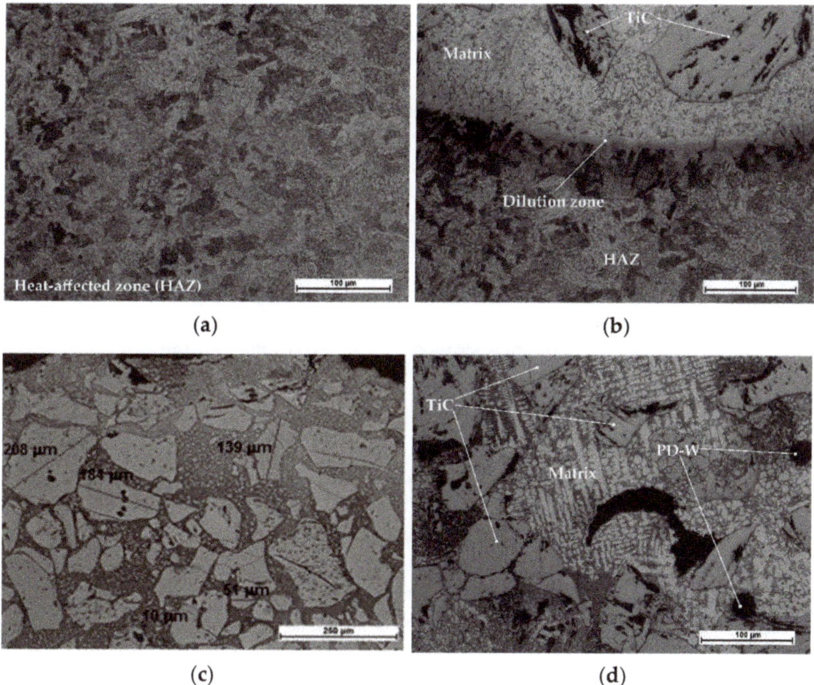

Figure 10. Microstructure of the surface layer (PPTAW metal deposition method, Co3+TiC+PD-W composite powder) deposited on structural low-alloy steel AISI 4715: (**a**) heat affected zone (HAZ), (**b**) dilution zone, (**c**) size and distribution of the hard phase in the middle of the cladding and (**d**) structure of the matrix of the solid solution near the padding weld.

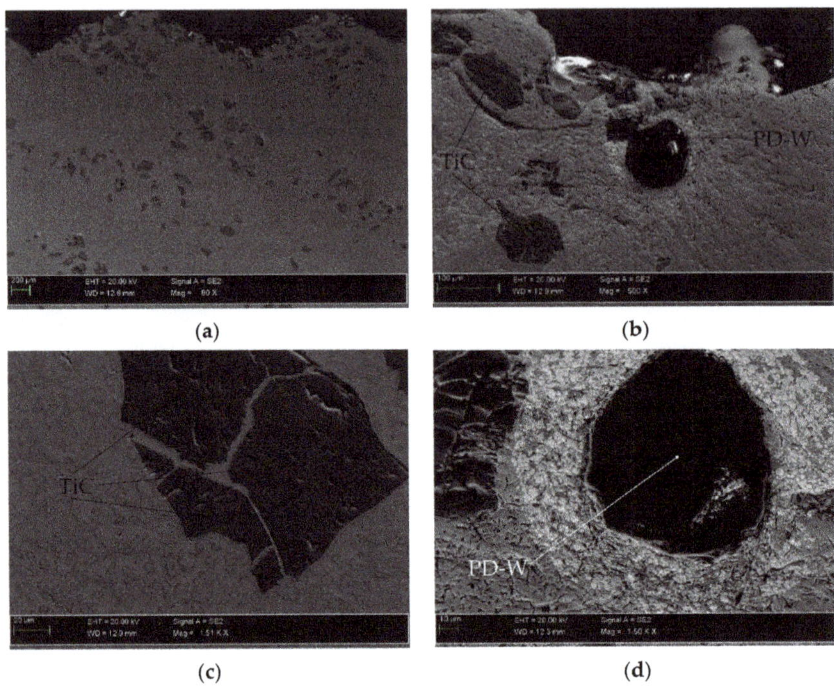

Figure 11. Microscopic image (SEM) of the surface layer (PPTAW metal deposition method, Co3+TiC+PD-W composite powder) deposited on structural low-alloy steel AISI 4715: (**a**) distribution of ceramic particles in the matrix, (**b**) agglomerate of ceramic particles of crushed sharp-edged titanium carbide (TiC) and the spherical particle of the synthetic metal–diamond composite (PD-W), (**c**) single particle of titanium carbide (TiC) and (**d**) the single particle of the synthetic metal–diamond composite (PD-W) in the lagging of tungsten that partly passed to the solid solution of the cobalt alloy.

The tests involving the use of light microscopy revealed that the microstructure of the metal matrix of the surface layer was dendritic, multi-directional and contained numerous inclusions of ceramic particles of titanium carbide (TiC) as well as single particles of the synthetic metal–diamond composite. The subsurface zone of the cladding contained agglomerates of hard phase particles characterised by smaller dimensions, whereas the dilution zone contained relatively larger particles. Titanium carbide (TiC) is characterised by favourable mechanical properties, yet, similar to most ceramic materials, it is also known for its brittleness. Plasticity limited by strong bonds translates into susceptibility go catastrophic cracking. Reference publications do not provide information concerning the crack resistance of titanium carbides or nitrides. It is probable that the first phase of the cooling of the liquid metal in the melt pool is accompanied by the cracking of the carbide, triggered by the concentration of tensile stresses in carbide defects (Figure 11c). Tests concerning tensile stresses generated during the laser-aided alloying of steel with cobalt alloys (using various laser process parameters) are discussed, among others, in publications [61,62]. As regards the particles of synthetic metal–diamond composite coated with tungsten, only some part of tungsten passes to the solution (Figure 12d). The remaining amount of tungsten reacts with diamond particles and creates the coating with tungsten carbide (WC), characterised by favourable thermal stability and abrasive wear resistance. In addition, even the thin layer of the tungsten coating provides the cohesion of the particle, increases the strength of the diamond and improves the thermal conductivity of the alloy (thus extending the service life of drilling tools). Many properties of cobalt alloys result from the crystalline structure of this chemical element. Alloying agents such as Ni, Fe and C stabilise regular face-centred structure A1 of cobalt, which, above a temperature of 417 °C, transforms into crystals of hexagonal close-packed lattice A3 stabilised by Cr, W and Mo. At ambient temperature, cobalt alloys, instead of the hexagonal phase, often contain the metastable regular face-centred phase. The above-named phase, referred to as cobalt (alloy) austenite, is a solid solution: Cr, Ni, Fe, W, Mo or Mn in cobalt [63].

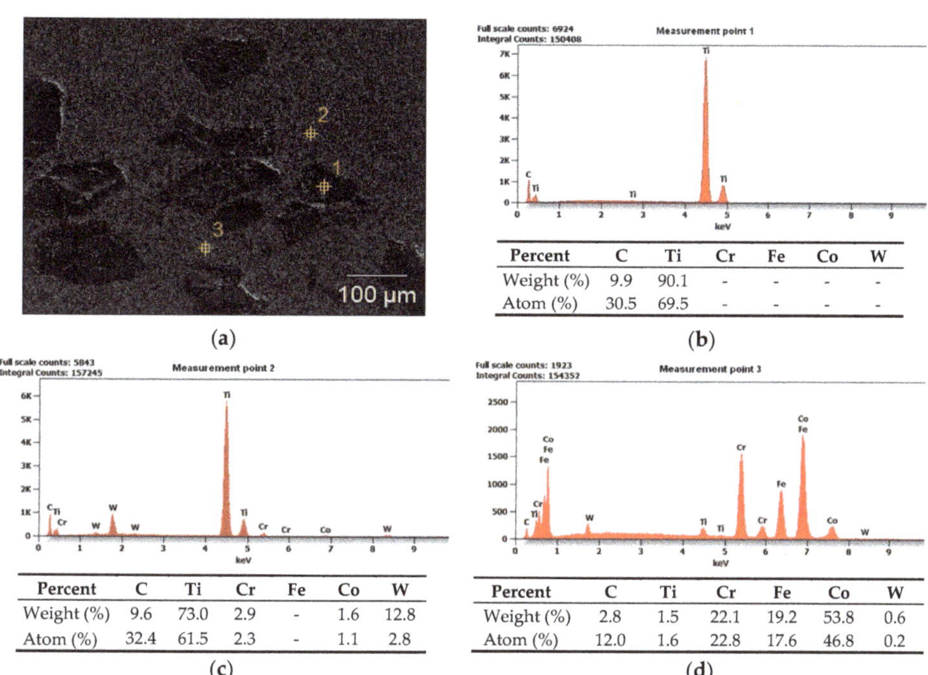

(a)

Percent	C	Ti	Cr	Fe	Co	W
Weight (%)	9.9	90.1	-	-	-	-
Atom (%)	30.5	69.5	-	-	-	-

(b)

Percent	C	Ti	Cr	Fe	Co	W
Weight (%)	9.6	73.0	2.9	-	1.6	12.8
Atom (%)	32.4	61.5	2.3	-	1.1	2.8

(c)

Percent	C	Ti	Cr	Fe	Co	W
Weight (%)	2.8	1.5	22.1	19.2	53.8	0.6
Atom (%)	12.0	1.6	22.8	17.6	46.8	0.2

(d)

Figure 12. Results of the microanalysis of the chemical composition of the surface layer (PPTAW metal deposition method, Co3+TiC+PD-W composite powder) deposited on structural low-alloy steel AISI 4715: (**a**) area subjected to analysis, (**b–d**) diagram of the energy of scattered X-radiation with energy lines present in the area of components (chemical elements (ceramic particle of TiC and metal matrix)) subjected to analysis.

Because of the dilution of the weld metal in the layer, the chemical composition of the layer differed from the chemical compositions of the powders used in the cladding process. The composite powder contained up to 5% Fe, whereas the layer contained the increased amount of this chemical element (Fe = 19.2%, see Figure 12d). The passage of Fe from the base material to the weld deposit was affected by the method and the parameters of the cladding process. The microstructure of the matrix was chemically inhomogenous. The dendritic area was formed of cobalt austenite, solution-hardened with chromium, tungsten or molybdenum. The interdendritic eutectics were rich in chromium, tungsten, silicon and carbides. In the designed alloy, Cr was "tasked with" providing corrosion resistance and reinforcing the solid solution by forming M_7C_3 and $M_{23}C_6$ carbides. According to Madadi et al. (2011) [64], the significant amount of tungsten in the layer (amounting up to 13%) can reinforce the solid solution and favour the formation of MC and M_6C carbides as well as intermetallic phases.

The X-ray diffraction phase and quantitative analysis were performed to identify phases present in the layer. The XRD pattern is presented in Figure 13, whereas the results of the X-ray qualitative phase analysis are presented in Table 4. The analysis revealed the presence of approximately 31% of γ-Co (cobalt austenite), nearly 57% of regularly-structured titanium carbide (TiC) and nearly 12% of hexagonal tungsten carbide (WC).

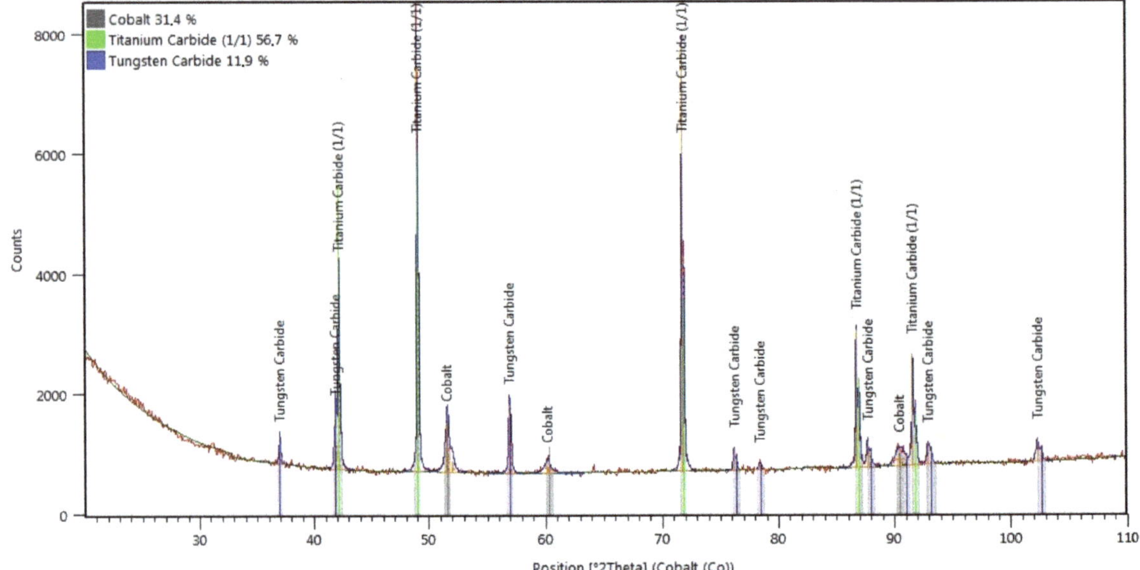

Figure 13. XRD pattern of the surface layer (PPTAW metal deposition method, Co3+TiC+PD-W composite powder) deposited on structural low-alloy steel AISI 4715 with the standard lines of identified crystalline phases.

Table 4. Results of the X-ray qualitative phase analysis of the surface layer (PPTAW metal deposition method, Co3+TiC+PD-W composite powder) deposited on structural low-alloy steel AISI 4715.

ICSD Card No.	Phase Name	Chemical Formula	Percentage (%)	Crystalline Structure
98-015-1365	Titanium carbide (1/1)	TiC	56.7	Regular (F m -3 m)
98-026-0166	Tungsten carbide (1/1)	WC/(PD-W) [1]	11.9	Hexagonal (P -6 m 2)
98-062-2439	Cobalt	Co	31.4	Regular (F m -3 m)

3.3. Deposited Layer Density and Porosity

The performance of calculations concerning the specific density of the composite and the parameters related to its porosity and absorbability involved the sampling of the surface layer for test specimens. The results of related measurements and calculations are presented in Table 5.

Table 5. Results of the μCT analysis and calculations concerning the surface layer (PPTAW metal deposition method, Co3+TiC+PD-W composite powder) deposited on structural low-alloy steel AISI 4715.

Physical Quantity	Average Value of Measured Quantity
Density ρ (measured using the Archimedes method), g/cm^3	5.7785
Absorbability A, %	1.0219
Open porosity P_o, %	5.6467
Closed porosity P_c, %	0.2526
Apparent density ρ_a, g/cm^3	5.4787
Total porosity P_c, %	5.8993

The composite layer was not porosity-free. The total porosity of the layer amounted to approximately 6%. In spite of the foregoing, the layer appears promising as regards further tests performed on an in situ basis using a drilling tool. Owing to its slight porosity, the layer can keep lubricant in its pores and thus reduce friction between interacting surfaces.

3.4. Hardness Test Results

The results of the hardness measurements concerning the external surface and the cross-section of the surface layer are presented in Table 6 and Figure 14, respectively.

Table 6. Hardness (HV10) measurement results concerning the external surface of the layer (PPTAW metal deposition method, Co3+TiC+PD-W composite powder) deposited on structural low-alloy steel AISI 4715 and of the surface of the reference material (abrasion-resistant steel AR400).

Specimen Designation	Specimen Number	Hardnesess, (HV10)					Average Hardness of the Tested Samples	Average Hardness of the Tested Materials
		Measurement Point Number						
		1	2	3	4	5		
Co3+TiC+PD-W	C 01	673	733	657	733	657	690.6	688.7
	C 02	657	675	675	733	694	686.8	
AR400 Steel	S 01	430	421	420	429	424	424.8	424.1
	S 02	421	424	422	421	429	423.4	

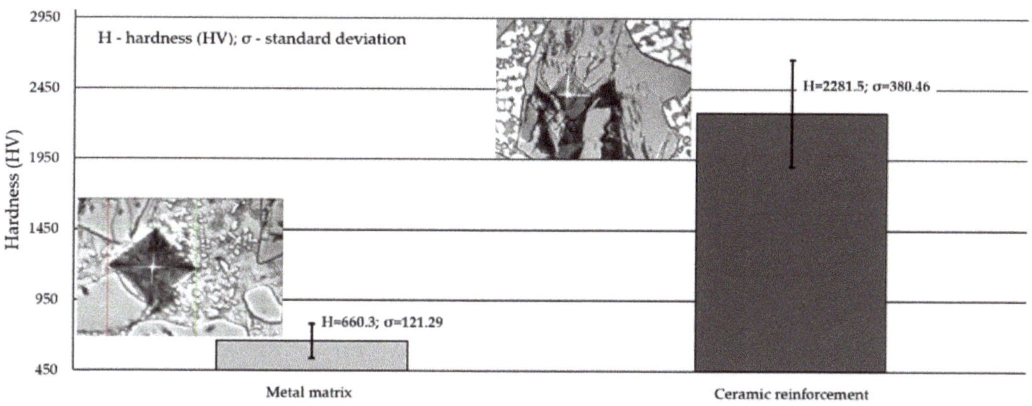

Figure 14. Hardness (HV0.5) measurement results concerning the cross-section of the deposited layer (PPTAW metal deposition method, Co3+TiC+PD-W composite powder) deposited on structural low-alloy steel AISI 4715.

The hardness of the surface layer over the entire external area was relatively uniform and, on the average, amounted to approximately 689 HV10 (59.7 HRC). In turn, the hardness of the reference material used in the abrasive wear resistance tests (abrasion-resistant steel AR400) was lower by more than 265 HV10. The hardness measurements involving the cross-section of the surface layer revealed that the average microhardness of the cobalt alloy-based interdendritic areas amounted to 660 HV0.5 and was lower by approximately 4% than the hardness measured on the surface of the deposited layer. The results concerning the microhardness of the tested layer revealed that the microhardness increased along with the distance between the measurement point and the fusion line (growing towards the surface of the layer). As is known, the microstructure of the layer in the area adjacent to the HAZ differed from the microstructure of the surface layer of the cladding as regards the cladding crystallisation manner [65]. The higher hardness could also result from the significant amount of the hard phase particles near the surface of the cladding. The average hardness of the particles of the hard phase reinforcing the matrix exceeded 2280 HV0.5. It should also be noted that the measurements concerning the hardness of titanium carbide (TiC) did not pose difficulties, yet the measurements concerning the hardness of the synthetic metal–diamond composite appeared problematic. The measurements results confirmed the effect of the base material on microhardness and reflected the chemical and structural homogeneity of the deposited layer.

3.5. Abrasive Wear Test Results

The tests concerning the metal-mineral abrasive wear resistance of the surface layer referred to the abrasive wear resistance of the plate made of popular abrasion-resistant steel AR400 (made by a Swedish manufacturer). As a result, it was possible to determine the relative abrasive wear of the surface layer (Table 7). The nature of the abrasive wear of the surface layer was assessed on the basis of visual tests (Figure 15) and observations involving the use of a confocal microscope (Figure 16).

Table 7. Results of the mineral-metal abrasive wear resistance tests concerning the surface layer (PPTAW metal deposition method, Co3+TiC+PD-W composite powder) deposited on structural low-alloy steel AISI 4715 in comparison with the abrasive wear resistance of abrasion-resistant steel AR400.

Specimen Designation	Spec. Number	Mass Before Test, g	Mass After Test, g	Mass Loss, g	Average Mass Loss, g	Material Density, g/cm^3	Average Volume Loss, mm^3	Relative [1] Abrasive Wear Resistance
				Composite coating				
Co3+TiC+PD-W	C 01	149.7652	149.7560	0.0092	0.0093	5.7997	1.6035	139.64
	C 02	149.4113	149.4019	0.0094				
				Reference material				
AR400 Steel	S 01	123.9290	122.2067	1.7223	1.7429	7.7836	223.9195	1
	S 02	121.7386	119.9752	1.7634				

Note: [1] relative abrasive wear resistance in relation to abrasion-resistant steel AR400.

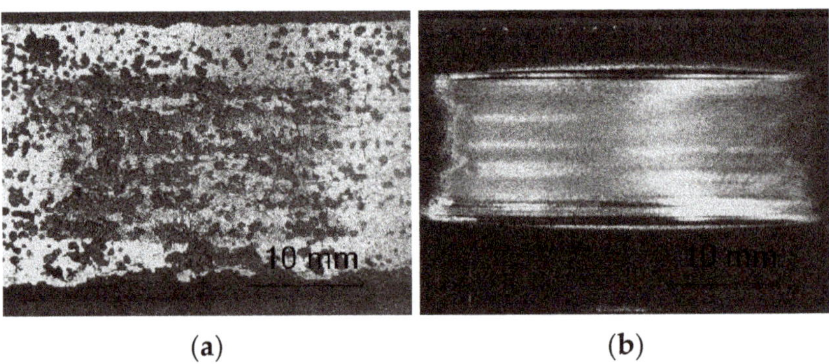

(a)   (b)

Figure 15. Surface of the composite layer and the surface of abrasion-resistant steel AR400 after the metal-mineral abrasive wear resistance test ASTM G64: (**a**) magnified area of layer abrasion and (**b**) magnified area of steel abrasion after the abrasive wear resistance test.

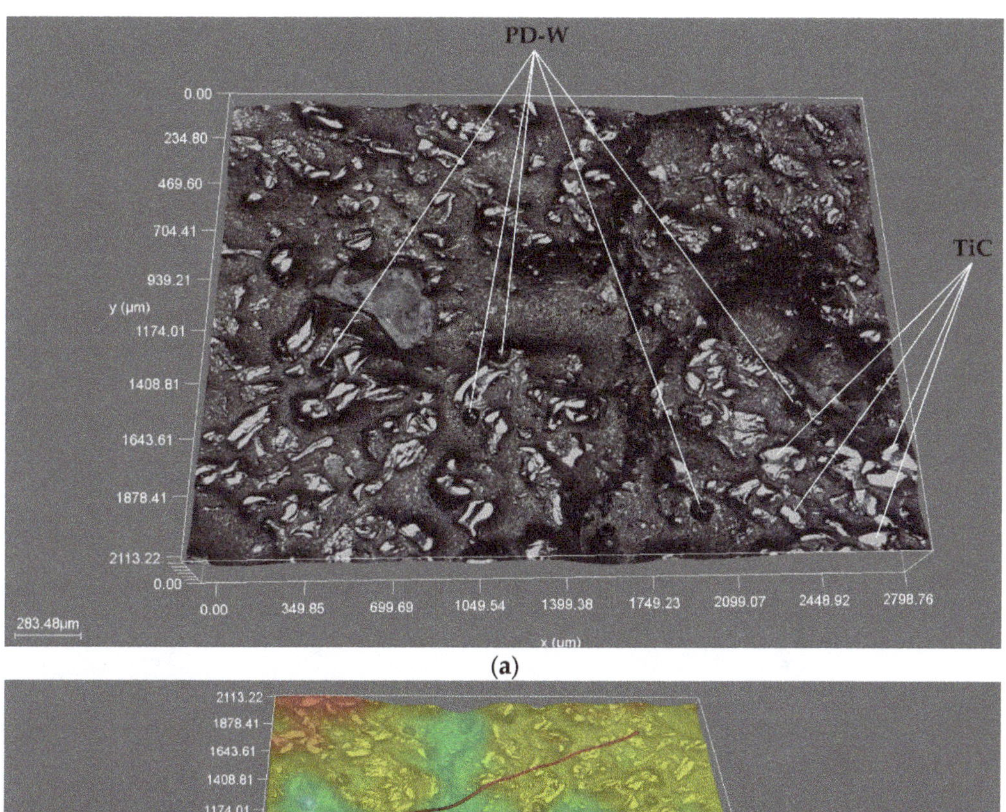

(**a**)

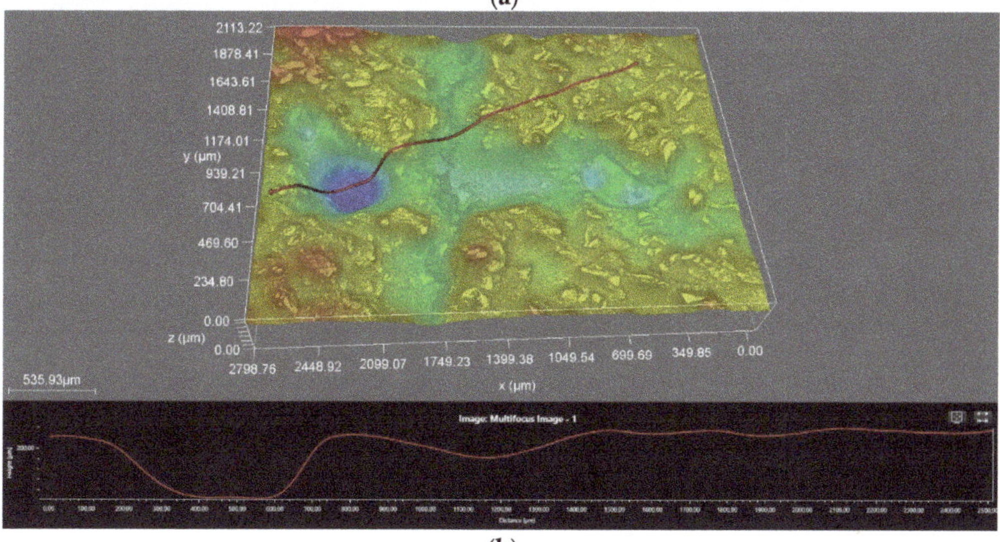

(**b**)

Figure 16. *Cont.*

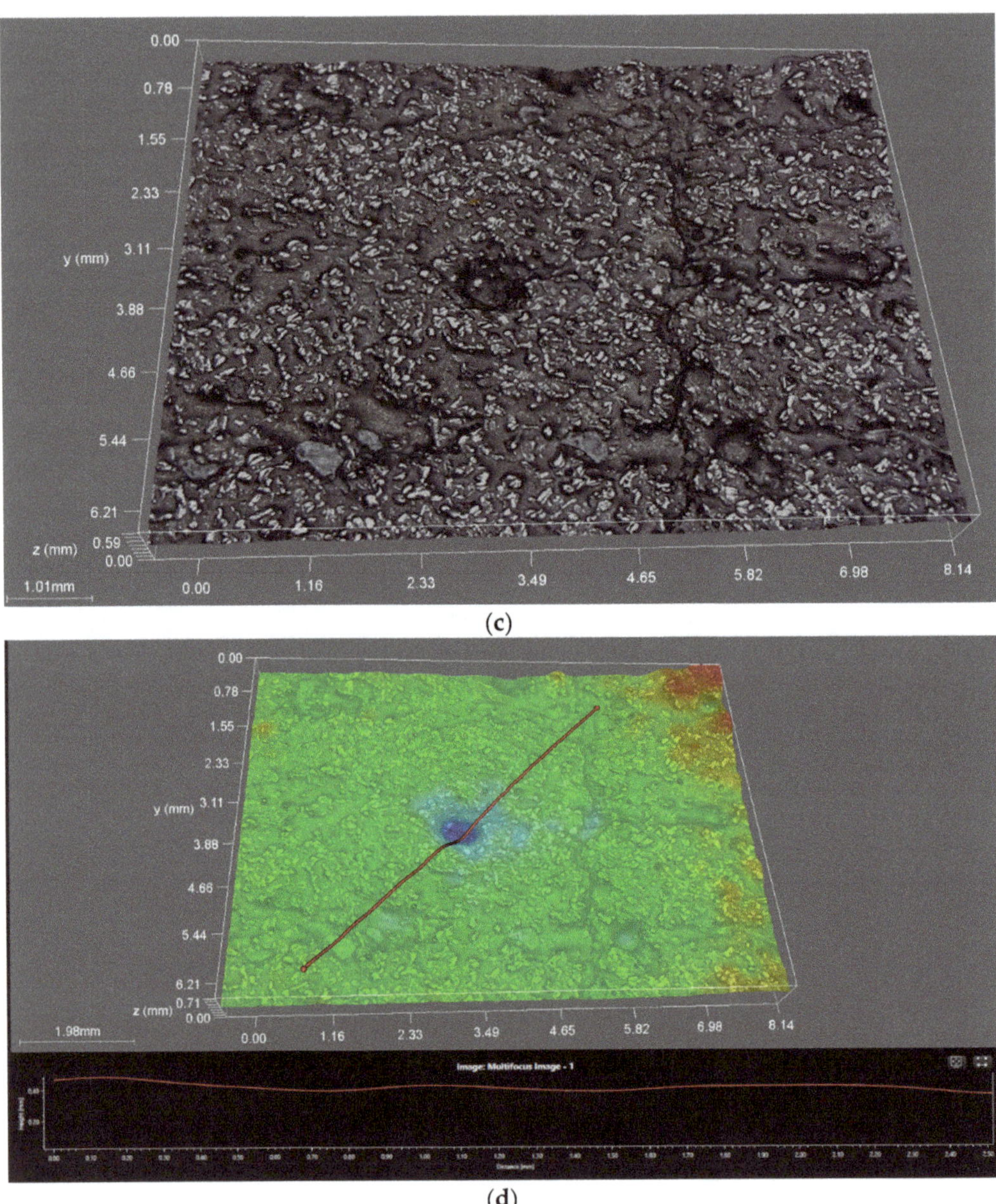

Figure 16. Surface of the composite layer after the metal-mineral abrasive wear resistance test observed using the confocal microscope: (**a,c**) main view of specimen wear, (**b,d**) measurement of defects (single craters).

The intensity of abrasive wear affecting the surface layer was medium as, after interaction with the layer, the grains of sand were partly crushed. As expected, the relative abrasive wear resistance of the deposited layer was higher than that of abrasion-resistant steel AR 400 (almost 140 times). The loss of mass after the test amounted to a mere 0.0093 g. The analysis of the surface condition after test ASTM G65, Procedure A revealed the abrasive mechanism of wear. The hard phase, in the form of titanium carbide (TiC) and the particles of the synthetic metal–diamond composite uniformly distributed in the cobalt matrix, provided a natural barrier to the abrasive. The dominant wear mechanism affecting the surface layer was micro-cutting manifested by continuous micro-scratches and, to a significantly lesser extent, micro-ridging. The grains of the abrasive created slight micro-scratches on the surface of the composite layer. Locally, the course of the micro-scratches deviated from the rectilinear direction, which indicated the effectiveness of the base material reinforcement and confirmed the presence of hard phases in the structure. The width of the micro-traces of wear amounted to approximately 15 μm, whereas the average size of the sand grains used in the abrasive wear tests amounted to 250 μm. Detailed examination involving the use of confocal microscopy revealed the presence of cracks and single craters (having a depth of up to 250 μm) inside the layer (Figure 16b,d). The small ceramic particles uniformly distributed in the layer structure constituted an effective barrier to the grains of the abrasive. The abrasive wear process was intensified by the additional effect of loose particles of the hard phase torn out of the matrix and moving between the (interacting) surface of the specimen and that of the counterspecimen ("rubber wheel"). The aforesaid situation slightly increased the abrasive wear of the surface layer. The freely rolling sharp-edged particles of titanium carbide (TiC) were primarily responsible for the formation of micro-scratches on the counterspecimen surface or the plastic deformation of matrix fragments, manifested by characteristic micro-ridges. In publication [63] concerning their research work, the Authors also confirmed that abrasive wear was a dominant factor responsible for damage to the surface layer of Co-Cr-W-Mo alloys. In relation to data contained in previous publications [31] and individual research [30] concerning the abrasive wear resistance of surface layers made of cobalt alloys reinforced only with TiC particles, it was possible to notice the favourable abrasion-resistant effect of synthetic metal–diamond composite particles constituting approximately 20 wt% of the entire matrix reinforcement. In cases of abrasion-resistant cobalt-based alloys containing ceramic particles, abrasive wear resistance increased along with a growth in the volume fraction of the hard phase. It should be noted that under in situ conditions, the effect of natural factors including the ground structure, texture and presence of stresses, as well as hydrogeological conditions and humidity, abrasive wear resistance may differ from the test results presented in the article.

3.6. Impact Resistance Test Results

The condition of the surface layer during the individual stages of impact resistance tests is presented in Figure 17.

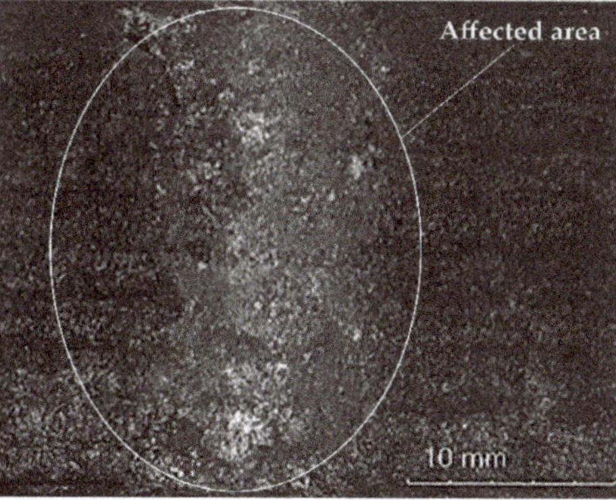

Figure 17. Macrostructure of the affected area. Condition of the surface of the composite layer (PPTAW metal deposition method, Co3+TiC+PD-W composite powder) deposited on structural low-alloy steel AISI 4715 after impact resistance tests.

The deposited composite layer was characterised by favourable resistance to moderate dynamic impact loads. After a cycle of twenty (ram) strokes affecting the surface of the layer with a potential energy of 200 J and the visual tests concerning the affected area, no damage in the form of visible cracks or chips of the weld deposit was observed. The only visible deformation was the plastic distortion of the deposited layer surface having a depth of approximately 0.3 mm, indicating the effective reinforcement of the cobalt-based alloy by cold work hardening.

4. Conclusions

The research-related tests aimed to assess the metallographic structure and to identify the metal-mineral abrasive wear resistance and the impact resistance of powder plasma transferred arc welding (PPTAW), which are innovative in terms of the chemical composition and the type of the hard phase reinforcing the matrix. The layer is intended for contact surfaces of inserts in drilling tools used in the extraction industry. The analysis of the above-presented test results justified the formulation of the following conclusions:

1. The chemical composition, type, amount and the size of the particles of the hard and metallic phases of the cobalt alloy-based composite powder enable its highly accurate and repeatable feeding, ensure its excellent melting and provide good weldability when using PPTAW metal deposition systems.
2. The cladding with powder plasma transferred arc welding (PPTAW) method (i.e., with the composite powder fed directly to the melt pool) favours the maintaining of the structural and thermal stability of the particles of the ceramic reinforcement of the matrix (having the form of tungsten-coated synthetic metal–diamond composite). During the solidification of the liquid metal in the melt pool, some of the reinforcement particles (TiC) underwent brittle cracking. The cracking process was probably triggered by the concentration of tensile stresses in carbide defects.
3. The composite layer was characterised by high hardness, very high metal-mineral abrasive wear resistance, relatively low internal porosity and advantageous resistance to moderate dynamic impact loads.
4. The mechanical and tribological features of the PPTAW deposited layer made using the innovative Co3+TiC+PD-W composite powder appear promising as regards the use of the layer on the contact surfaces of inserts in drilling tools applied in the used in the extraction industry.

The second part of the article will contain test results concerning the structural and the mechanical properties of a nickel-based layer reinforced with particles of tungsten carbide (WC) and synthetic metal–diamond composite.

Funding: The research was funded from the Silesian University of Technology Rector's pro-quality grant 10/050/RGJ_20/0080.

Data Availability Statement: The data are not publicly available due to initiation of a patent procedure No. P435997.

Conflicts of Interest: The author declares no conflict of interests.

References

1. Lisiecki, A. Tribology and Surface Engineering. *Coatings* **2019**, *9*, 663. [CrossRef]
2. Mele, C.; Bozzini, B. Localised corrosion processes of austenitic stainless steel bipolar plates for polymer electrolyte membrane fuel cells. *J. Power Sources* **2010**, *195*, 3590–3596. [CrossRef]
3. Tomków, J.; Rogalski, G.; Fydrych, D.; Łabanowski, J. Advantages of the Application of the Temper Bead Welding Technique during Wet Welding. *Materials* **2019**, *12*, 915. [CrossRef] [PubMed]
4. Bremerstein, T.; Potthoff, A.; Michaelis, A.; Schmiedel, C.; Uhlmann, E.; Blug, B.; Amann, T. Wear of abrasive media and its effect on abrasive flow machining results. *Wear* **2015**, *342–343*, 44–51. [CrossRef]
5. Kenny, P.; Johnson, S. An investigation of the abrasive wear of mineral-cutting tools. *Wear* **1976**, *36*, 337–361. [CrossRef]
6. Gharahbagh, E.A.; Qiu, T.; Rostami, J. Evaluation of Granular Soil Abrasivity for Wear on Cutting Tools in Excavation and Tunneling Equipment. *J. Geotech. Geoenviron. Eng.* **2013**, *139*, 1718–1726. [CrossRef]

7.   Dewangan, S.; Chattopadhyaya, S.; Hloch, S. Wear Assessment of Conical Pick used in Coal Cutting Operation. *Rock Mech. Rock Eng.* **2015**, *48*, 2129–2139. [CrossRef]

8.   Dewangan, S.; Chattopadhyaya, S. Characterization of Wear Mechanisms in Distorted Conical Picks after Coal Cutting. *Rock Mech. Rock Eng.* **2016**, *49*, 225–242. [CrossRef]

9.   Amoun, S.; Sharifzadeh, M.; Shahriar, K.; Rostami, J.; Azali, S.T. Evaluation of tool wear in EPB tunneling of Tehran Metro, Line 7 Expansion. *Tunn. Undergr. Space Technol.* **2017**, *61*, 233–246. [CrossRef]

10.   Nahak, S.; Dewangan, S.; Chattopadhyaya, S.; Somnath, B.; Krolczyk, G.; Hloch, S. Discussion on importance of tungsten carbide—Cobalt (Wc-Co) cemented carbide and its critical characterization for wear mechanisms based on mining applications. *Arch. Min. Sci.* **2018**, *63*, 229–246.

11.   Richerson, D.W.; Lee, W.E. *Modern Ceramic Engineering: Properties, Processing, and Use in Design*, 4th ed.; CRC Press Taylor & Francis Group: Boca Raton, FL, USA, 2018.

12.   Davis, J.R. *ASM Specialty Handbook: Tool Materials*; ASTM International: West Conshohocken, OH, USA, 2001.

13.   Sousa, V.F.C.; Silva, F.J.G. Recent Advances on Coated Milling Tool Technology—A Comprehensive Review. *Coatings* **2020**, *10*, 235. [CrossRef]

14.   Klimpel, A.; Dobrzański, L.; Lisiecki, A.; Janicki, D. The study of properties of Ni–W2C and Co–W2C powders thermal sprayed deposits. *J. Mater. Process. Technol.* **2005**, *164–165*, 1068–1073. [CrossRef]

15.   Medvedovski, E.; Jiang, J.R.; Robertson, M. Tribological properties of boride based thermal diffusion coatings. *Adv. Appl. Ceram.* **2014**, *113*, 427–437. [CrossRef]

16.   Aboua, K.A.M.; Umehara, N.; Kousaka, H.; Tokoroyama, T.; Murashima, M.; Mabuchi, Y.; Higuchi, T.; Kawaguchi, M. Effect of Carbon Diffusion on Friction and Wear Behaviors of Diamond-Like Carbon Coating Against Germanium in Boundary Base Oil Lubrication. *Tribol. Lett.* **2019**, *67*, 65. [CrossRef]

17.   Khor, K.; Yu, L.; Sundararajan, G. Formation of hard tungsten boride layer by spark plasma sintering boriding. *Thin Solid Films* **2005**, *478*, 232–237. [CrossRef]

18.   Medvedovski, E.; Chinski, F.A.; Stewart, J. Wear- and Corrosion-Resistant Boride-Based Coatings Obtained through Thermal Diffusion CVD Processing. *Adv. Eng. Mater.* **2014**, *16*, 713–728. [CrossRef]

19.   Bartkowska, A.; Pertek, A. Laser production of B–Ni complex layers. *Surf. Coat. Technol.* **2014**, *248*, 23–29. [CrossRef]

20.   Latka, L.; Michalak, M.; Jonda, E. Atmospheric plasma spraying of Al2O3 + 13% TiO2 coatings using external and internal injection system. *Adv. Mater. Sci.* **2019**, *19*, 5–17. [CrossRef]

21.   Czupryński, A. Properties of Al_2O_3/TiO_2 and ZrO_2/CaO flame-sprayed coatings. *Mater. Tehnol.* **2017**, *51*, 205–212. [CrossRef]

22.   Chmielewski, T.; A Golański, D. New method of in-situ fabrication of protective coatings based on Fe–Al intermetallic compounds. *Proc. Inst. Mech. Eng. Part B J. Eng. Manuf.* **2011**, *225*, 611–616. [CrossRef]

23.   Andolfi, A.; Mammoliti, F.; Pineschi, F.; Catastini, R. Advanced laser cladding application for oil and gas components. *Technol. Insights* **2012**, 164–173.

24.   Mele, C.; Lionetto, F.; Bozzini, B. An Erosion-Corrosion Investigation of Coated Steel for Applications in the Oil and Gas Field, Based on Bipolar Electrochemistry. *Coatings* **2020**, *10*, 92. [CrossRef]

25.   Khaskin, V.; Korzhyk, V.; Tkachuk, V.; Peleshenko, S.; Voitenko, O.; Oleinychenko, T. The process of laser and laser-plasma cladding. *Am. Sci. J.* **2016**, *2*, 74–78. Available online: https://www.researchgate.net/publication/318463056_THE_PROCESS_OF_LASER_AND_LASER-PLASMA_CLADDING (accessed on 1 May 2021).

26.   Brunner-Schwer, C.; Kersting, R.; Graf, B.; Rethmeier, M. Laser-plasma-cladding as a hybrid metal deposition-technology applying a SLM-produced copper plasma nozzle. *Procedia CIRP* **2018**, *74*, 738–742. [CrossRef]

27.   Brunner-Schwer, C.; Petrat, T.; Graf, B.; Rethmeier, M. Highspeed-plasma-laser-cladding of thin wear resistance coatings: A process approach as a hybrid metal deposition-technology. *Vacuum* **2019**, *166*, 123–126. [CrossRef]

28.   Xia, D.; Xu, B.S.; Lv, Y.H.; Jiang, Y.; Liu, C.L. Microstructure and Properties of Ni-Base Coatings Obtained by Micro-Plasma Arc Cladding Processes. *Adv. Mater. Res.* **2010**, *154–155*, 1371–1374. [CrossRef]

29.   Gao, G.; Li, K.; Chen, W.; Zhang, H.; Yang, X.; Zhang, H.; Li, J. Physical Characteristics of Plasma Cladding Fe-Cr-Nb-Si-Mo Alloy Cladding Layers on Different Substrates. *J. Wuhan Univ. Technol. Sci. Ed.* **2020**, *35*, 820–824. [CrossRef]

30.   Czupryński, A. Comparison of Properties of Hardfaced Layers Made by a Metal-Core-Covered Tubular Electrode with a Special Chemical Composition. *Materials* **2020**, *13*, 5445. [CrossRef] [PubMed]

31.   Cherepova, T.; Dmitrieva, G.; Tisov, O.; Dukhota, O.; Kindrachuk, M. Research on the Properties of Co-Tic and Ni-Tic Hip-Sintered Alloys. *Acta Mech. Autom.* **2019**, *13*, 57–67. [CrossRef]

32.   Adamiec, J.; Łyczkowska, K. Remelting of Inconel 713C alloy by laser and plasma arc. *Weld. Technol. Rev.* **2017**, *89*, 11–16. Available online: http://www.pspaw.wip.pw.edu.pl/index.php/pspaw/article/view/757 (accessed on 1 May 2021).

33.   Czupryński, A.; Wyględacz, B. Comparative Analysis of Laser and Plasma Surfacing by Nickel-Based Superalloy of Heat Resistant Steel. *Materials* **2020**, *13*, 2367. [CrossRef]

34.   Afzal, M.; Ajmal, M.; Khan, A.N.; Hussain, A.; Akhter, R. Surface modification of air plasma spraying WC–12%Co cermet coating by laser melting technique. *Opt. Laser Technol.* **2014**, *56*, 202–206. [CrossRef]

35.   Hou, Q.; Gao, J.; Zhou, F. Microstructure and wear characteristics of cobalt-based alloy deposited by plasma transferred arc weld surfacing. *Surf. Coat. Technol.* **2005**, *194*, 238–243. [CrossRef]

36. Sawant, M.S.; Jain, N. Investigations on wear characteristics of Stellite coating by micro-plasma transferred arc powder deposition process. *Wear* **2017**, *378–379*, 155–164. [CrossRef]

37. Chen, J.; Li, X.; Bell, T.; Dong, H. Improving the wear properties of Stellite 21 alloy by plasma surface alloying with carbon and nitrogen. *Wear* **2008**, *264*, 157–165. [CrossRef]

38. Yang, J.; Liu, F.; Miao, X.; Yang, F. Influence of laser cladding process on the magnetic properties of WC–FeNiCr metal–matrix composite coatings. *J. Mater. Process. Technol.* **2012**, *212*, 1862–1868. [CrossRef]

39. Sahin, F.C.; Apak, B.; Akin, I.; Kanbur, H.E.; Genckan, D.H.; Turan, A.; Goller, G.; Yucel, O. Spark plasma sintering of B4C–SiC composites. *Solid State Sci.* **2012**, *14*, 1660–1663. [CrossRef]

40. Cliche, G.; Dallaire, S. Synthesis and deposition of TiC-Fe coatings by plasma spraying. *Surf. Coat. Technol.* **1991**, *46*, 199–206. [CrossRef]

41. Bober, M.; Grześ, J. The structure of Ni-TiC composite coatings deposited by PPTAW method. *Compos. Theory Pract.* **2015**, *15*, 72–77. Available online: https://kompozyty.ptmk.net/pliczki/pliki/3_2015_t2_BoberGrzes.pdf (accessed on 2 May 2021).

42. Kindrachuk, M.; Shevchenko, A.; Kryzhanovskyi, A. Improvement of the quality of tic-co system plasma coating by laser treatment. *Aviation* **2016**, *20*, 155–159. [CrossRef]

43. Onuoha, C.C.; Jin, C.; Farhat, Z.N.; Kipouros, G.J.; Plucknett, K.P. The effects of TiC grain size and steel binder content on the reciprocating wear behaviour of TiC-316L stainless steel cermets. *Wear* **2016**, *350–351*, 116–129. [CrossRef]

44. Sakamoto, T.; Kurishita, H.; Matsuo, S.; Arakawa, H.; Takahashi, S.; Tsuchida, M.; Kobayashi, S.; Nakai, K.; Terasawa, M.; Yamasaki, T.; et al. Development of nanostructured SUS316L-2%TiC with superior tensile properties. *J. Nucl. Mater.* **2015**, *466*, 468–476. [CrossRef]

45. Jung, S.-A.; Kwon, H.; Suh, C.-Y.; Oh, J.-M.; Kim, W. Preparation of a fine-structured TiC–Co composite by high-energy milling and subsequent heat treatment of a Ti–Co alloy. *Ceram. Int.* **2015**, *41*, 14326–14331. [CrossRef]

46. Karantzalis, A.E.; Lekatou, A.; Evaggelidou, M. Microstructure and sliding wear assessment of Co–TiC composite materials. *Int. J. Cast Met. Res.* **2013**, *27*, 73–79. [CrossRef]

47. Anasori, B.; Barsoum, M.W. Energy damping in magnesium alloy composites reinforced with TiC or Ti2AlC particles. *Mater. Sci. Eng. A* **2016**, *653*, 53–62. [CrossRef]

48. Cai, B.; Tan, Y.-F.; He, L.; Tan, H.; Gao, L. Tribological properties of TiC particles reinforced Ni-based alloy composite coatings. *Trans. Nonferrous Met. Soc. China* **2013**, *23*, 1681–1688. [CrossRef]

49. Jones, D.R.; Ashby, M.F. *Engineering Materials 1: An Introduction to Properties, Applications and Design*, 5th ed.; Elsevier Ltd.: Amsterdam, The Netherlands, 2019.

50. *EN 14700. Welding Consumables. Welding Consumables for Hard-Facing*; CEN: Brussels, Belgium, 2014.

51. *ISO 17637. Non-Destructive Testing of Welds—Visual Testing of Fusion-Welded Joints*; ISO: Geneva, Switzerland, 2016.

52. *ISO 3452. Non-Destructive Testing—Penetrant Testing—Part 2: Testing of Penetrant Materials*; ISO: Geneva, Switzerland, 2013.

53. *ASTM D792. Standard Test Methods for Density and Specific Gravity (Relative Density) of Plastics by Displacement*; American Society for Testing and Materials: West Conshohocken, PA, USA, 2020.

54. *ISO 6507. Metallic materials—Vickers hardness test—Part 1: Test method*; ISO: Geneva, Switzerland, 2018.

55. *ASTM G65-00. Standard Test Method for Measuring Abrasion Using the Dry Sand/Rubber Wheel Apparatus*; American Society for Testing and Materials: West Conshohocken, PA, USA, 2015.

56. Archard, J.F.; Hirst, W. The wear of metals under unlubricated conditions. *Proc. R. Soc. London. Ser. A Math. Phys. Sci.* **1956**, *236*, 397–410. [CrossRef]

57. Przestacki, D.; Majchrowski, R.; Marciniak-Podsadna, L. Experimental research of surface roughness and surface texture after laser cladding. *Appl. Surf. Sci.* **2016**, *388*, 420–423. [CrossRef]

58. Berger, L.-M. Application of hardmetals as thermal spray coatings. *Int. J. Refract. Met. Hard Mater.* **2015**, *49*, 350–364. [CrossRef]

59. *ISO 15614. Specification and Qualification of Welding Procedures for Metallic Materials—Welding Procedure Test—Part 7: Overlay Welding*; ISO: Geneva, Switzerland, 2016.

60. *ISO 5817. Welding—Fusion-Welded Joints in Steel, Nickel, Titanium and Their Alloys (Beam Welding Excluded)—Quality Levels for Imperfections*; ISO: Geneva, Switzerland, 2014.

61. Gireń, B.G.; Szkodo, M.; Steller, J. The influence of residual stresses on cavitation resistance of metals — an analysis based on investigations of metals remelted by laser beam and optical discharge plasma. *Wear* **1999**, *233–235*, 86–92. [CrossRef]

62. Grum, J.; Šturm, R. A new experimental technique for measuring strain and residual stresses during a laser remelting process. *J. Mater. Process. Technol.* **2004**, *147*, 351–358. [CrossRef]

63. Balagna, C.; Spriano, S.; Faga, M. Characterization of Co–Cr–Mo alloys after a thermal treatment for high wear resistance. *Mater. Sci. Eng. C* **2012**, *32*, 1868–1877. [CrossRef]

64. Madadi, F.; Shamanian, M.; Ashrafizadeh, F. Effect of pulse current on microstructure and wear resistance of Stellite6/tungsten carbide claddings produced by tungsten inert gas process. *Surf. Coat. Technol.* **2011**, *205*, 4320–4328. [CrossRef]

65. Lancaster, J.F. *Metallurgy of Welding*, 6th ed.; Woodhead Publishing: Sawston Cambridge, UK, 1999.

Article

Microstructure and Abrasive Wear Resistance of Metal Matrix Composite Coatings Deposited on Steel Grade AISI 4715 by Powder Plasma Transferred Arc Welding Part 2. Mechanical and Structural Properties of a Nickel-Based Alloy Surface Layer Reinforced with Particles of Tungsten Carbide and Synthetic Metal–Diamond Composite

Artur Czupryński 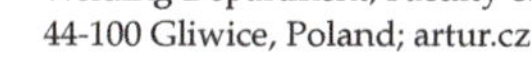

Welding Department, Faculty of Mechanical Engineering, Silesian University of Technology, Konarskiego 18A, 44-100 Gliwice, Poland; artur.czuprynski@polsl.pl

Abstract: The article is the continuation of a cycle of works published in a Special Issue of MDPI entitled "Innovative Technologies and Materials for the Production of Mechanical, Thermal and Corrosion Wear-Resistant Surface Layers and Coatings" related to tests concerning the microstructure and mechanical properties of innovative surface layers made using the Powder Plasma Transferred Arc Welding (PPTAW) method and intended for work surfaces of drilling tools and machinery applied in the extraction industry. A layer subjected to tests was a metal matrix composite, made using powder based on a nickel alloy containing spherical fused tungsten carbide (SFTC) particles, which are fused tungsten carbide (FTC) particles and spherical particles of tungsten-coated synthetic metal–diamond composite (PD-W). The layer was deposited on the substrate of low-alloy structural steel grade AISI 4715. The results showed that the chemical composition of the metallic powder as well as the content of the hard phase constituting the matrix enabled the making of a powder filler material characterised by very good weldability and appropriate melting. It was also found that the structure of the Ni-WC-PD-W layer was complex and that proper claddings (characterised by the uniform distribution of tungsten carbide (WC)) were formed in relation to specific cladding process parameters. In addition, the structure of the composite layer revealed the partial thermal and structural decomposition of tungsten carbide, while the particles of the synthetic metal–diamond composite remained coherent. The deposited surface layer was characterised by favourable resistance to moderate dynamic impact loads with a potential energy of 200 J, yet at the same time, by over 12 times lower metal–mineral abrasive wear resistance than the previously tested surface layer made of cobalt-based composite powder, the matrix of which contained the hard phase composed of TiC particles and synthetic metal–diamond composite. The lower abrasive wear resistance could result from a different mechanism responsible for the hardening of the spherical particles of the hard phase susceptible to separation from the metal matrix, as well as from a different mechanism of tribological wear.

Keywords: PPTAW; cladding; deposition; abrasion; impact load; tungsten carbide; synthetic metal–diamond composite

Citation: Czupryński, A. Microstructure and Abrasive Wear Resistance of Metal Matrix Composite Coatings Deposited on Steel Grade AISI 4715 by Powder Plasma Transferred Arc Welding Part 2. Mechanical and Structural Properties of a Nickel-Based Alloy Surface Layer Reinforced with Particles of Tungsten Carbide and Synthetic Metal–Diamond Composite. *Materials* **2021**, *14*, 2805. https://doi.org/10.3390/ma14112805

Academic Editor: Hendra Hermawan

Received: 18 April 2021
Accepted: 21 May 2021
Published: 25 May 2021

Publisher's Note: MDPI stays neutral with regard to jurisdictional claims in published maps and institutional affiliations.

1. Introduction

Most plasma cladding applications involve the use of a powder filler material enabling the obtainment of deposited layers characterised by various chemical compositions, structure and properties [1,2]. In deposited layers that have the structure of composite materials (used to improve the abrasive wear resistance of drilling tools or machinery), the matrix is usually composed of cobalt, nickel or iron-based alloys containing particles of high-melting phases, e.g., carbides of transition metals found in groups IVB–VIB of the periodic table [3,4]. High hardness, high melting points and high thermodynamic stability constitute the primary and very desirable features of the aforesaid carbides [5]. The combination of hard carbide phases with the relatively plastic metal matrix significantly improves the functional properties of surface layers exposed to various types of abrasion. Composite layers are characterised by particularly high abrasive wear resistance as well as resistance to moderate impact loads, unobtainable in cases of typical metallic layers. Under operational conditions of drilling tools (discussed in detail in the first part of the cycle of publications), it appears particularly reasonable to use composite layers reinforced with tungsten carbide particles [6,7].

Deposited composite layers are characterised by the repeatable structure providing high operational properties, particularly when particles of the hardening phase are uniformly distributed in the metal matrix. There are many factors affecting the formation of deposited composite layers and the dispersion of particles in the hard reinforcing phase. The most important of the aforementioned factors include the nature and the intensity of the effect of the liquid metal matrix containing the particles of the hardening phase (indicated by the wetting degree). According to Bober et al. (2016) [8], good wettability is observed when both the liquid and the solid phase are characterised by the same type of atomic bonds and when appropriate temperature conditions and surface activation conditions are satisfied. The carbides of the transition metals of the periodic table such as Ti, Zr, Nb, Cr, Hf, Nb, Ta, Mo and W are characterised by the complexity of chemical bonds. Typically, the above-named carbides contain mixed bonds, i.e., metallic bonds with the covalent-type effect or, in certain cases, with the ionic effect [9]. During the cladding process, when the particles of the hard phase characterised by a high melting point are wetted appropriately by the liquid metallic matrix, proper welds are formed and high yield (degree of carbide powder use) is obtained. In cases of insufficient wettability, the particles of the reinforcement (hardening) are expelled from the melt pool and the yield of hardening phase particles is low. The above-named situation may result in the formation of many welding imperfections including cracks, gas pores or the internal porosity of the deposited layer. In addition to providing composite components with good wettability, the difference in density between the hard phase and the matrix favourably affects the formation of composite claddings. Regarding the metal of the matrix and that of the hardening phase, significantly varying mass densities may result in the non-uniform distribution or the agglomeration of the hard phase particles in the volume of the cladding metal. Carbides characterised by high specific density tend to settle to the bottom of the melt pool, whereas the particles of the hard phase with lower density concentrate in the upper zone, slightly below the cladding weld. Many research works are concerned with the thorough investigation of the structure and the abrasive wear resistance of deposited composite layers that have the nickel alloy-based matrix containing the addition of tungsten carbide [10,11], chromium carbide [12,13] and titanium carbide [4,14,15]. However, significantly fewer publications focus on composite claddings reinforced with particles of the remaining carbides of transition metals, e.g., ZrC, HfC, NbC, TaC and MoC, or synthetic metal–diamond sinters. As the aforementioned reinforcement particles are also characterised by advantageous properties, their addition could significantly improve the abrasive wear resistance of deposited layers [16]. In light of research results presented in the first part of the cycle of publications, the use of the hard phase particles in the form of synthetic metal–diamond composite seems very promising. Powder filler materials that could be used in the PPTAW metal deposition of composite layers on selected structural materials are presented in Table 1.

Metal matrix composite (MMC) layers reinforced with, e.g., TiC, WC, B_4C, Cr_3C_2, NbC, etc. particles, combine the properties of the plastic, abrasive wear and corrosion resistant matrix with the properties of hard carbide ceramics [27]. Surface layers of the aforesaid type can be used where it is necessary to ensure high abrasive wear resistance combined with resistance to dynamic impact loads. Such requirements cannot be satisfied by commonly used metallic layers applied using welding methods or by hard and brittle ceramic layers. Therefore, it seems favourable to use MMC layers deposited on work surfaces of drilling tools. The addition of hard carbide particles characterised by the various degree of dispersion additionally reinforces the matrix. According to Deuis et al. (1997) metal matrix composites can transfer higher compressive and tensile stresses than monolithic alloys. This is because applied loads are transferred from the plastic matrix to the particles of the hardening phase, which is possible if there are appropriate bonds between individual components of a given composite [26]. During the deposition of composite layers, the nature of the interphase boundary between the metal matrix and the hard reinforcing phase depends on the time and the temperature of cladding formation as well as on the chemical composition of the matrix. The growing popularity of these materials also results from their advantageous structural ratio, i.e., the proportion of strength to mass. The acceptable quality and the highly satisfactory structural and mechanical properties of the composite layer with the composite powder based on a cobalt alloy containing titanium carbide (TiC) and synthetic metal–diamond composite [14] inspired the author's research on the material characteristics of the layer made of composite powder based on nickel with the addition of hard reinforcing phase composed of fused tungsten carbide (WC) and synthetic metal–diamond sinter. There is no scientific information and material data related to the abrasion tests concerning metal matrix surface layers reinforced with synthetic metal–diamond composite and obtained using the Powder Plasma Transferred Arc Welding (PPTAW) metal deposition method. The results of previous individual research led to the general conclusion that it was possible to obtain a nickel-based alloy surface layer reinforced with particles of tungsten carbide and synthetic metal–diamond composite that can characterize a good resistance to abrasive wear.

Table 1. Powder filler materials usable in the PPTAW metal deposition of composite layers on selected structural materials.

Base Material	Powder Filler Material		Powder Grain (μm)	References
	Matrix	Reinforcement		
Structural steel	Ni–Cr–B–Si	69% WC/Co	53–106	[17]
Structural steel	Ni–base alloy	30% Cr_3C_2	75–185	[18]
Structural steel	Ni–base alloy	NbC	50–150	[19]
Structural steel	Fe–Cr–C–Ni	Chromium (II) carbide Cr_3C_2, Cr_7C_3, and $Cr_{23}C_6$	70–140	[20]
Structural steel	Fe–C–B–Mn–Si	20% B_4C	50–150	[21]
Structural steel	Co–Cr–W–C	60% TiC+PD-W	60–250	[22]
Structural steel	Co–Cr–W–C	30% Cr_3C_2	60–145	[23]
Stainless steel	Co–Cr–W–C	50% Cr_3C_2, 20% WC, 50% TiC, 40% NbC	53–180	[24]
Stainless steel	Co–Mo–Cr–Si–Fe–Ni	35% WC 35% (WC−12% Cr)	53–180	[25]
Aluminium	Al–Ni	Al_2O_3, SiC, TiC	70	[26]
Titanium	Ti	50% NbC	80–120	[16]

2. Experimental Section

2.1. Objective of the Study

The study aimed to assess the metallographic structure as well as to identify the metal–mineral abrasive wear resistance and the resistance to moderate impact loads of a composite layer deposited using the PPTAW metal deposition method and "newly developed" nickel alloy-based powder with the addition of the hard reinforcing phase containing two types of tungsten carbide ($WC-W_2C$) and synthetic metal–diamond composite (PD-W). It was assumed that the spherical shape of the hard phase particles would result in their firm deposition in the metallic matrix, thus significantly improving the abrasive wear resistance parameters of the deposited layer protecting work surfaces of drilling tools used in the extractive industry. The results of the relative abrasive wear resistance of the nickel alloy-based composite layer were compared with those concerning the cobalt-based surface layer [22]. Criteria specifying the requirements related to the making and the acceptance of the deposited layer are presented in detail in the first part of the cycle of publications.

2.2. Materials and Methods

The surface layer was deposited on the substrate of low-alloy structural steel AISI 4715 (Table 2) having the form of a flat bar (75 mm × 25 mm × 10 mm). The surface of the substrate was cleaned directly before the cladding process. The filler material used in the process was a powder mixture, the matrix of which was a nickel alloy having a chemical composition corresponding to that of the group of Ni3 alloys (in accordance with EN 147000:2014) [28] (Table 2). The powder mixture also contained a hard phase composed of spherical fused tungsten carbide (SFTC) particles (Figure 1a), fused tungsten carbide (FTC) particles (Figure 1b) and spherical particles of tungsten-coated synthetic metal–diamond composite (PD-W) (Harmony Industry Diamond, Zhengzhou, China) (Figure 1c). The particles of the $WC-W_2C$ tungsten carbide and synthetic metal–diamond composite (PD-W) were mixed with the matrix powder in the 60/35 wt% ratio using a mixer-shaker and ceramic balls.

Table 2. Chemical composition of AISI 4715 low-alloy structural steel according to manufacturer data (TimkenSteel Ltd., Canton, OH, USA) and chemical composition of Ni3+ $WC-W_2C$-Co+PD-W powder.

Chemical Composition of AISI 4715 Low-Alloy Structural Steel, wg. %								
C	Mn	S	P	Si	Cr	Mo	Ni	Fe
0.12–0.18	0.65–0.95	$\leq$0.015	$\leq$0.015	0.15–0.35	0.40–0.70	0.45–0.60	0.65–1.00	Bal.

Chemical composition of the Ni3 alloy, wg. %							Ceramic reinforcement of the matrix, wg. %		
C	Si	Mn	Cr	B	Fe	Ni	SFTC	FTC	PD-W
$\leq$0.05	2.4	0.5	2.0	$\leq$1.4	$\leq$0.5	Bal.	70	10	20

Carbide to matrix ratio, 60/35 (wg. %); bulk density of the powder determined by pycnometry, 11.64 g/cm^3.

The composite layer was deposited using an industrial welding station provided with a modern robotic hardfacing plasma coating machine for cladding with PPTAW Durweld 300T PTA (Durum Verschleiss-Schutz GmbH, Willich, Germany) having a maximal current of 300 A and parameters determined during initial cladding tests (Figure 2, Table 3). The deposited layer was obtained by weave-bead technique with an overlap of 33%.

The cladding parameters identified as optimum were those ensuring the uniform distribution of the powder over the entire liquid metal area in the weld metal pool, uniform and shallow penetration having depth of g = 1.2 mm, the height of the layer in one run h $\geq$ 2 and the dilution of the base material in the cladding D below 4.5%.

2.3. Testing Methodology

The testing methodology and equipment are discussed in detail in the first part of the cycle of publications. The tests involving the analysis of the morphology and the size of the composite powder (MMC) particles as well as the assessment of the quality of the layer were based on non-destructive tests including visual tests (VT), penetrant tests (PT) and tests concerning the roughness (Ra) of the composite layer. The analysis of the structure and of the surface properties of the composite layer was based on macro and microscopic metallographic test results, chemical composition analysis results, X-ray diffraction results, weld deposit hardness and roughness measurement results, as well as results of tests concerning metal–mineral abrasive wear resistance and resistance to moderate impact loads.

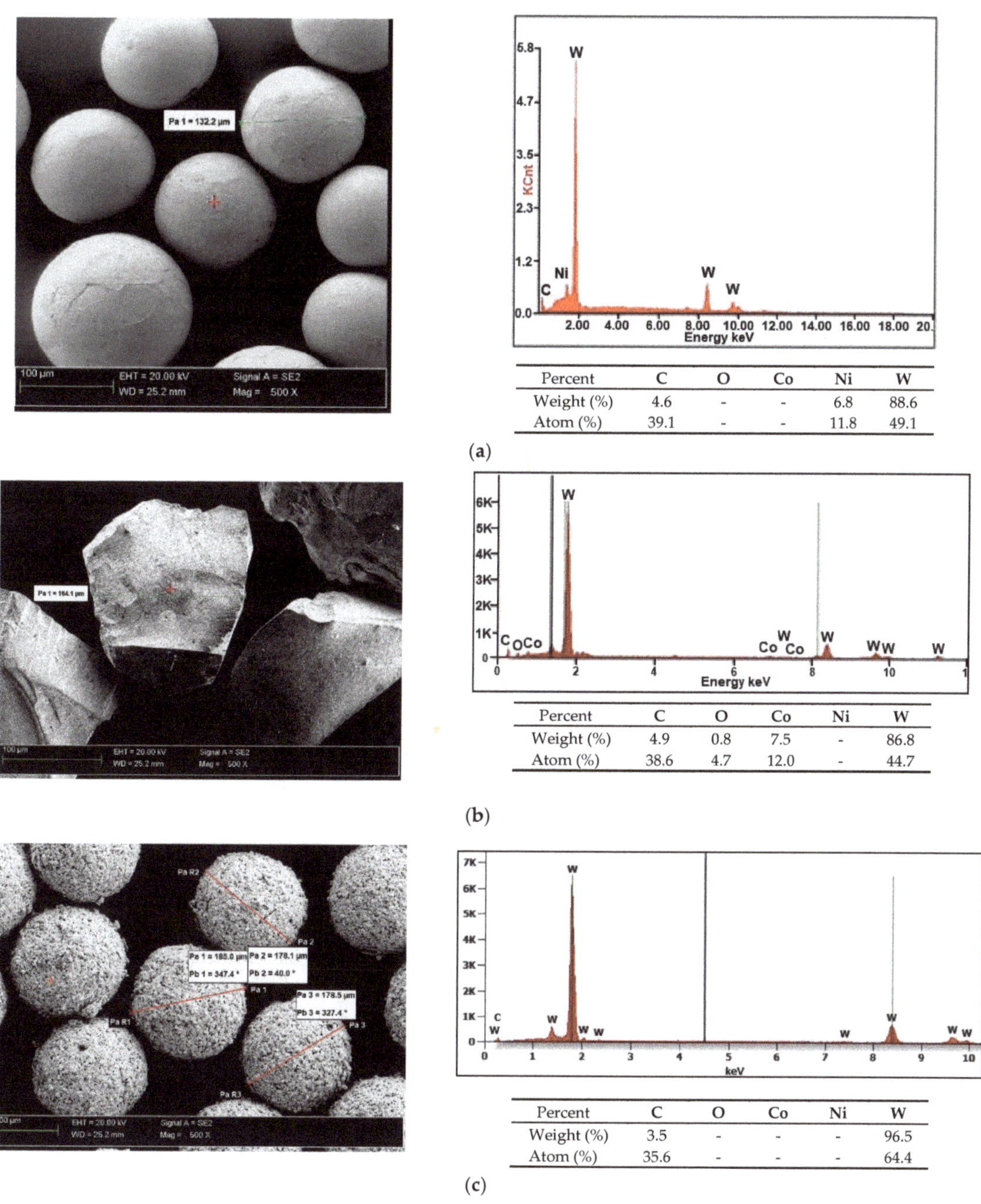

Percent	C	O	Co	Ni	W
Weight (%)	4.6	–	–	6.8	88.6
Atom (%)	39.1	–	–	11.8	49.1

(**a**)

Percent	C	O	Co	Ni	W
Weight (%)	4.9	0.8	7.5	–	86.8
Atom (%)	38.6	4.7	12.0	–	44.7

(**b**)

Percent	C	O	Co	Ni	W
Weight (%)	3.5	–	–	–	96.5
Atom (%)	35.6	–	–	–	64.4

(**c**)

Figure 1. Components of the hard ceramic phase in the Ni3+WC-W$_2$C+PD-W powder: (**a**) spherical fused tungsten carbide (SFTC), (**b**) fused tungsten carbide (FTC) and (**c**) tungsten-coated synthetic metal–diamond composite (PD-W).

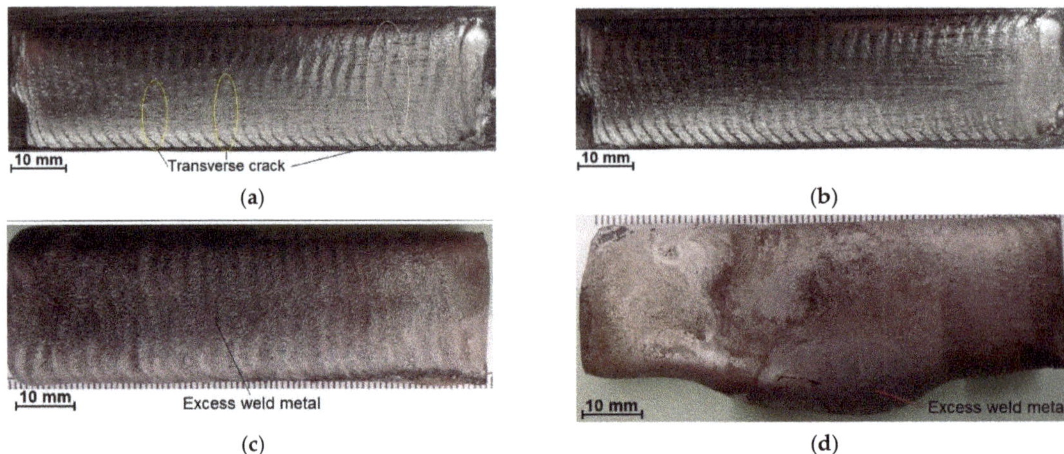

Figure 2. Optimization process parameters of the PPTAW metal deposition of the surface layer (Ni3+WC-W$_2$C+PD-W composite powder) on steel AISI 4715: (**a**) too low heat input (275 J/mm); (**b**) the right heat input (400 J/mm); (**c**) too much heat input (520 J/mm); (**d**) too much heat input (580 J/mm) and amount of supplies powder (22 g/min).

Table 3. Welding parameters of the plasma transfer arc welding (PTAW) metal deposition of the surface layer (Ni3+WC-W$_2$C+PD-W composite powder) on steel AISI 4715.

Process Parameters	Value of Parameter
Current, I (A)	80
Voltage, U (V)	25
Travel speed, S (mm/s)	3
Powder feed rate, q (g/min)	18
Heat input, E$_u$[1] (J/mm)	400

Notes: Argon 5.0 (99.999%) acc. ISO 14175—I1: 2009 was used as plasma gas (flow rate = 1.6 L/min), Argon/Hydrogen 5% H2, Ar (welding mixture ISO 14175-R1-ArH-5) was used as shielding gas (flow rate = 12 L/min) and carrier gas (flow rate = 4 L/min), [1] calculated acc. to the formula: $E_u = k \cdot (U \times I)/S$. The thermal efficiency coefficient for plasma transferred arc k = 0.6 was used.

3. Results and Discussion

3.1. Composite Powder Morphology

The morphology and the size of the Ni3+WC-W$_2$C+PD-W powder particles are presented (in the form of SEM images) in Figure 3. The image of the powder mixture components was made in the contrast of back-scattered electrons (BSE); the bright particles represent the tungsten carbide particles, whereas the darker particles are the nickel alloy particles. The tests revealed that the size of the powder particles was restricted within the range of 30 μm to 170 μm (median Q$_{50}$ = 152 μm) and that the powder was a homogenous mixture containing primarily spherical components and a small fraction of irregularly-shaped particles. The components of the powder were mixed in a laboratory powder mixer-shaker, using ceramic balls. The powder mixer-shaker of a container with two perpendicular rotational axes was used. Main rotational axis was driven with constant rotational velocity of 46 rpm. The secondary axis was performing rocking motions in the range 0–180°. The powder ingredients were mixed for 1 minute. The particle bulk were moving across the vessel through a three-dimensional unsteady periodical behaviour.

The applied manner of mixing favourably limited the agglomeration of components and enabled the obtainment of the powder filler material characterised by good flowability. Microanalysis results concerning the chemical composition of the powder are presented in Figure 4 (in the form of photographs and diagrams of scattered X-radiation). The tests were performed on the surface of the powder particles using point or micro-area-based analysis.

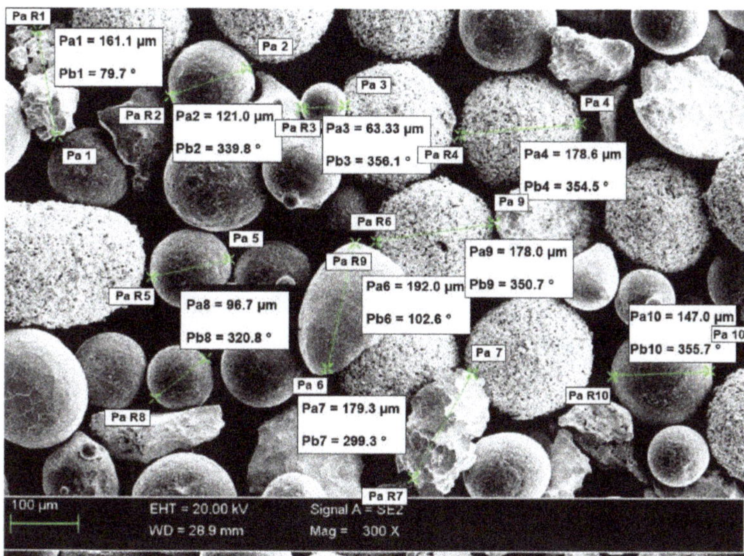

Figure 3. Scanning electron microscopic (SEM) image of the morphology of the Ni3+WC-W$_2$C+PD-W powder particles along with measurement values.

3.2. Non-Destructive Tests—Visual Test Results

The visual and the penetrant tests of the surface of the deposited surface layer only revealed the presence of apparent welding imperfections or single gas pores (Figure 5b). The above-named tests did not reveal shape and dimension-related imperfections. The average surface roughness (Ra) amounted to 12 µm. Roughness tests were performed in five test lines on the untreated surface. Comparing the obtained test results with the results presented in [22], it should be stated that the average surface roughness of the composite layer on the matrix of the nickel alloy was slightly lower (about 2 µm) than the roughness of the layer on the matrix of the cobalt alloy. The overlapping beads were characterised by appropriate symmetry, which translated into the uniform distribution of the composite powder on the specimen surface (Figure 5a).

In view of the fact that the surface layer is intended for work surfaces of drilling tools, the presence of single small gas pores on the surface of the deposited layer can be regarded as acceptable.

3.3. Metallographic Test Results and Results of the XRD Analysis

The results of microscopic metallographic observations made it possible to identify the structure of the matrix as well as the type, distribution and the dimensions of the surface layer reinforcement. The results of the microstructural observations of the surface layer cross-section are presented in Figure 6. The microstructure and the results of the qualitative point analysis, identifying the individual chemical elements present in the surface layer (made using the PPTAW method), are presented in Figure 7.

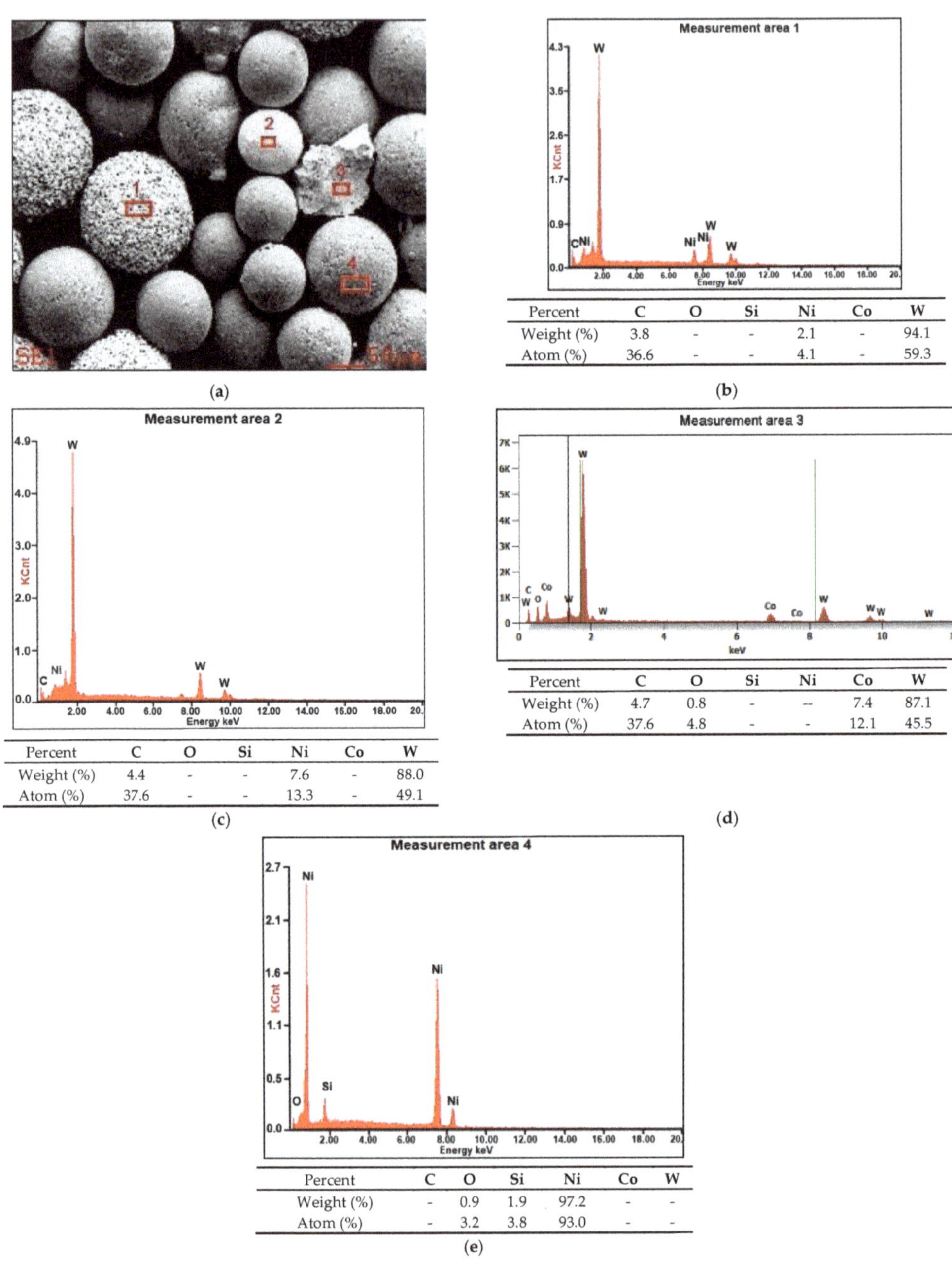

(a)

Measurement area 1 (b)

Percent	C	O	Si	Ni	Co	W
Weight (%)	3.8	-	-	2.1	-	94.1
Atom (%)	36.6	-	-	4.1	-	59.3

Measurement area 2 (c)

Percent	C	O	Si	Ni	Co	W
Weight (%)	4.4	-	-	7.6	-	88.0
Atom (%)	37.6	-	-	13.3	-	49.1

Measurement area 3 (d)

Percent	C	O	Si	Ni	Co	W
Weight (%)	4.7	0.8	-	--	7.4	87.1
Atom (%)	37.6	4.8	-	-	12.1	45.5

Measurement area 4 (e)

Percent	C	O	Si	Ni	Co	W
Weight (%)	-	0.9	1.9	97.2	-	-
Atom (%)	-	3.2	3.8	93.0	-	-

Figure 4. Morphology, the size of the powder mixture particles and the microanalysis results related to the chemical composition of the Ni3+WC-W$_2$C+PD-W composite powder: (**a**) SEM image of the morphology of the powder particles with the area subjected to analysis and the diagrams of scattered X-radiation energy with energy lines present in the area of components (chemical elements) subjected to analysis; (**b**) tungsten-coated synthetic metal–diamond composite (PD-W); (**c**) spherical fused tungsten carbide (SFTC); (**d**) fused tungsten carbide (FTC); (**e**) nickel matrix.

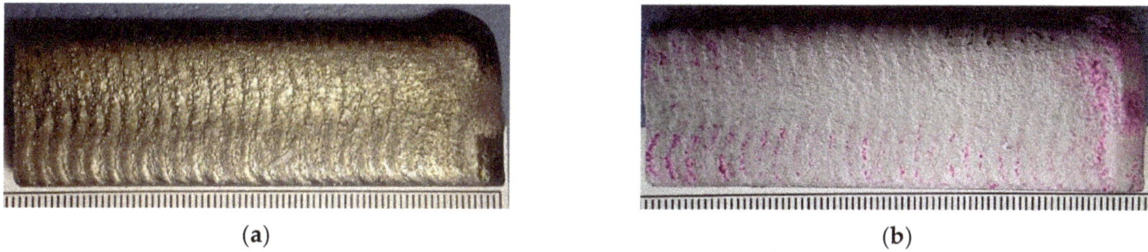

Figure 5. Surface layer made using the Ni3+WC-W$_2$C+PD-W composite powder: (**a**) layer after the visual tests (VT); (**b**) layer after the penetrant tests (PT).

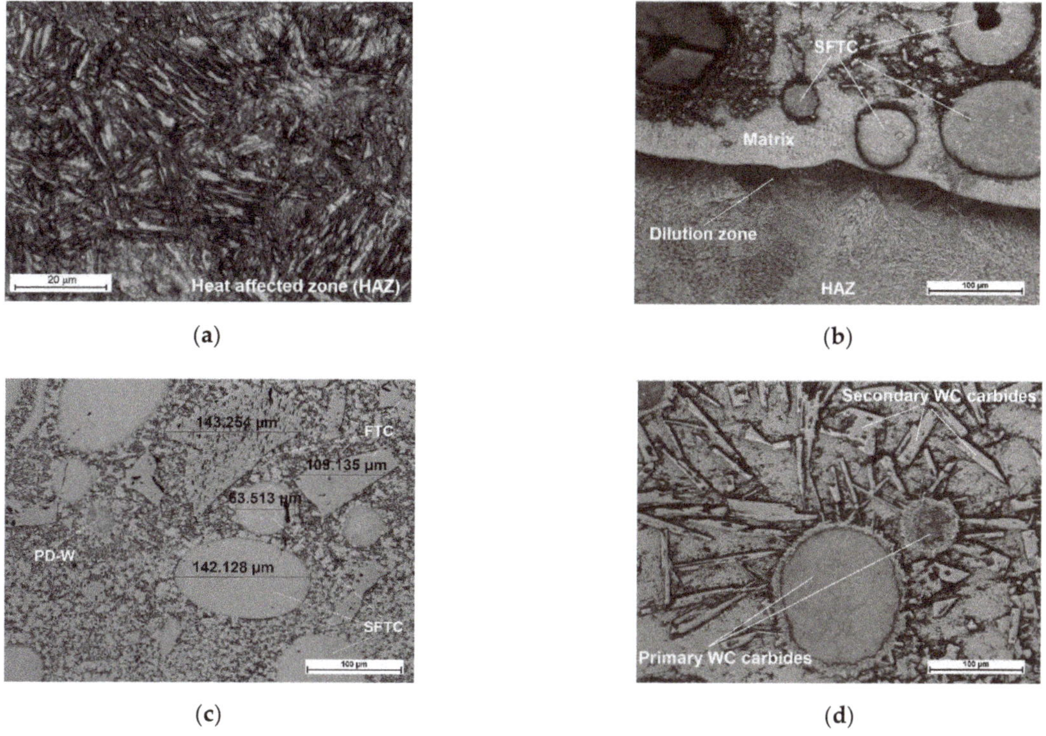

Figure 6. Microstructure of the surface layer (PPTAW metal deposition method, Ni3+WC-W$_2$C+PD-W composite powder) deposited on structural low-alloy steel AISI 4715: (**a**) heat-affected zone (HAZ); (**b**) dilution zone; (**c**) size and distribution of the hard phase in the middle of the cladding; (**d**) distribution of the reinforcing phase near the padding weld.

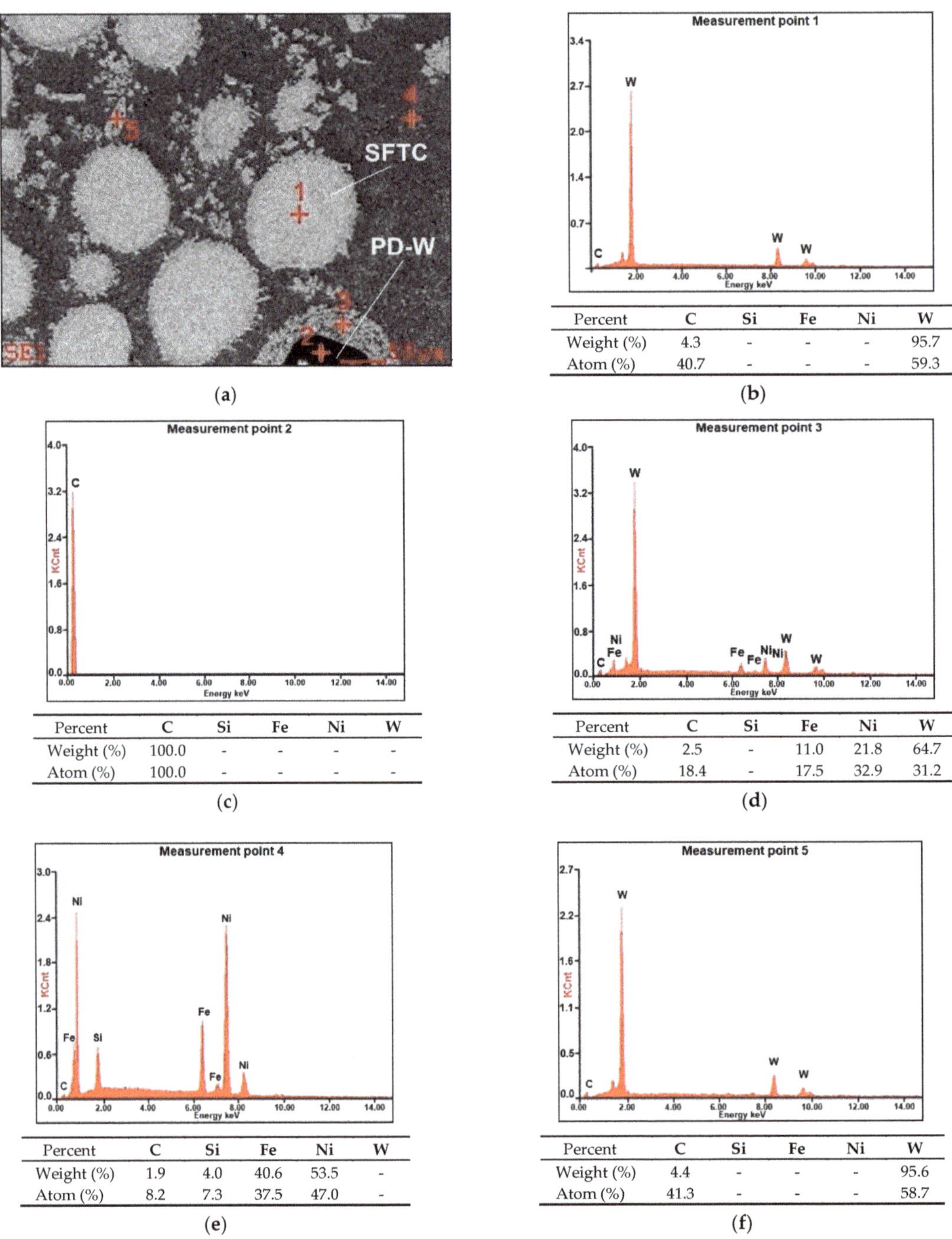

(a)

(b)

Percent	C	Si	Fe	Ni	W
Weight (%)	4.3	-	-	-	95.7
Atom (%)	40.7	-	-	-	59.3

(c)

Percent	C	Si	Fe	Ni	W
Weight (%)	100.0	-	-	-	-
Atom (%)	100.0	-	-	-	-

(d)

Percent	C	Si	Fe	Ni	W
Weight (%)	2.5	-	11.0	21.8	64.7
Atom (%)	18.4	-	17.5	32.9	31.2

(e)

Percent	C	Si	Fe	Ni	W
Weight (%)	1.9	4.0	40.6	53.5	-
Atom (%)	8.2	7.3	37.5	47.0	-

(f)

Percent	C	Si	Fe	Ni	W
Weight (%)	4.4	-	-	-	95.6
Atom (%)	41.3	-	-	-	58.7

Figure 7. Microstructure and the results of the qualitative point analysis in the carbide–matrix area of the surface layer (PPTAW metal deposition method, Ni3+WC-W$_2$C+PD-W composite powder) deposited on structural low-alloy steel AISI 4715: (**a**) area subjected to analysis; (**b–f**)diagram of the energy of scattered X-radiation with energy lines present in the area of components (chemical elements (ceramic particle of WC, synthetic metal–diamond composite and metal matrix)) subjected to analysis.

The microstructure of the layer was composed of spherical particles of primary tungsten carbides in the nickel alloy matrix. The carbides were present in the entire area of the deposited layer. The carbides were mostly uniformly distributed in the layer, without forming agglomerates typical of ceramic composite castings with the metal matrix [29]. The uniform distribution of carbides in the surface layer resulted from the proper blending of the powder mixture and the fact that the density of tungsten carbide was only twice that of the matrix. During the cladding process, the use of mixtures composed of ingredients characterised by significantly varying density may lead to the formation of agglomerates of the hard reinforcing phase, the reduction of their wettability by the matrix alloy and to the formation of welding imperfections [3,11]. The microstructure of the substrate in the heat-affected zone was composed of martensite or tempered martensite (Figure 6a). In turn, the microstructure of the matrix contained dendritic grains, on the boundary of which it was possible to observe eutectics (Figure 6c). The quantitative analysis performed in measurement points marked in Figure 7a in the carbide–matrix area revealed that the chemical composition in measurement points 1 and 5 corresponded to that of tungsten carbide. Measurement point 2 contained synthetic metal–diamond composite, the tungsten coating of which (measurement point 3), after partial melting, was responsible for the transfer of tungsten to the solution of the composite matrix (alloy Ni-C-Fe-Si). The presence of iron in the composite matrix resulted from the dilution of the base metal in the layer, amounting to approximately 4.5%. The cladding process was accompanied by the complete melting of the nickel alloy powder ($T_{melt.}$ of approximately 1300 °C) and the partial melting of the primary tungsten carbides ($T_{melt.}$ = 2870 °C), potentially leading to the formation of complex secondary carbides on the carbide–matrix boundary [11]. The secondary carbides were responsible for the diffusive bond of the primary carbides with the matrix. According to Bober et al. (2011) [3] and Poloczek et al. (2019) [10], the above-named mechanism of the bonding of carbides with the matrix should ensure their stable deposition. The partial melting of the primary carbides led to the partial saturation of the matrix with tungsten and carbon (Figure 7b).

The structural X-ray diffraction analysis aimed to identify phases present in the composite layer. An exemplary XRD pattern is presented in Figure 8, whereas the results of the X-ray qualitative phase analysis are presented in Table 4. The analysis revealed the presence of the γ-Ni solid solution and of the γ-Ni/Ni$_3$B eutectic phase, which was consistent with information provided in related reference publications [30,31]. In addition, the structure also contained hexagonal carbide having the W$_2$C structure as well as hexagonal carbide having the WC structure.

Table 4. Results of the X-ray qualitative phase analysis of the surface layer (PPTAW metal deposition method, Ni3+WC-W$_2$C+PD-W composite powder) deposited on structural low-alloy steel AISI 4715.

ICSD Card No	Phase Name	Chemical Formula	Crystalline Structure
98-007-7568	Tungsten carbide (2/1)	W$_2$C	Hexagonal (P 63/m m c)
98-026-0166	Tungsten carbide (1/1)	WC	Hexagonal (P $\bar{6}$ m 2)
98-026-0172	Nickel	Ni	Regular (F m $\bar{3}$ m)
98-002-4306	Nickel boride (3/1)	Ni$_3$B	Orthorhombic (P n m a)

3.4. Density and Porosity of the Deposited Layer

The density, porosity and absorbability of the composite was determined with the Archimedes method (according to ISO ASTM-D-792) on the basis of measurements concerning the mass of a specimen sampled from the surface layer. The apparent density was determined using a pycnometer. The pycnometer was used in ISO 1183–1:2004 standard. The results of related measurements and calculations are presented in Table 5. The surface layer porosity was analysed using a μCT microtomography. The images obtained as a result of the μCT analysis of the composite layer are presented in Figure 9.

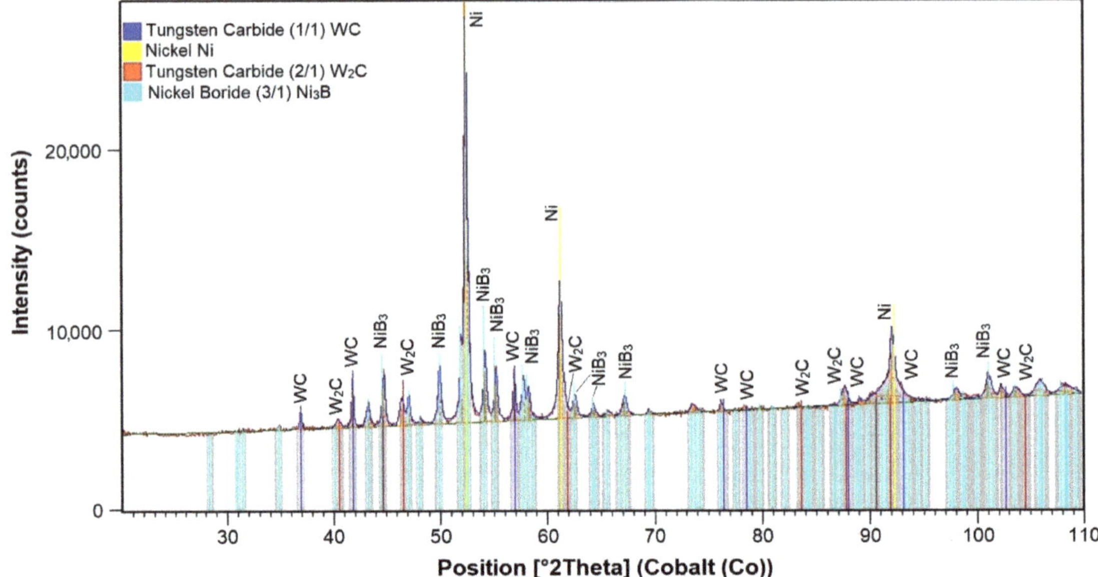

Figure 8. XRD pattern of the surface layer (PPTAW metal deposition method, Ni3+WC-W$_2$C+PD-W composite powder) deposited on structural low-alloy steel AISI 4715 with the standard lines of identified crystalline phases.

Table 5. Density-related measurement results and calculations concerning the porosity of the surface layer (PPTAW metal deposition method, Ni3+WC-W$_2$C+PD-W composite powder) deposited on structural low-alloy steel AISI 4715.

Physical Quantity	Average Value of the Measured Quantity
Density ρ (measured using the Archimedes method), g/cm^3	9.3425
Absorbability A, %	1.3856
Open porosity P$_o$, %	11.2421
Closed porosity P$_c$, %	0.0082
Apparent density ρ$_a$, g/cm^3	8.9729
Total porosity P$_c$, %	11.2503

The composite layer was characterised by a relatively high total porosity of more than 11% and a density of 9.34 g/cm^3. The total porosity of the tested composite layer was almost twice as high as the porosity of the layer made of cobalt alloy reinforced with TiC and synthetic metal–diamond composite particles [22]. According to Bober et al. [19] (2018), this may be due to the poor wettability of the hard reinforcement phase by the nickel alloy matrix, especially in the case of bigger grains. In respect of the fact that the composite layer is intended for work surfaces of drilling tools, such high porosity can appear problematic and requires the performance of in situ tests involving the use of an actual tricone bite.

3.5. Hardness Measurements Test Results

The results of hardness measurements concerning the external surface and the cross-section of the deposited layer are presented in Table 6 and Figure 10, respectively.

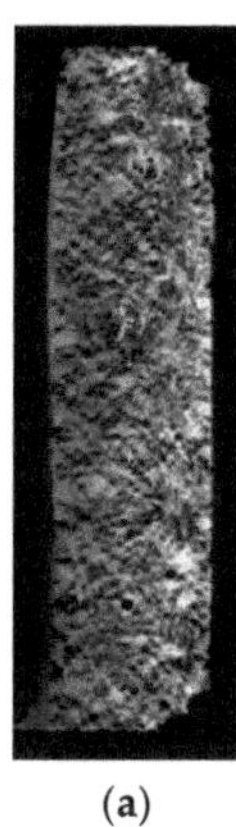 

(a)   (b)

Figure 9. Images obtained as a result of the μCT analysis of the surface layer (laser metal deposition method, Ni3+WC-W$_2$C+PD-W composite powder) deposited on structural low-alloy steel AISI 4715: (**a**) specimen porosity in cross-section surface; (**b**) specimen porosity in longitudinal section surface.

Table 6. Hardness (HV10) measurement results concerning the external surface of the surface layer (PPTAW metal deposition method, Ni3+WC-W$_2$C+PD-W composite powder) deposited on structural low-alloy steel AISI 4715 and the surface of the reference material (abrasion-resistant steel AR400).

Specimen Designation	Specimen Number	Measurement Point Number					Average Hardness of the Tested Samples	Average Hardness of the Tested Materials
		1	**2**	**3**	**4**	**5**		
Ni3+WC-W$_2$C+PD-W	N 01	634	746	883	545	733	708.2	702.5
	N 02	615	686	773	746	664	696.8	
AR400 Steel	S 01	430	421	420	429	424	424.8	424.1
	S 02	421	424	422	421	429	423.4	

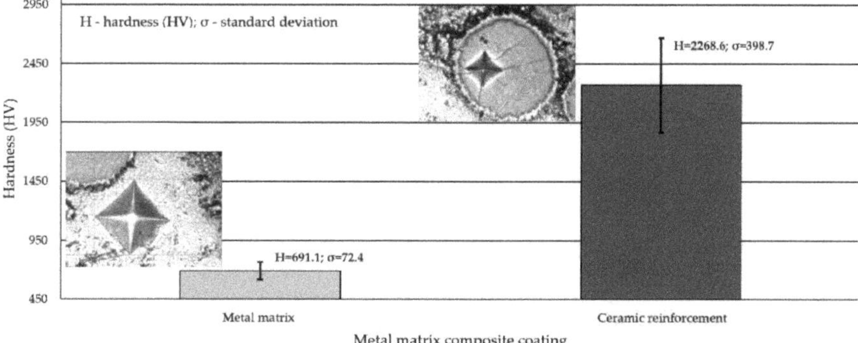

Figure 10. Hardness (HV0.5) measurement results concerning the cross-section of the surface layer (PPTAW metal deposition method, Ni3+WC-W$_2$C+PD-W composite powder) deposited on structural low-alloy steel AISI 4715.

Hardness measured on the external surface of the deposited layer was slightly above 700 HV10 (approximately 60 HRC) and was approximately 208 HV10 higher than the hardness of the reference material used in abrasive wear tests, i.e., abrasion-resistant steel AR400. The average microhardness of the matrix of the composite cladding (obtained using the powder plasma transferred arc cladding method) measured on the cross-section of the layer at 10 measurement points amounted to 691 HV0.5 and was nearly the same as the hardness measured on the surface. The profile of the microhardness of the cross-section was sharp, which resulted from the composite structure of the cladding. The average hardness of ceramic reinforcement amounted to approximately 2270 HV0.5. The high value of the standard deviation resulted from the large discrepancy of the results between the hardness particles of the fused tungsten carbide (about 1400 HV0.5) and spherical fused tungsten carbide (about 2600 HV0.5). As the particles of synthetic metal–diamond composite has ultrahigh hardness compared to the hardest carbides currently used in the industry, the hardness of this matrix reinforcement was difficult to quantify. It was found that the length of the Vickers indentation on the compact surface of the polished synthetic metal–diamond composite was too small to be measured, even if we increased the load force to 9.8 N with a dwelling time of 15 s. Yahiaoui et al. (2016) [32] describe similar research difficulties when assessing hardness of graded polycrystalline diamond compact cutters. It was revealed that the hardness of the matrix increased along with the distance between the measurement point and the fusion line (growing towards the surface of the layer). Because of the dilution of the weld metal in the layer and due to the different cladding crystallisation manner, the microstructure of the layer in the area adjacent to the HAZ differed from the microstructure of the subsurface layer.

3.6. Abrasive Wear Test Results

The tests concerning the metal–mineral abrasive wear resistance of the surface layer were performed in accordance with ASTM G65, Procedure A, and referred to the abrasive wear resistance of a plate made of popular abrasion-resistant steel AR400 (made by a Swedish manufacturer) (SSAB AB, Stockholm, Sweden). The tests enabled the determination of the relative abrasive wear resistance of the surface layer (Table 7). The test results were compared with those obtained in relation to the previously tested Co3+TiC+PD-W composite layer. The nature of the abrasive wear of the surface layer was assessed on the basis of visual tests as well as observations involving the use of a stereoscopic microscope (Figure 11) and a confocal microscope (Figure 12).

Table 7. Results of the metal–mineral abrasive wear resistance tests concerning the surface layer (PPTAW metal deposition method, Ni3+WC-W$_2$C+PD-W composite powder) deposited on structural low-alloy steel AISI 4715 in comparison with the abrasive wear resistance of abrasion-resistant steel AR400.

Specimen Designation	Spec. Number	Mass Before Test, g	Mass After Test, g	Mass Loss, g	Average Mass Loss, g	Clad Layer Density, g/cm^3	Average Volume Loss, mm^3	Relative [1] Abrasive Wear Resistance
Composite Coating								
Ni3+WC-W$_2$C +PD-W	N 01	173.2112	173.0283	0.1829	0.1894	9.3425	20.2730	11.05
	N 02	162.8753	162.6794	0.1959				
Reference Material								
AR400 Steel	S 01	123.9290	122.2067	1.7223	1.7429	7.7836	223.9195	1
	S 02	121.7386	119.9752	1.7634				

Note: [1] relative abrasive wear resistance to Abrasion-Resistant Steel type 400.

The abrasive wear process, performed using the medium pressure of the counterspecimen affecting the surface of the deposited layer, led to the partial crushing of abrasive particles. The relative abrasive wear resistance of the deposited layer was higher than that of abrasion-resistant steel AR400 (slightly over 11 times). The average loss of surface mass after the test amounted to 0.1894 g. The assessment of the surface after the test revealed the abrasive mechanism of wear. The hard phase, having the form of spheroidal particles of tungsten carbide (WC) and synthetic metal–diamond composite (PD-W), did not reveal sufficient embedment in the nickel alloy-based matrix and was relatively easily peeled by the abrasive medium. The dominant wear mechanism affecting the surface layer was micro-cutting, manifested by the presence of continuous micro-scratches along the traces of wear and, to a significantly lesser extent, by the micro-ridging of the surface. The most intensive wear was observed in the overlap area (Figure 12a). A similar mechanism of increased wear of the layer deposited with nickel alloy in the area of overlapping seams was confirmed by Katsich et al. (2006) [6]. The authors state that the wear behaviour of WC-W$_2$C reinforced Ni-based MMCs strongly depends on the formation of hard phase structures during manufacturing process. The surfacing parameters and the cladding bead trajectory are of particular importance. The abrasive grains uniformly affected the surface of the layer, where it was possible to notice small single craters after removing particles of the hard reinforcing phase. The micro-scratches were parallel in relation to the longer side of the specimen (which confirmed the thesis about the insufficient embedment of the particles of the carbide ceramics). Detailed examination involving the use of confocal microscopy revealed the presence of single crack (Figure 12a) and single crater (having a depth of up to 100 μm) inside the layer (Figure 12b,c). The abrasive wear process was significantly intensified by the effect of loose particles of the hard carbide phase torn out of the matrix and moving between the surfaces of the specimen and of the "rubber wheel". The aforesaid situation could significantly increase the abrasive wear of the surface layer. The freely rolling spheroidal tungsten carbides (WC) and synthetic metal–diamond composite particles were primarily responsible for the plastic deformation of matrix fragments, manifested by the presence of characteristic micro-ridges. The aforementioned observation was confirmed by results obtained by Cheng et al. (2013) [33] who, in their work, mentioned such a possibility. The thesis that the partial surface melting of the primary tungsten carbides could translate into a sufficiently high increase in the diffusive force of the bond between the carbides and the matrix was not confirmed. In relation to data contained in previous publications [34] and individual research [12] concerning the abrasive wear resistance of surface layers made of nickel alloys reinforced only with the hard WC layer (SFTC), it was possible to notice the highly favourable abrasion-resistant effect of synthetic metal–diamond composite particles (PD-W), constituting approximately 20 wt% of the entire matrix reinforcement. The comparison of the abrasive wear resistance of the layer made using the Ni3+WC-W$_2$C+PD-W composite powder with that of the layer made using the Co3+TiC+PD-W composite powder revealed that the abrasive wear resistance of the cobalt-based layer was more than 14 times higher (Figure 13).

The obtained test results concerning the abrasive wear resistance of the composite layer should be confirmed under actual conditions affecting the operation of drilling tools, including the ground structure, texture, the presence of stresses as well as hydrogeological conditions and humidity.

(a)

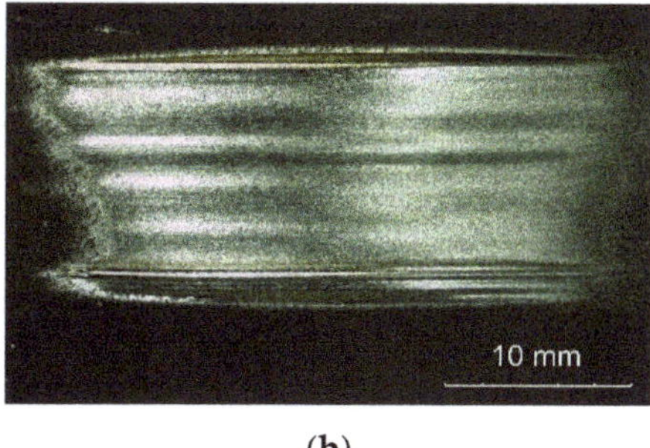

(b)

Figure 11. Surface of the composite layer and the surface of abrasion-resistant steel AR400 after the metal–mineral abrasive wear resistance test ASTM G64: (**a**) magnified area of layer abrasion; (**b**) magnified area of steel abrasion after the abrasive wear resistance test.

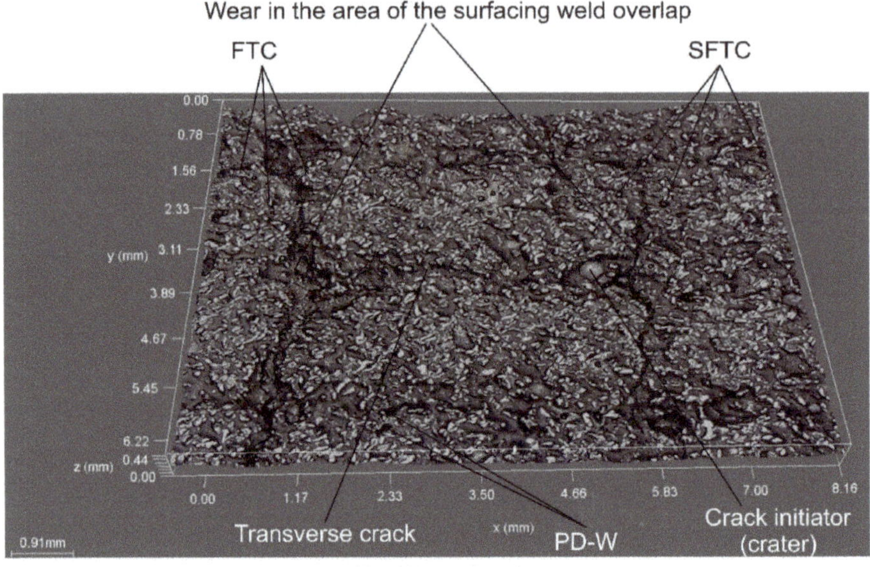

(**a**)

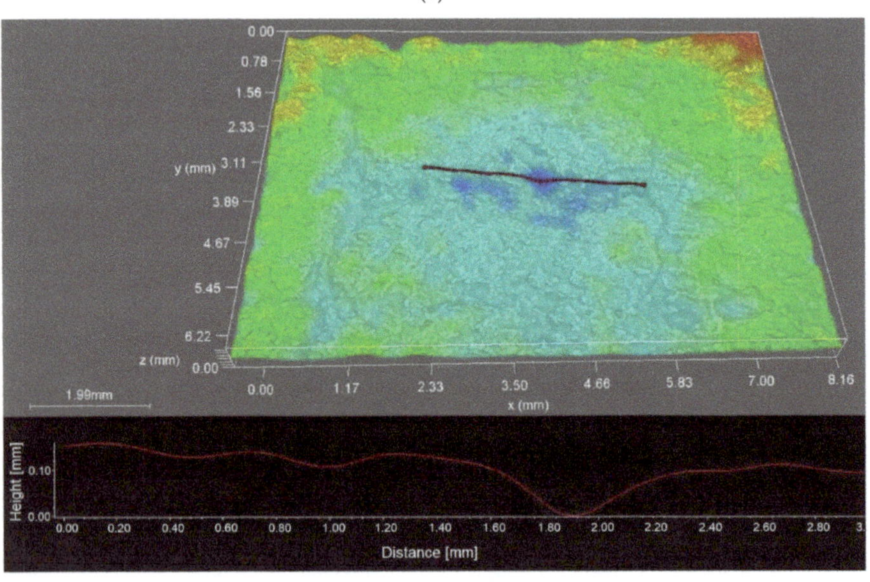

(**b**)

Figure 12. *Cont.*

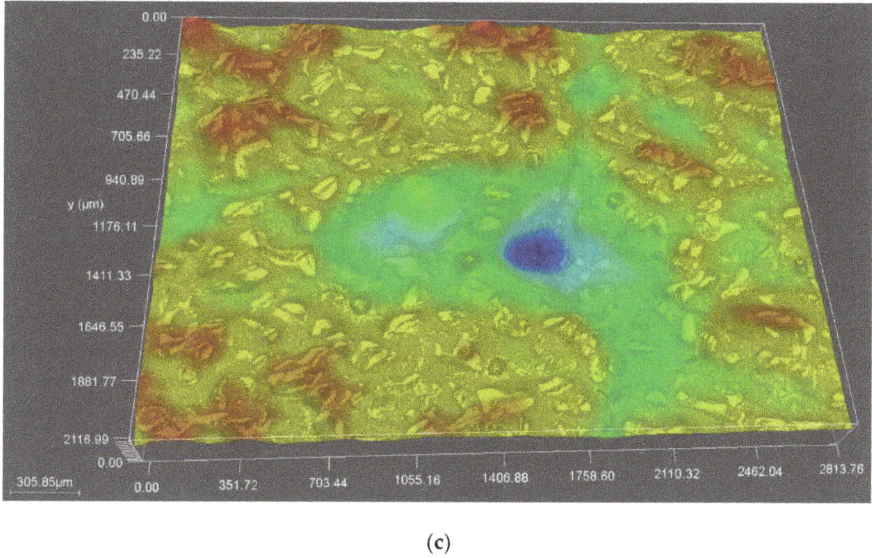

(c)

Figure 12. Surface of the composite layer after the metal–mineral abrasive wear resistance test observed using the confocal microscope: (**a,c**) main view of specimen wear; (**b**) measurement of defects (single craters).

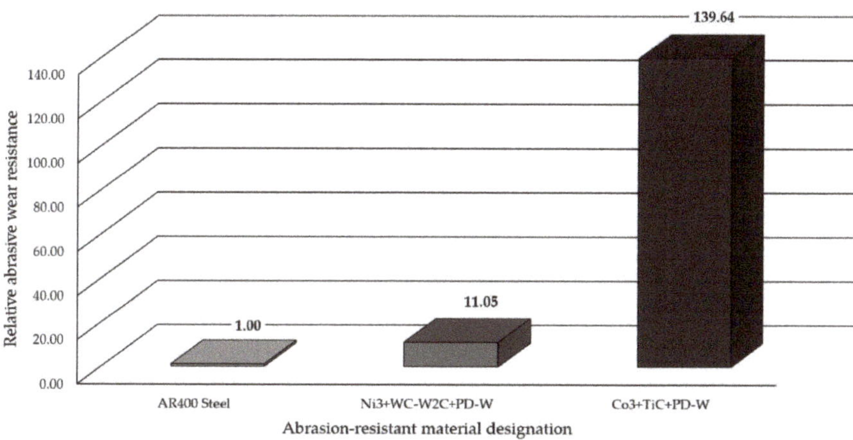

Figure 13. Relative metal–mineral abrasive wear resistance (ASTM 65-00, Procedure A) of the abrasive wear resistant deposited layers containing innovative cobalt and nickel alloys compared with the abrasive wear resistance of abrasion-resistant steel AR400.

3.7. Impact Resistance Test Results

The condition of the deposited layer surface during the individual stages of tests concerning resistance to moderate impact loads is presented in Figure 14.

The deposited layer made using the Ni3+WC-W$_2$C+PD-W composite powder was characterised by highly favourable resistance to moderate dynamic impact loads. After a cycle of 20 (ram) strokes affecting the surface of the layer with a potential energy of 200 J, no damage in the form of visible cracks or chips of the weld deposit was observed. The only visible deformation was the plastic distortion of the deposited layer, particularly near the edges of the specimen.

Figure 14. Condition of the surface of the composite layer (PPTAW metal deposition method, Ni3+WC-W$_2$C+PD-W composite powder) deposited on structural low-alloy steel AISI 4715-after impact resistance tests.

4. Conclusions

The research-related tests aimed to assess the metallographic structure as well as to identify the metal–mineral abrasive wear resistance and the impact resistance of the innovative composite surface layer obtained using the Powder Plasma Transferred Arc Welding (PPTAW) metal deposition method and the Ni3+WC-W$_2$C+PD-W powder. The analysis of the above-presented test results justified the formulation of the following conclusions:

1. The obtained layer was characterised by the classical composite structure and the uniform distribution of reinforcement composed of primary tungsten carbides (WC-W$_2$C) and the particles of synthetic metal–diamond sinter in the matrix composed of the γ-Ni solid solution and the γ-Ni/Ni$_3$B eutectic phase.

2. The applied powder plasma transferred arc (PPTAW) metal deposition method favours the maintaining of the structural and thermal stability of the particles of the ceramic reinforcement of the matrix having the form of tungsten-coated synthetic metal–diamond composite (PD-W). The partial surface melting of the primary spherical tungsten carbides (WC) did not significantly increase the force of the diffusive bond between the hard phase and the matrix. During the abrasion test, the spherical particles of the carbide reinforcement (WC) underwent peeling, likely because of the insufficient wetting of the particle surface with the metal of the matrix.

3. The absolute porosity of the composite layer slightly exceeded 11%, whereas its specific density amounted to 9.34 g/cm^3. The average hardness of the composite matrix amounted to 691 HV0.5, whereas tungsten carbides were characterised by a hardness of approximately 2268 HV0.5.

4. The relative metal–mineral abrasive wear resistance of the deposited composite layer obtained using the Ni3+WC-W$_2$C+PD-W powder was more than 11 times higher than that of abrasion-resistant steel AR400 and more than 14 times lower than the abrasive wear resistance of the layer obtained using the Co3+TiC+PD-W powder [22].

5. The very high resistance of the composite layer to moderate dynamic impact loads appears very promising as regards to its application as the preventive protection of work surfaces of drilling tools used in the extractive industry.

The third part of the article will contain comparative test results concerning the brittle fracture resistance of iron, nickel and cobalt-based composite layers.

Funding: The research was founded by Silesian University of Technology Rector's habilitation grant 10/050/RGH_20/1006.

Data Availability Statement: The data are not publicly available due to initiation of a patent procedure No. P435997.

Conflicts of Interest: The author declares no conflict of interest.

References

1. Yildiz, T.; Kaya Gür, A. Microstructural characteristic of N2 shielding gas in coating FeCrC composite to the surface of AISI 1030 steel with PTA method. *Arch. Metall. Mater.* **2011**, *56*, 723–729. [CrossRef]
2. Brezinová, J.; Viňáš, J.; Guzanová, A.; Živčák, J.; Brezina, J.; Sailer, H.; Vojtko, M.; Džupon, M.; Volkov, A.; Kolařík, L.; et al. Selected Properties of Hardfacing Layers Created by PTA Technology. *Metals* **2021**, *11*, 134. [CrossRef]
3. Bober, M. Composite coatings deposited by the plasma transferred arc—Characterization and coating formation. *Weld. Technol. Rev.* **2011**, *83*, 43–47.
4. Cherepova, T.; Dmitrieva, G.; Tisov, G.; Dukhota, O.; Kindrachuk, M. Research on the properties of Co-TiC and Ni-TiC HIP-sintered alloys. *Acta Mech. Autom.* **2019**, *13*, 57–67. [CrossRef]
5. Toth, L.E. *Transition Metal Carbides and Nitrides*; Academic Press: New York, NY, USA; London, UK, 1971.
6. Katsich, C.; Badisch, E. Effect of carbide degradation in a Ni-based hardfacing under abrasive and combined impact/abrasive conditions. *Surf. Coat. Technol.* **2011**, *206*, 1062–1068. [CrossRef]
7. Kulu, P.; Surzhenkov, A.; Tarbe, R.; Saarna, M.; Tarraste, M.; Viljus, M. Hardfacings for Extreme Wear Applications. In Proceedings of the XXVIII International Conference on Surface Modification Technologies, Tampere, Finland, 16–18 June 2014.
8. Bober, M.; Senkara, J. Study of the structure of composite coatings Ni-WC deposited by plasma transferred arc. *Weld. Technol. Rev.* **2016**, *88*, 67–70.
9. Oyama, S.T. *The Chemistry of Transition Metal Carbides and Nitrides*; Balckie Academic & Professional, Chapman & Hall: London, UK, 1996.
10. Poloczek, T.; Czupryński, A.; Żuk, M.; Chruściel, M. Structure and tribological properties of wear-resistant layers produced in process of plasma powder surfacing. *Weld. Technol. Rev.* **2019**, *91*, 35–41. [CrossRef]
11. Bober, M.; Senkara, J. Formation of nickel-based weld overlays strengthened with transition metal carbides. *Inst. Weld. Bull.* **2010**, *54*, 103–107.
12. Czupryński, A. Comparison of Properties of Hardfaced Layers Made by a Metal-Core-Covered Tubular Electrode with a Special Chemical Composition. *Materials* **2020**, *13*, 5445. [CrossRef]
13. Sobolev, A.; Mirzoev, A. Structure and stability of (Cr, Fe)7C3 ternary carbides in solid and liquid state. *J. Alloys Compd.* **2019**, *804*, 566–572. [CrossRef]
14. Chukwuma, C.; Onuoha, X.; Chenxin, J.; Zoheir, N.F.; Georges, J.K.; Kevin, P.P. The effects of TiC grain size and steel binder content on the reciprocating wear behaviour of TiC-316L stainless steel cermets wear. *Wear* **2016**, *350*, 116–129.
15. Sakamoto, T.; Kurishita, H.; Matsuo, S.; Arakawa, H.; Takahashi, S.; Tsuchida, M.; Kobayashi, S.; Nakai, K.; Terasawa, M.; Yamasaki, T.; et al. Development of nanostructured SUS316L-2%TiC with superior tensile properties. *J. Nucl. Mater.* **2015**, *466*, 468–476. [CrossRef]
16. Hung, F.Y.; Yan, Z.Y.; Chen, L.H.; Lui, T.S. Microstructural characteristics of PTA-overlayed NbC on pure Ti. *Surf. Coat. Technol.* **2006**, *200*, 6881–6887. [CrossRef]
17. Smirnov, A.; Kozlov, E.; Radchenko, M.; Knyaz'kov, K.; Knyaz'kov, V. State of Ni–Cr–B–Si–Fe/WC coatings after plasma powder surfacing with nanopowder modifier. *Steel Transl.* **2016**, *46*, 251–255. [CrossRef]
18. Ozel, S.; Kurt, B.; Somunkiran, I.; Orhan, N. Microstructural characteristic of NiTi coating on stainless Steel by plasma transferred arc process. *Surf. Coat. Technol.* **2008**, *202*, 3633–3637. [CrossRef]
19. Bober, M. Formation and structure of the composite coatings Ni-NbC deposited by plasma transferred arc. *Weld. Technol. Rev.* **2018**, *90*, 65–69.
20. Liu, Y.F.; Han, J.M.; Li, R.H.; Li, W.J.; Xu, X.Y.; Wang, J.H.; Yang, S.Z. Microstructure and dry-sliding wear resistance of PTA cald (Cr, Fe)7C3/γ-Fe ceramal composite coating. *Appl. Surf. Sci.* **2006**, *252*, 7539–7544. [CrossRef]
21. Wang, X. The metallurgical behavior of B4C in the iron-based surfacing alloy during PTA powder surfacing. *Appl. Surf. Sci.* **2005**, *252*, 2021–2028. [CrossRef]
22. Czupryński, A. Microstructure and Abrasive Wear Resistance of Metal Matrix Composite Coatings Deposited on Steel Grade AISI 4715 by Powder Plasma Transferred Arc Welding Part 1. Mechanical and Structural Properties of a Cobalt-Based Alloy Surface Layer Reinforced with Particles of Titanium Carbide and Synthetic Metal–Diamond Composite. *Materials* **2021**, *14*, 2382. [CrossRef]
23. Aoh, J.N.; Jeng, Y.R.; Chu, E.L.; Wu, L.T. On the wear behavior of surface clad layers under high temperature. *Wear* **1999**, *225*, 1114–1122. [CrossRef]
24. Wu, J.B.C.; Redman, J.E. Hardfacing with Cobalt and Nickel Alloys. *Weld. J.* **1994**, *73*, 63–68.
25. Durejko, T.; Łazińska, M.; Dworecka-Wójcik, J.; Lipiński, S.; Varin, R.A.; Czujko, T. The Tribaloy T-800 Coatings Deposited by Laser Engineered Net Shaping (LENSTM). *Materials* **2019**, *12*, 1366. [CrossRef]
26. Deuis, R.L.; Subramanian, C.; Yellup, J.M. Abrasive wear of composite coatings in a saline sand slurry environment. *Wear* **1997**, *203*, 119–128. [CrossRef]

27. Mele, C.; Lionetto, F.; Bozzini, B. An Erosion-Corrosion Investigation of Coated Steel for Applications in the Oil and Gas Field, Based on Bipolar Electrochemistry. *Coatings* **2020**, *10*, 92. [CrossRef]

28. EN 14700. *Welding Consumables. Welding Consumables for Hard-Facing*; CEN: Brussels, Belgium, 2014.

29. Zygmuntowicz, J.; Miazga, A.; Konopka, K.; Jędrysiak, K.; Kaszuwara, W. Alumina matrix ceramic-nickel composites formed by centrifugal slip casting. *Process. Appl. Ceram.* **2015**, *9*, 199–202. [CrossRef]

30. Yao, S.H. Tribological behaviour of NiCrBSi–WC(Co) coatings. *Mater. Res. Innov.* **2014**, *18*, 332–337. [CrossRef]

31. Mele, C.; Bozzini, B. Localised corrosion processes of austenitic stainless steel bipolar plates for polymer electrolyte membrane fuel cells. *J. Power Sources* **2010**, *195*, 3590–3596. [CrossRef]

32. Yahiaoui, M.; Paris, J.-Y.; Delbé, K.; Denape, J.; Gerbaud, L.; Colin, C.; Ther, O.; Dourfaye, A. Quality and wear behavior of graded polycrystalline diamond compact cutters. *Int. J. Refract. Met. Hard Mater.* **2016**, *56*, 87–95. [CrossRef]

33. Cheng, H.; Yi, J.; Fang, Z.; Dai, S.; Zhao, X. Tribology Property of Laser Cladding Crack Free Ni/WC Composite Coating. *Mater. Trans.* **2013**, *54*, 50–55. [CrossRef]

34. Wu, P.; Du, H.M.; Chen, X.L.; Li, Z.Q.; Bai, H.L.; Jiang, E.Y. Influence of WC particle behavior on the wear resistance properties of Ni-WC composite coatings. *Wear* **2004**, *257*, 142–147. [CrossRef]

Article

Matrix Composite Coatings Deposited on AISI 4715 Steel by Powder Plasma-Transferred Arc Welding. Part 3. Comparison of the Brittle Fracture Resistance of Wear-Resistant Composite Layers Surfaced Using the PPTAW Method

Artur Czupryński * and Marcin Żuk

Welding Department, Faculty of Mechanical Engineering, Silesian University of Technology, Konarskiego 18A, 44-100 Gliwice, Poland; rmt5@polsl.pl
* Correspondence: artur.czuprynski@polsl.pl; Tel.: +48-322371443

Abstract: This article is the last of a series of publications included in the MDPI special edition entitled *"Innovative Technologies and Materials for the Production of Mechanical, Thermal and Corrosion Wear-Resistant Surface Layers and Coatings"*. Powder plasma-transferred arc welding (PPTAW) was used to surface metal matrix composite (MMC) layers using a mixture of cobalt (Co3) and nickel (Ni3) alloy powders. These powders contained different proportions and types of hard reinforcing phases in the form of ceramic carbides (TiC and WC-W$_2$C), titanium diboride (TiB$_2$), and of tungsten-coated synthetic polycrystalline diamond (PD-W). The resistance of the composite layers to cracking under the influence of dynamic loading was determined using Charpy hammer impact tests. The results showed that the various interactions between the ceramic particles and the metal matrix significantly affected the formation process and porosity of the composite surfacing welds on the AISI 4715 low-alloy structural steel substrate. They also affected the distribution and proportion of reinforcing-phase particles in the matrix. The size, shape, and type of the ceramic reinforcement particles and the surfacing weld density significantly impacted the brittleness of the padded MMC layer. The fracture toughness increased upon decreasing the particle size of the hard reinforcing phase in the nickel alloy matrix and upon increasing the composite density. The calculated mean critical stress intensity factor K_{Ic} of the steel samples with deposited layers of cobalt alloy reinforced with TiC and PD-W particles was 4.3 MPa·m$^{\frac{1}{2}}$ higher than that of the nickel alloy reinforced with TiC and WC-W$_2$C particles.

Keywords: PPTAW; cladding; deposition; impact strength; brittle fracture strength; tungsten carbide; titanium carbide; titanium diboride; synthetic polycrystalline diamond

Citation: Czupryński, A.; Żuk, M. Matrix Composite Coatings Deposited on AISI 4715 Steel by Powder Plasma-Transferred Arc Welding. Part 3. Comparison of the Brittle Fracture Resistance of Wear-Resistant Composite Layers Surfaced Using the PPTAW Method. *Materials* **2021**, *14*, 6066. https://doi.org/10.3390/ma14206066

Academic Editor: Daolun Chen

Received: 15 September 2021
Accepted: 12 October 2021
Published: 14 October 2021

Publisher's Note: MDPI stays neutral with regard to jurisdictional claims in published maps and institutional affiliations.

1. Introduction

Composites of multi-component ceramics have been the subject of research and wider applications in materials engineering [1,2]. Among strengthening ceramics, WC, TiC, and TiB$_2$ are some of the most popular materials because of their large Young's modulus, good thermal stability, low density, and chemical compatibility with metals such as iron, nickel, cobalt, and titanium [3]. To improve the thermal conductivity and abrasive wear and erosion resistance of the working surface of drilling tools used in mining, Sue et al. [4] developed an innovative material suitable for surfacing with powder plasma-transferred arc welding (PPTAW) or laser metal deposition (LMD) techniques. The composite consisted of spherical fused tungsten carbide (SFTC) particles, with WC-W$_2$C evenly distributed in the Ni-Si-B matrix. Compared with conventional hard layers padded with composite powder containing WC-W$_2$C tungsten carbide particles and Ni-Cr-Si-B-Fe alloy, the developed material was characterized by a significantly higher thermal resistance and lower wear by abrasion and erosion. This improved the durability and increased the economic efficiency of the steel bits and polycrystalline diamond compact (PDC).

In recent years, research has also been conducted on composites using TiB_2-TiC as a hard reinforcing phase, which are promising materials for use in the surfacing of parts that are resistant to abrasive wear and high temperatures [5,6]. According to Baoshuai et al. [7], the hardness and wear resistance of the laser plane was significantly improved due to the presence of TiB_2 particles in an iron matrix. Similar observations were made by Tijo et al. (2018) [8], who investigated the mechanical properties of a composite coating on a Ti-6Al-4V alloy matrix containing TiC-TiB_2 particles, which was obtained via in situ plating using the TIG method. The coating showed high strength and high abrasion resistance during pin-on-disc tests. A prospective ceramic metal matrix reinforcement in overlaid composite layers is polycrystalline synthetic diamond (PD). The previously presented results of research on the tribological properties of composite layers welded with the PPTAW method showed that compared with the wear-resistant AR 400 steel, the addition of polycrystalline diamond particles to the nickel matrix increased the wear resistance of the metal-mineral type by more than 11 times [9]. In the case of the cobalt warp, the increase was almost 140 times [10]. The properties of ceramic materials used to reinforce metal matrixes in surfaced composite layers are presented in Table 1.

Table 1. Properties of ceramic materials used to reinforce the metal matrix in surfaced composite layers.

Composite Reinforcement Properties	TiC	WC	TiB$_2$	PD [1]
Density, g/cm^3	4.93	15.8	4.52	3.51
Decomposition temperature, °C	3065	2870	2980	1450
Thermal expansion coefficient at 298 K, K^{-1}	7.74×10^{-6}	6×10^{-6}	3.5×10^{-6}	1.3×10^{-6}
Thermal conductivity at 298 K, W/(m·K)	50	90–110	96	550
Young's modulus, GPa	410–510	550	510–570	1050
Compressive Strength, GPa	1.8	2–2.5	1.8	4.2
Flexural Strength, MPa	240–390	350	400–450	850
Fracture toughness K_{Ic}, MPa·m$^{\frac{1}{2}}$	4–5.5	4.5	4–6	8.5
Hardness, GPa	28–35	30	25–35	75
Other features	High chemical resistance, very high abrasion resistance, relatively good resistance to oxidation	High chemical resistance, very high abrasion resistance, weaker resistance to oxidation	Very high abrasion resistance, good chemical resistance, very good resistance to liquid metals, resistance to oxidation (in the air) to 1000 °C	Low friction coefficient, high strength, very good chemical resistance, biocompatibility, poor resistance to oxidation around 780 °C
Reference	[11]	[12]	[7]	[10]

[1] Polycrystalline synthetic diamond made by high pressure and high temperature (HPHT) technology.

The mechanical and tribological properties of the composite surfacing layers are particularly important for preventive protection and the regeneration of contact surfaces of the inserts of drilling tools. In recent years, drilling in the oil and gas mining sector has predominantly used cutter drill bits with polycrystalline synthetic diamond compact (PDC) blades. Among diamond tools, they reduce the working load of the drilling machine. In terms of construction, they are characterized by a steel or matrix body and segmented, ribbed, or winged arrangement of their blades (Figure 1).

The steel body of a drill bit is made of one piece of heat-treated alloy steel and reinforced with inserts on the outer peripheral surface made of cemented carbide or polycrystalline synthetic diamond. The matrix body of the bit is manufactured by infiltrating tungsten carbide particles, macrocrystalline WC, or fused-and-crushed WC-W_2C, or a mixture of them, with a Cu-Ni-Zn-Mn alloy [4]. The steel body is more resistant to shock loading than the matrix body. The main disadvantage of the steel body is that it is quite susceptible to abrasive wear and erosion due to contact with the rock material to be mined. To protect the steel core of the rock cutting core from damage caused by tribological wear during operation, a coating is usually sprayed onto its surface or a layer more resistant to abrasion and erosion is deposited on its surface. The overarching goal of producing a durable steel body is to provide a hardfacing filler that has abrasion, erosion, and impact resistance equal to or better than that of the matrix body.

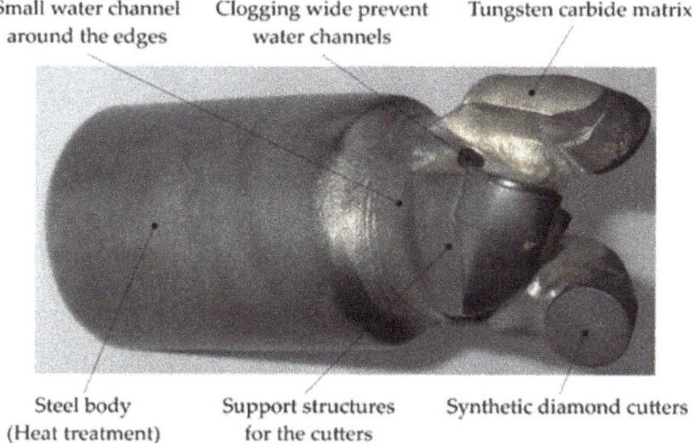

Figure 1. PDC small-diameter bit used in underground drilling.

So far, the commonly used methods to protect the surface of the PDC steel body against wear include flame surfacing with powders based on the Ni-Cr-Si-B-Fe alloy matrix with the addition of irregular particles of angular fused-and-crushed (FTC) WC-W$_2$C with a diameter of 10–160 µm. Large particles of spherical fused tungsten carbide (SFTC) WC-W$_2$C with a diameter of 750–1200 µm have also been used. According to Badish et al. [13], these layers ensure the correct operation of the PDC drill for over 10 years. Examples of the chemical compositions of filler materials used for the hardfacing of tools working in the extraction, mining, and cement industries are presented in Table 2.

Table 2. Examples of chemical compositions of filler materials used to protect the working surface of drilling tools.

Hard-Facing Alloy Composition, wt.%		Carbide-to-Matrix Ratio	Reference
Carbide Reinforcement	Matrix		
WC-CoCr (FTC)	Ni–17Cr–4Fe–4Si–3.5B–1C	70/30	[14]
WC-W$_2$C (FTC)	Ni–0.2C–3.5–Si–3Fe–2.3B	60/40	[15]
WC-W$_2$C (FTC)	Ni–7.5Cr–3Fe–3.5Si–1.5B-0.3C	55/45	[4]
WC-W$_2$C (SFTC)	Ni–0.1C–3Si–3B–2Fe	60/40	[16]
WC-W$_2$C (SFTC)	Ni–9.5Cr–3Fe–3Si–1.6B–0.6C	68/32	[4]
WC-W$_2$C (FTC)	Co–27Mo–16.9Cr–3Si–0.5Fe–0.8Ni	35/65	[17]
WC-W$_2$C (SFTC)	Ni– < 3Si– < 1B	40–90/60–10	[4]
TiB$_2$	Ti–6Al–4V	50/50	[18]

In recent years, the geological conditions of the exploitation of mineral, oil, and natural gas deposits have become more difficult. The filler materials and surfacing technologies currently used in most drilling applications do not sufficiently protect the steel body of the PDC drill bit; therefore, research was undertaken to develop a next-generation composite (MMC), dedicated to plasma surfacing that effectively protects the working surface of the PDC steel drill body against abrasive wear, erosion, high temperatures, and impact loading [9,10].

Powders based on cobalt and nickel belonging to Co3 and Ni3 alloys (in accordance with EN 147000) [19] with a hard strengthening phase containing, among others, WC, TiC, TiB_2 particles, and of tungsten-coated synthetic polycrystalline diamond (PD-W). Powders based on these alloys often contain non-metallic elements such as B and Si. Higher contents of silicon and boron in the cladding powder resulted in more efficient melting of matrix components and improved the wetting of the reinforcing phase particles in composites. At the same time, when Si >1%, the plastic properties of the padded layer decrease, and with higher boron contents, the tendency to form austenite grains increases. The content of silicon and boron in the produced PPTAW hardfacing powders were optimized individually for each of the alloys depending on the type, shape, and size of particles included in the reinforcing components of the matrix metal. At the stage of preliminary preparation of the filler material, the content of silicon and boron in the metallic powders was changed from 0 to 2.5% in 0.5% incremental steps. Plasma arc melting of small volumes of composite powders allowed to select the filler material with the best weldability properties. In general, the silicon and boron contents were slightly lower than that of commercial hardfacing alloys. The use of spherical ceramic particles in the form of fused tungsten carbide WC-W_2C and polycrystalline synthetic diamond determined the stable feeding of the filler material to the weld pool and the continuity of the surfacing process. Moreover, these particles in combination with a matrix of alloys of the Ni3 and Co3 groups had the effect of high hardness and stress reduction in the composite layer. Due to the larger surface of the carbide, there was a higher proportion of WC-W_2C and TiC particles with irregular shapes in the hard reinforcing phase of the composite. This required a slightly higher content of fluxing components in the powder mixture, which enabled the production of powders for the plasma surfacing of steel body bits and tools with a "matrix-type armor" for drilling applications. In materials science, it is important to properly qualify materials for specific engineering applications. Usually, this is done based on the results of mechanical tests, in which material parameters such as Young's modulus, proportionality limit, yield point, or temporary strength are determined; however, these are not the only mechanical properties used in engineering practice. The parameter defined within the framework of linear-elastic fracture mechanics is also taken into account, i.e., the stress intensity factor, K_I, which is used to determine the fracture toughness of a material. It is generally accepted that for composite coatings, the critical value of the stress intensity factor determines its mechanical resistance to cracking under an impact load; thus, the evaluation of the K_{Ic} parameter value is very important for ceramic-reinforced composite coatings produced in via in situ surfacing processes, where the percentage content of the composition varies depending on technological conditions. Determination of the critical values of fracture toughness K_{Ic}, crack tip opening displacement (CTOD), δ_{Ic}, or J_{Ic} integrals according to the *I* crack growth model in accordance with the standards [20,21] is difficult and burdensome and requires appropriate laboratory equipment. These difficulties result from the methods used to make and prepare test samples and the need to maintain appropriate test conditions and procedures [22]. For these reasons, the relationship between the results of standard tests of the mechanical properties of the material and the critical values of the fracture toughness parameters is constantly sought. Many research results on the fracture toughness of composite coatings are based on the indentation method and the formula proposed by Evans and Wilshaw [23]. Several researchers have used this method to evaluate the fracture toughness of a brittle composite coating developed using various methods and compared the results with the fracture toughness measured by standard methods [24,25]. These are most often the dependencies connecting the results of the work of impact breaking of samples with a Charpy V–KV sharp notch (expressed in Joule units) with the critical values K_{Ic} and δ_{Ic}. There are also papers that discuss other methods for determining these values, e.g., determining the K_{Ic} value for plasma-sprayed coatings based on measuring the length of cracks around the impression formed during hardness measurements using the Vickers method [26]. Such alternative approaches for determining the fracture toughness of a material are generally applicable to industrial conditions. This article presents the analytical evaluation of the resistance to brittle fracture under a dynamic load of abrasion-resistant metal matrix composite layers for protection against wear of working surfaces of drilling tools used in the oil and natural gas mining sector. There is no information related to tests concerning the brittle fracture resistance of the MMC layers reinforced with metal-diamond composite, obtained using the powder plasma transferred arc welding or the laser metal deposition methods. The analysis of related reference publications and the results of personal research led to the conclusion that it is possible to obtain a ceramic reinforcement-metal matrix composite surface layer characterized by microstructure and brittle fracture resistance similar to the classic metal surfacing weld. The original article achievement is, by selecting the right research methodology and analytical tools, obtaining information about the shock load resistance of innovative composite layers.

2. Materials and Methods

For PPTAW, four different powders with a proprietary chemical composition were used—cobalt Co3 and nickel Ni3 matrix composites (chemical compositions in accordance with EN 147,000 [19]) containing super hard phases in the form of ceramic particles with crushed sharp-edged TiC, spherical fused tungsten carbide particles (SFTC) $WC-W_2C$, fused tungsten carbide (FTC) $WC-W_2C$, fine particles of titanium diboride TiB_2, and spherical particles of polycrystalline synthetic diamond with a tungsten coating PD-W (Harmony Industry Diamond, Zhengzhou, China). The chemical composition of the hardfacing powders is given in Table 3.

Table 3. Chemical composition of powders.

Filler Material	Chemical Composition of the Matrix, wt. (%)										Ceramic Reinforcement of the Matrix, wt. (%)				
	C	Si	Mn	Cr	B	Mo	W	Fe	Ni	Co	TiB_2	TiC	SFTC	FTC	PD-W
C1	2.5–3	≤1	≤2	24–28	-	≤1	12–14	<5	≤3	Bal.	-	90	-	-	10
C2	≤0.05	<2.4	0.5	2.0	≤1.4	-	-	≤0.5	Bal.	-	-	60	30	10	-
C3	≤0.05	<2.4	0.5	2.0	≤1.4	-	-	≤0.5	Bal.	-	-	-	70	10	20
C4	≤0.05	<2.4	0.5	2.0	≤1.4	-	-	≤0.5	Bal.	-	10	-	90	-	-

Carbide-to-matrix ratio: 60/40 (acc. to %). Bulk density of the powder determined by pycnometry: C1 = 7.47 g/cm^3, C2 = 11.03 g/cm^3, C3 = 11.64 g/cm^3, C4 = 13.80 g/cm^3.

A SEM image of the micromorphology of the powder components used for PPTAW cladding is shown in Figure 2. The density of the powders was tested by the volumetric pycnometric method using an AccuPyc II 1340 density analyzer (Micromeritics Instrument Corporation, Norcross, GA, USA) with a measurement accuracy of 0.03%. The volume of the samples was determined as part of a previously marked measuring chamber that was not filled with gas (helium pressure 134,447.77 Pa). The obtained values of five measurements for each type of powder indicate the repeatability of its density. The powders were dried before use in an S 60/03 chamber dryer (LAC, Židlochovice, Czech Republic) in a 6-h cycle at 150 °C.

The particle sizes of the powders constituting the additional material for cladding were measured by laser diffraction using an Analysette 22 MicroTec plus laser particle size meter (Fritsch GmbH, Idar-Oberstein, Germany) equipped with two semiconductor laser sources: green (λ = 532 nm, 7 mW) and infrared (λ = 940 nm, 9 mW). Measurements were made using a wet dispersion unit with an ultrasonic exciter.

The surfacing tests were carried out on an automated welding station equipped with a Eutronic GAP 2501 DC power source, EP2 powder feeder, and E52 universal plasma torch (Castolin Eutectic, Gliwice, Poland), as shown in Figure 3. Single-pass padding was deposited on samples with dimensions of 75 × 10 × 8 mm, made of low-alloy AISI 4715 structural steel (Table 4). Surfacing was performed with constant technological parameters, which were the same for each type of powder (Table 5).

Figure 2. SEM image of the micromorphology of the powder components used for PPTAW cladding: (**a**) C1 filler material; (**b**) C2 filler material; (**c**) C3 filler material; (**d**) C4 filler material (Table 3).

Figure 3. Gap automated unit robotic coating system.

Table 4. Chemical composition of low-alloy structural steel AISI 4715 according to the manufacturer's data (TimkenSteel Ltd., Canton, OH, USA).

Chemical Composition, wt.%								
C	Mn	S	P	Si	Cr	Mo	Ni	Fe
0.12–0.18	0.65–0.95	≤0.015	≤0.015	0.15–0.35	0.40–0.70	0.45–0.60	0.65–1.00	Bal.

Notes: the average density of steel at a temperature of 25 °C, $\rho = 7.865$ g/cm^3.

Table 5. Cladding parameters of the plasma transfer arc welding (PTAW) metal deposition of the surface layers on AISI 4715 steel.

Process Parameters	Value of Parameter
Current, I (A)	80
Voltage, U (V)	25
Travel speed, S (mm/s)	2.7
Powder feed rate, q (g/min)	18
Heat input, E_u [1] (J/mm)	444

Notes: Argon 5.0 (99.999%) acc. ISO 14175—I1: 2009 [27] was used as the plasma gas (flow rate = 1.6 L/min), argon/hydrogen 5% H$_2$, Ar (welding mixture ISO 14175-R1-ArH-5) was used as shielding gas (flow rate = 12 L/min) and carrier gas (flow rate = 4 L/min), [1] calculated acc. to the formula: $E_\mathrm{u} = k \cdot (U \times I)/S$. The thermal efficiency coefficient for plasma-transferred arc $k = 0.6$ was used.

Fracture toughness tests of the parent material and steel samples with wear-resistant composite surfacing welds were carried out in accordance with ISO 148 [28] using a standard Charpy SUNPOC JB-300B impact hammer (Sunpoc, Guiyang City, China) with a hammer with a rounding radius in an impact point of 2 mm.

The apparent density, porosity, and water absorption of the composite layers were determined by Archimedes' method in accordance with the ASTM D792-00 standard [29]. An AS 220.R2 Plus analytical balance (Radwag, Radom, Poland) with a reading accuracy of ±0.1 mg was used to measure the mass. The analytical balance was equipped with a set KIT 85 for determining the density of solids using the hydrostatic weighing method.

The metallographic examination of the structures of the wear-resistant surfacing layers and the parent material was carried out in accordance with ISO 17639 [30]. Metallographic specimens were prepared in a standard manner and were collected perpendicularly and parallel to the cladding direction. The polished specimens were etched in a mixture of concentrated hydrochloric acid and concentrated nitric acid in a volume ratio of 3:1. The etching time was selected experimentally, individually for each of the layer materials. Observation and registration of microstructure images were performed with an Olympus GX 71 inverted metallographic microscope (Olympus Corporation, Tokyo, Japan). The surface topography of impact fractures was studied using a Zeiss Supra 25 system (Carl Zeiss AG, Oberkochen, Germany) using the secondary electron (SE) detector, with an accelerating voltage of 20 kV.

3. Results

3.1. Particle Size Distribution in Powders

The particle size of the powders of the additional material for surfacing was measured in distilled water. Five measurements were made for each of the tested powders. Each test powder was ultrasonically dispersed for 5 s before and during the test. The overall measurement range was 0.08–2000 µm. The obtained results are presented in the form of particle size distributions (Figure 4), and the selected statistical values are collected in Table 6.

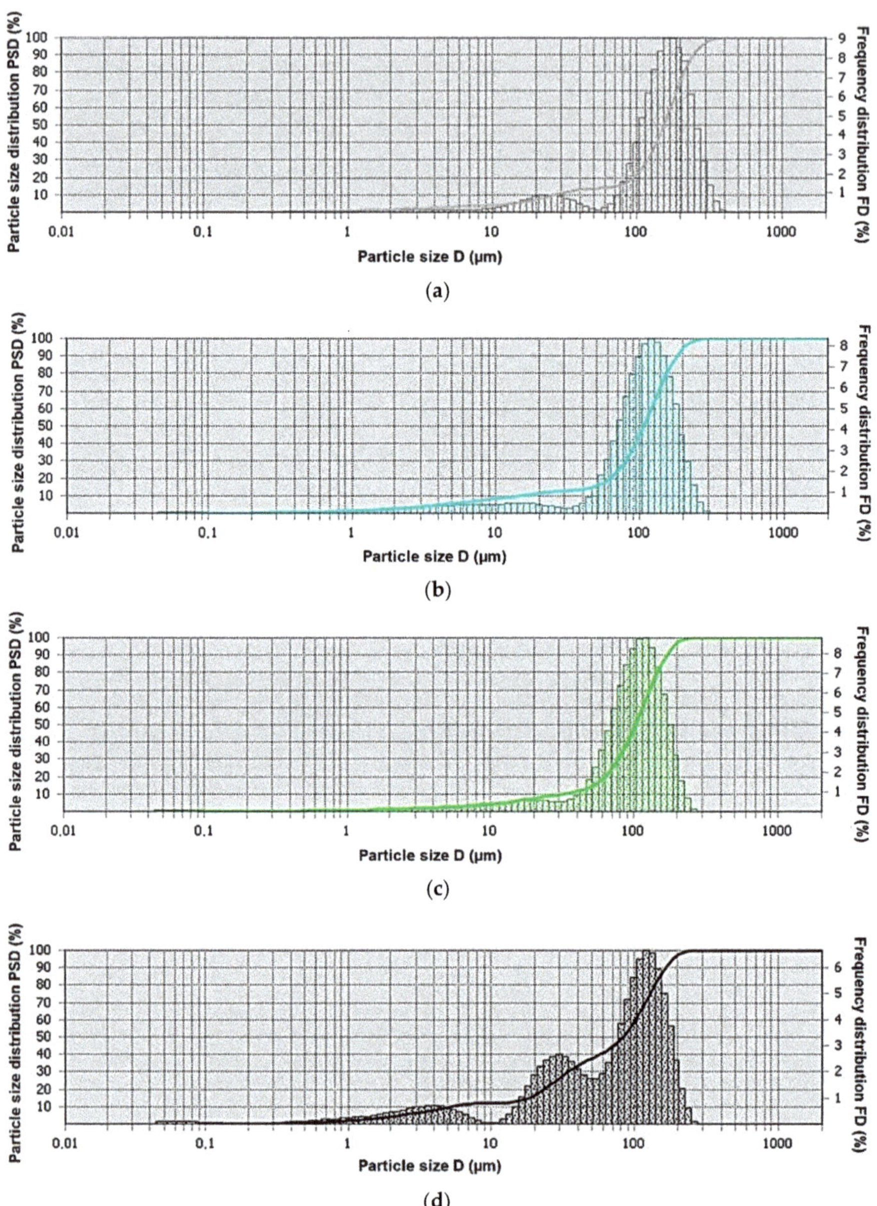

Figure 4. Graphs of cumulative percent passing versus the logarithmic particle size: (**a**) C1 filler material; (**b**) C2 filler material; (**c**) C3 filler material; (**d**) C4 filler material.

Table 6. Statistical mean values of the particle size distribution of the powders used for surfacing (PPTAW) of composite layers.

Average Statistical Value	Sample Determination			
	C1	**C2**	**C3**	**C4**
Quantile Q_{10} (µm)	26.60	14.17	31.22	5.13
Median Q_{50} (µm)	152.41	105.17	100.94	82.07
Quantile Q_{90} (µm)	247.55	180.10	167.32	162.88

The test results showed that the cobalt alloy powder with a hard reinforcing phase in the form of particles of crushed titanium carbide TiC and spherical polycrystalline synthetic diamond sintered PD-W (C1 filler material) had the highest mean particle size (152 µm). On the other hand, the C4 filler material—a powder based on a nickel alloy matrix with a hard reinforcing phase in the form of particles made of titanium diboride TiB_2 and spherical tungsten carbide WC-W_2C—was characterized by the smallest average particle size, amounting to 82 µm.

3.2. Density and Porosity of Composite Surfacing Welds

Based on the average of three mass measurements of test samples taken from the weld metal of each surfacing weld, calculations were performed to characterize the physical properties of composite layers, i.e., apparent density, open porosity, and water absorption. Samples cut from the welds were dried in a laboratory dryer at $110 \pm 5\,°C$ to a constant mass and then cooled in a desiccator. After determining the dry mass of the tested materials, the samples were deaerated in a vacuum device, and then saturated with liquid. The apparent mass of the sample immersed in the liquid and the mass of the sample saturated with liquid and weighed in the air were determined. The results of measurements and calculations are presented in Table 7. An example view of the longitudinal section of the macrostructure of selected surfacing welds is shown in Figure 5.

Table 7. Density-related measurement results and calculations of the porosity of the surface layers deposited (PPTAW) on structural low-alloy AISI 4715 steel.

Physical Quantity	Average			
	C1	**C2**	**C3**	**C4**
Density ρ (g/cm^3)	5.7785	8.1582	9.3425	9.6013
Standard deviation σ_ρ	0.2117	0.6267	0.3150	0.2349
Absorbability A (%)	1.0219	0.6368	1.3856	0.1315
Open porosity P_o (%)	5.6467	0.6368	11.2421	1.2346
Closed porosity P_c (%)	0.2526	5.7292	0.0082	1.4318
Apparent density ρ_a (g/cm^3)	5.4787	7.7335	8.9729	9.3400
Total porosity P_c (%)	5.8993	6.0366	11.2503	2.6778

The results of density measurements and calculations of the porosity of composite layers deposited with the PPTAW method showed that the weld metal obtained from powder based on a nickel alloy matrix with a hard reinforcing phase in the form of particles made of titanium diboride TiB_2 and spherical tungsten carbide WC-W_2C (C4 filler material) has the highest density (9.6 g/cm^3) and the lowest porosity (2.7%). The lowest density (5.78 g/cm^3) was found for the layer obtained from cobalt alloy powder with a hard reinforcing phase in the form of particles of crushed titanium carbide TiC and spherical polycrystalline synthetic diamond sintered PD-W (C1 filler material), and the highest porosity (11.2%) for the weld metal of the cladding weld made from nickel alloy powder with a hard reinforcing phase in the form of particles of WC-W_2C spherical and broken tungsten carbide and PD-W spherical polycrystalline synthetic diamond sinter (C3 filler material).

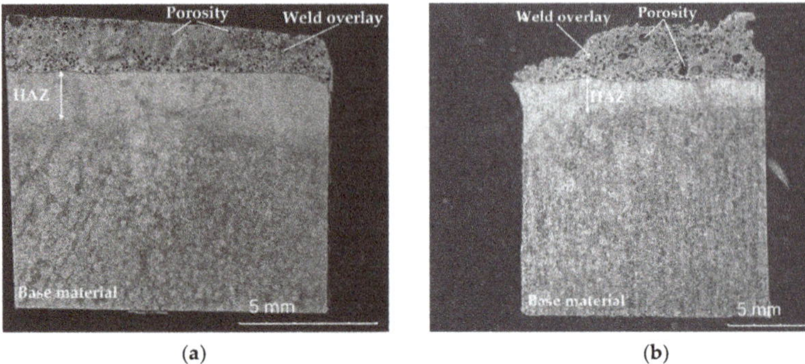

(a)

(b)

Figure 5. Exemplary macroscopic image of the longitudinal section of a composite surfacing weld: (**a**) C1 filler material; (**b**) C3 filler material.

3.3. Assessment of the Microstructure of Composite Surfacing Welds

The results of microscopic metallographic observations were used to determine the structure, type, distribution, and dimensions of the reinforcement phase and matrix of the composite wear-resistant layers in the near-surface, middle, and transitional zones of the surfacing weld. Light microscopy images of the structure were taken at a magnification of 200× (Figure 6). The microstructure and the results of the qualitative surface analysis in the polycrystalline synthetic diamond are shown in Figures 7 and 8.

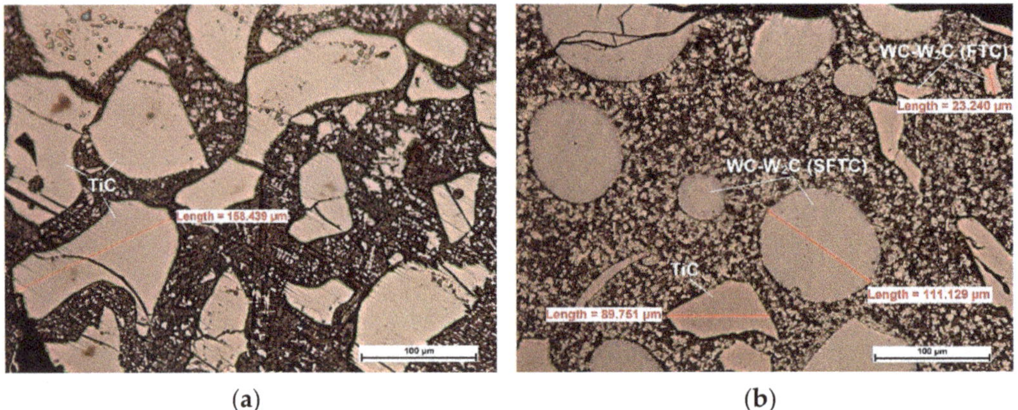

(a)

(b)

Figure 6. *Cont.*

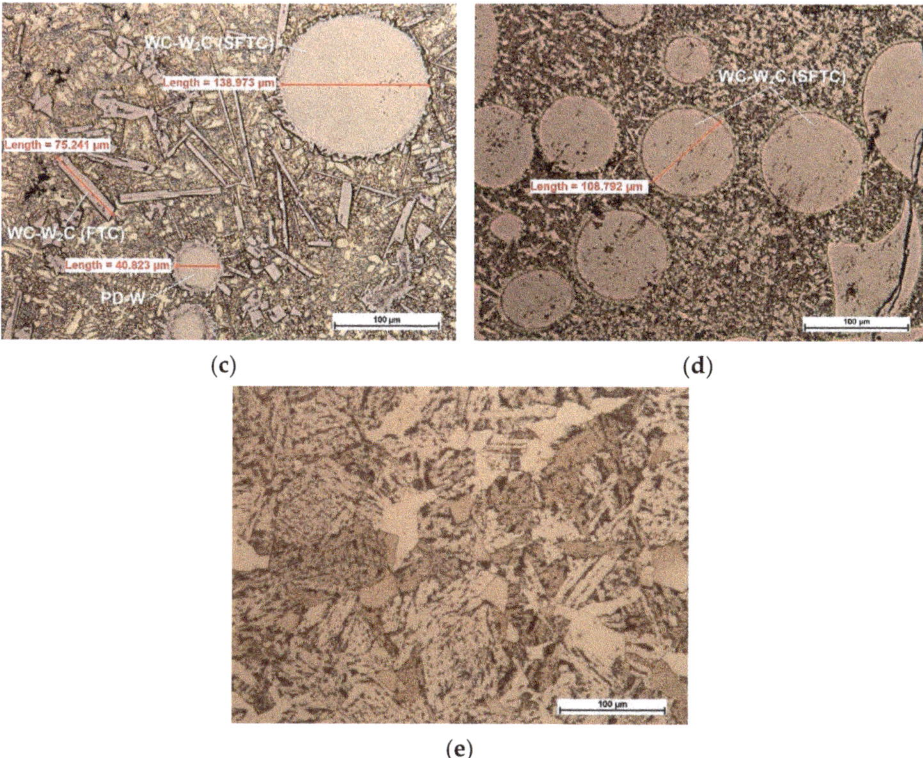

Figure 6. Light microscopy image of the structure of the composite layer padded using the PPTAW method with powder on the alloy matrix: (**a**) cobalt with a hard reinforcing phase in the form of crushed TiC and spherical PD-W particles (C1 filler material); (**b**) nickel with a hard reinforcing phase in the form of crushed TiC and spherical broken tungsten carbide WC-W2C particles (C2 filler material); (**c**) nickel with a hard reinforcing phase in the form of particles of WC-W$_2$C spherical and broken tungsten carbide and spherical polycrystalline synthetic diamond (C3 filler material); (**d**) nickel with a hard reinforcing phase in particles of spherical tungsten carbide WC-W$_2$C and titanium diboride TiB$_2$ (C4 filler material); (**e**) microstructure of the parent material of AISI 4715 non-alloy structural steel.

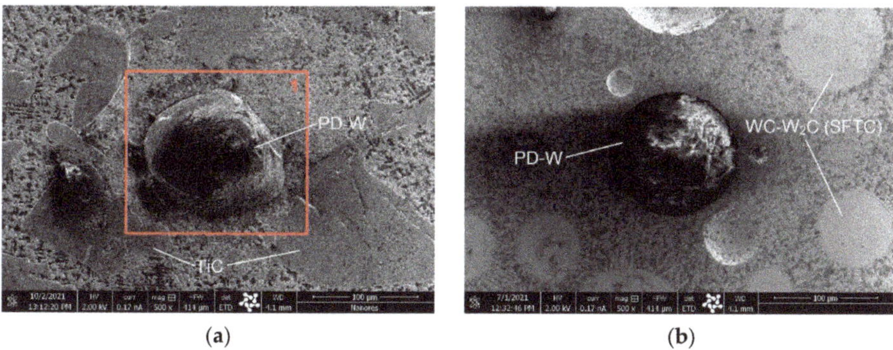

Figure 7. SEM image of the structure of selected composite layers padded using the PPTAW method with a powder on an alloy matrix: (**a**) cobalt with a hard reinforcing phase in the form of particles of crushed TiC and spherical polycrystalline synthetic diamond (C1 filler material); (**b**) nickel with a hard reinforcing phase in the form of particles of WC-W$_2$C spherical and broken tungsten carbide and PD-W spherical polycrystalline synthetic diamond (C3 filler material).

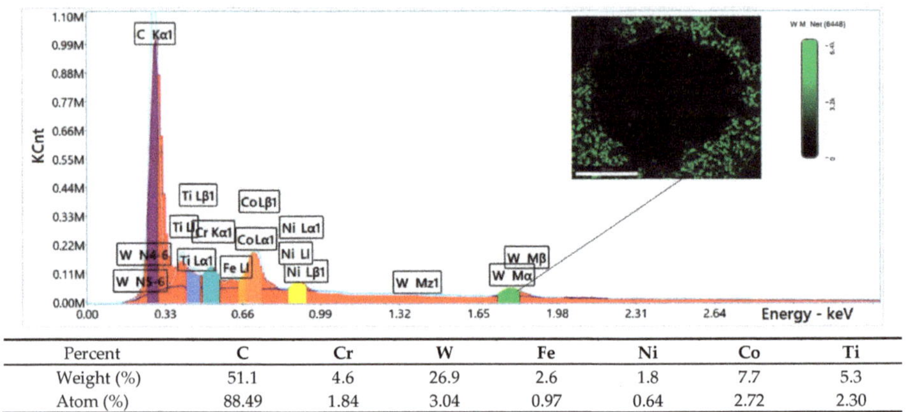

Percent	C	Cr	W	Fe	Ni	Co	Ti
Weight (%)	51.1	4.6	26.9	2.6	1.8	7.7	5.3
Atom (%)	88.49	1.84	3.04	0.97	0.64	2.72	2.30

Figure 8. Diagram of scattered X-radiation energy with energy lines present in area 1 (Figure 7a) of the analyzed components (chemical elements).

The tests involving the use of light microscopy revealed that the microstructure of the layer padded with a filler material C1 was dendritic, multidirectional, and contained numerous inclusions of ceramic particles of titanium carbide (TiC) as well as single particles of the synthetic metal-diamond composite (Figure 6a). The microstructure of the layer padded with a filler material C2 was composed of spherical particles of primary tungsten carbides and particles of crushed titanium carbide TiC in the nickel alloy matrix (Figure 6b). In turn in the microstructure of the layer padded with a filler material C3, morphologically diverse intermetallic phases consisting predominantly of spherical and crushed particles of primary tungsten carbide as well as smaller amounts of complex secondary carbides on the carbide–matrix boundary (Figure 6c) [9]. The secondary carbides were responsible for the diffusive bond of the primary carbides with the matrix. According to Bober et al. [3], the above-named mechanism of the bonding of carbides with the matrix should ensure their stable deposition. The partial melting of the primary carbides led to the partial saturation of the matrix with tungsten and carbon. In the structure of the padding weld made with C4 filler material, only different size particles of primary spherical tungsten carbide in the nickel alloy matrix were observed. The microstructure of alloy structural steel AISI 4715 is ferrite and lamellar perlite (Figure 6e).

3.4. Critical Parameters of the Material's Resistance to Cracking

3.4.1. Correlation Relationships Determining the Critical Parameters of Fracture Mechanics

Based on the work of the impact breaking KV (J) of the "Charpy V" type samples, the fracture toughness K_{Ic} (MPa·m$^{\frac{1}{2}}$) can be determined from the following relations [31]:

$$K_{Ic} = \sqrt{0.00022 \cdot E \cdot (KV)^{\frac{3}{2}}} \tag{1}$$

$$K_{Ic} = \sqrt{0.00137 \cdot E \cdot (KV)} \tag{2}$$

$$K_{Ic} = 14.5 \cdot \sqrt{(KV)} \tag{3}$$

$$K_{Ic} = 0.53 \cdot (KV) + 57.9 \tag{4}$$

where, in the absence of material data concerning the value of Young's modulus (*E*), Formula (3) is most often used.

On the other hand, the critical value of the fracture front opening, *CTOD* (units: mm) can be determined by Equation (5), as shown in [32]:

$$CTOD = \delta_{Ic} = 0.0024 \cdot KV \tag{5}$$

3.4.2. Test Results and Calculations of the Critical Values of K_{Ic} and CTOD (δ_{Ic})

The impact tests were performed to determine the brittleness threshold and the nature of the degradation of the AISI 4715 steel parent material and steel samples with wear-resistant composite surfacing welds. The work of the impact breaking *KV* was determined using a Charpy pendulum hammer with an initial energy of 300 J. The test was performed at 23 °C on standardized samples with a V-shaped notch with an angle of 45°, a depth of 2 mm, and a bottom rounding radius of 0.25 mm. A notch was cut from the underside of the sample on the side opposite to the surfacing weld (Figure 9). For each type of material, one test set was prepared that consisted of three samples with dimensions of 10 × 10 × 55 mm (full-sized sample). During the tests, the sample was adhered to the supports, the hammer impact was centered, the notch axis was in the plane of the hammer's motion, and the notch was directed so that the hammer hit the surfacing weld during the test. The averages of test results and calculated K_{Ic} and *CTOD* values are summarized in Table 8. Moreover, the test results were supplemented with morphology and chemical composition analyses of fractured micro-areas recorded with a stereoscopic microscope (Figure 10) and a scanning electron microscope (Figure 11).

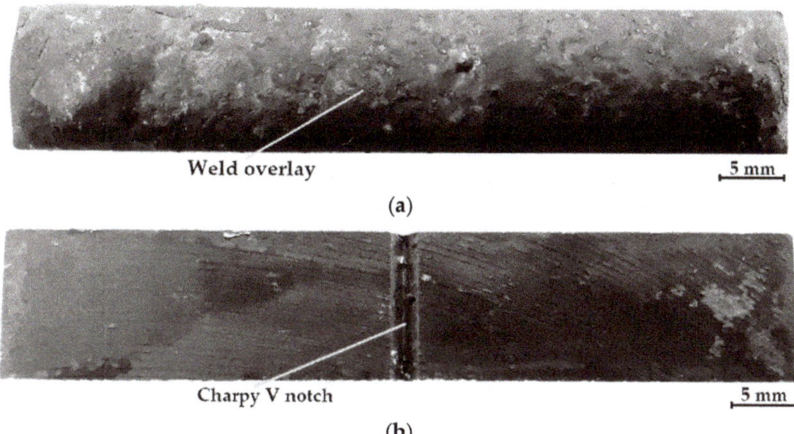

Figure 9. View of the sample before the Charpy impact test: (**a**) view from the weld overlay; (**b**) view from the Charpy V notch.

Table 8. Fracture toughness K_{Ic} and critical value of crack opening *CTOD* calculated based on the value of the average impact work of breaking *KV* at 23 °C of the AISI 4715 steel parent material and steel samples with wear-resistant composite surfacing welds.

Sample No. (Composite Weld Overlay)	*KV* (J) [1]	Standard Deviation σ (J)	K_{Ic} Equation (3) (MPa·m$^{\frac{1}{2}}$)	*CTOD* Equation (5) (mm)
C1 (Co3 + TiC + PD-W)	18.9	1.9	63.0	0.045
C2 (Ni3 + TiC + SFTC + FTC)	16.4	1.0	58.7	0.039
C3 (Ni3 + SFTC + FTC + PD-W)	22.3	1.4	68.5	0.054
C4 (Ni3 + SFTC + TiB$_2$)	24.1	2.2	71.2	0.058
C5 (Steel AISI 4715)	10.7	0.4	47.4	0.026

[1] Average value of three test results.

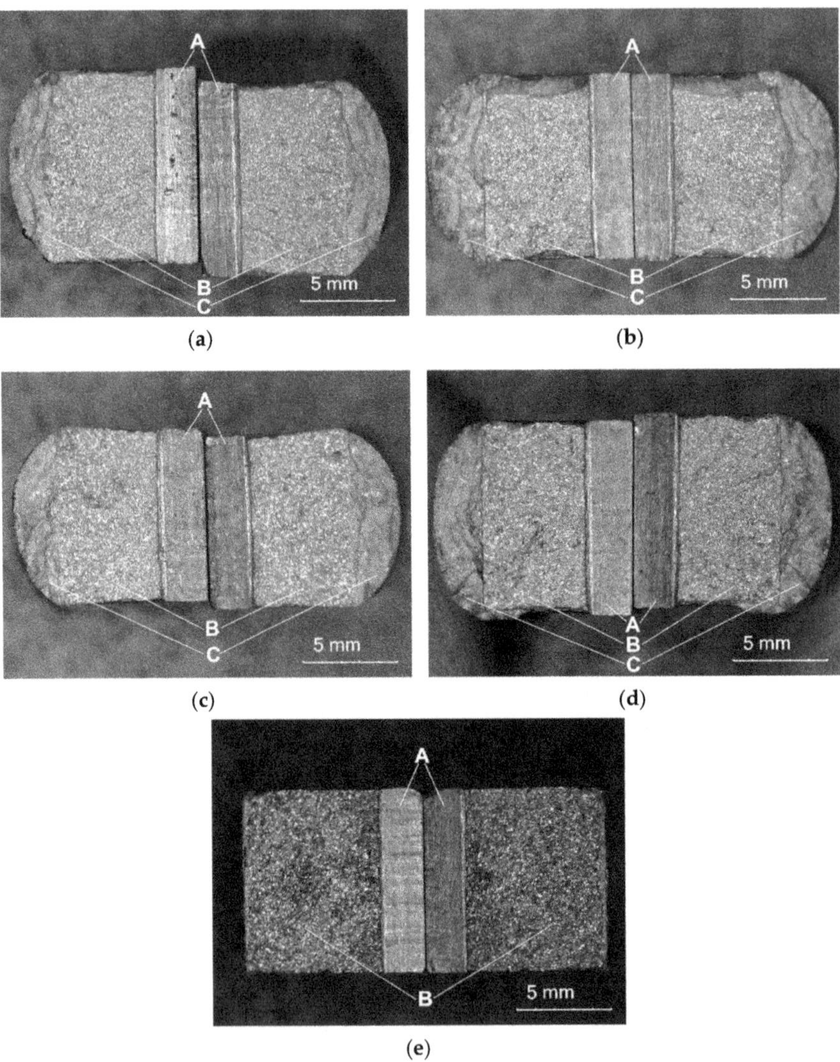

Figure 10. Fractures after impact tests at a temperature of +23 °C, A—notch area, B—area of the dynamic propagation zone of the fracture front in the native material, C—area of the dynamic propagation zone of the fracture front in the layer deposited with powder on a matrix of: (**a**) C1 filler material; (**b**) C2 filler material); (**c**) C3 filler material; (**d**) C4 filler material; (**e**) parent material.

The SEM images of fracture surfaces of all samples with composite layers (Figure 11a–d) showed an obvious brittle fracture mechanism. Also, in the base material (Figure 11e) no specimen exhibited the presence of ductile fracture areas with the formation of typical dimples, brittle intergranular fracture dominated.

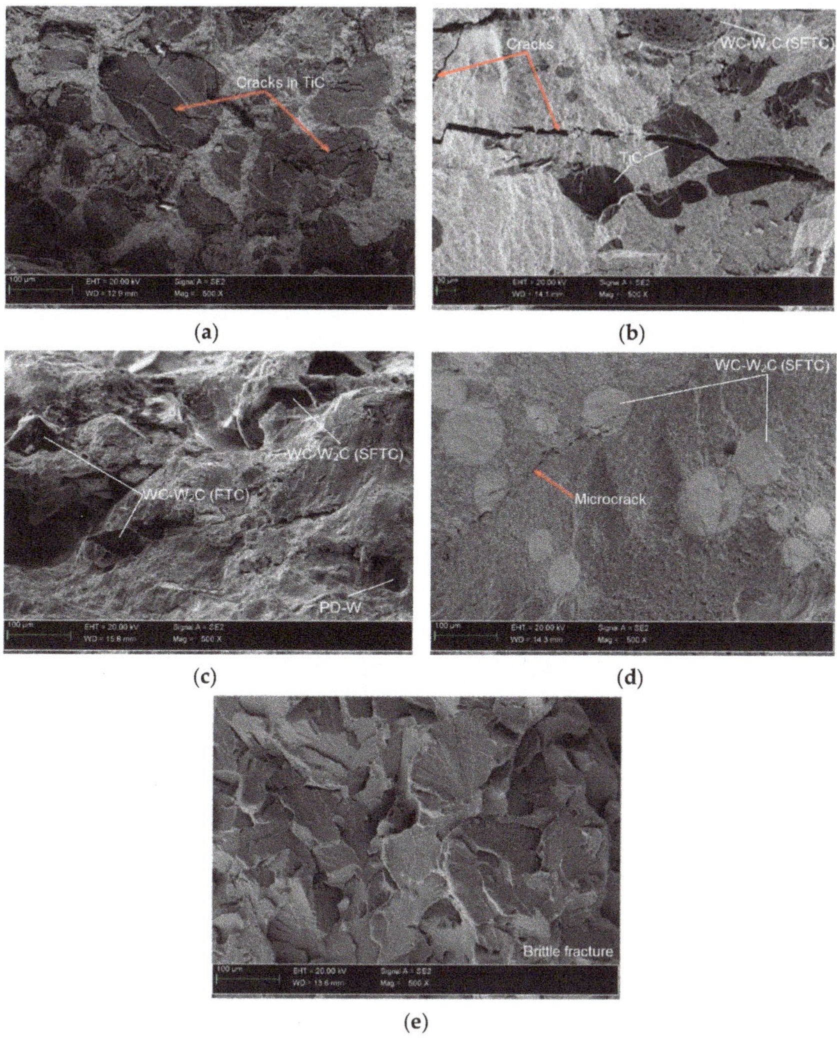

(a)
(b)
(c)
(d)
(e)

Figure 11. SEM images of the brittle fracture surface for samples: (**a**) C1; (**b**) C2; (**c**) C3; (**d**) C4; (**e**) base material.

4. Discussion

In the first publication of the series, we reported the structural and tribological properties of composite layers padded using PPTAW that contained Co-Cr-W-Mo alloys (C1 filler material) and Ni-Cr-B-Si (C2, C3, and C4 filler materials). These filler materials contained various combinations and volume fractions of hard ceramic particles (TiC, WC-W$_2$C, TiB$_2$, and PD-W) [9,10]. This earlier research showed that the surface plasma-welded layers made of these alloys displayed metal-mineral abrasion resistance, which made them suitable for the contact surfaces of the drilling tool inserts used in the oil and natural gas sectors. The current research was carried out to assess the resistance of the above-mentioned composite layers to impact loading.

4.1. Influence of Powder Particle Size on the Cracking of Surfaced Composite Layers under an Impact Load

Figure 2 shows the morphology of the components of the powders used for hardfacing the tested composite layers with the PPTAW method. The powders, apart from easily fusible matrix metal particles (Co3 and Ni3 alloys), whose diameter did not exceed 50 μm, also contained flame-retardant and super-hard phases in the form of ceramic particles from crushed, sharp-edged TiC (about 250 μm), spherical fused tungsten carbide WC-W_2C (about 160 μm), broken tungsten carbide WC-W_2C (about 80 μm), fine-grained titanium diboride TiB_2 (about 40 μm), and a spherical polycrystalline synthetic diamond with a tungsten coating (about 60 μm). Measurements of the particle sizes of powders (Table 6) in relation to the relevant calculations of the critical stress values K_{Ic} and the fracture front *CTOD* (Table 8) showed that the particle size of the hard ceramic phase was closely related to the fracture toughness of the MMC layers (Figure 12).

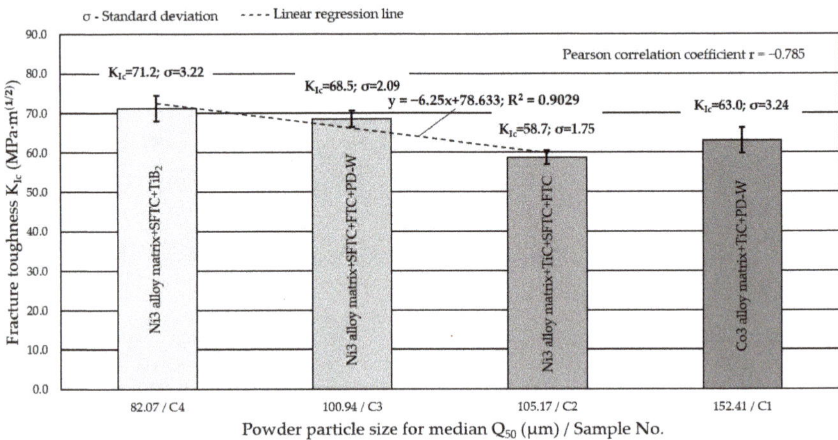

Figure 12. Influence of the particle size of the powder used for surfacing composite layers on the critical value of the stress intensity factor K_{Ic}.

The micrographs showed that in the composite layers on the Ni3 alloy matrix (C2, C3, C4 filler materials) large particles of sharply crushed TiC (Figures 6b and 11b) cracked more easily than spherical particles of WC-W_2C or PD-W (Figures 6c and 11d). Larger TiC particles were more susceptible to structural defects than smaller ones [33]. The larger the particle size of the hard ceramic phase, the larger the interfacial surface (matrix–ceramic particles) and the greater the stress transferred from the interfacial surfaces to the particles. Large ceramic particles fixed in a relatively plastic matrix showed less tendency to plastic deformation of the composite layer.

It was noticed that during crystallization of the liquid metal in the weld pool (C1 filler material), some TiC particles cracked due to tensile stress concentration in carbide defects (Figure 6a). The relatively low Young's modulus (Table 1) and the resulting reduction in the plasticity of TiC particles by strong intermolecular bonds indicates catastrophic crack propagation in the material. This thesis is confirmed by the research presented in refs. [34,35], which investigated the influence of tensile stress on the fracture of TiC particles during laser alloying of steel with cobalt alloys at different laser treatment parameters. The literature shows that the fracture toughness of TiC is 4–5.5 MPa·m$^{\frac{1}{2}}$ [36].

Based on SEM observations of fractures in the composite layers containing fused spherical WC-W_2C particles (Figure 11d), it was found that the initiation of a single microcrack occurred inside the WC-W_2C particles. The high temperature gradient resulting from the rapid heating and cooling of the composite layer welded with the PPTAW method, along with the 1/3 lower thermal expansion coefficient of the WC-W_2C particle (Table 1) lower than the matrix metal, led to the generation of thermal stress in the ceramic particles. When the value of the thermal stress exceeded the yield point of the polycrystalline WC-W_2C particle, microcracks formed on the surface of grain boundaries, spread, and then penetrated the particles. After exceeding the critical dimension, the microcrack propagated further from the WC-W_2C particle into the composite matrix, easily developing along the fracture-prone and brittle dendritic and interdendritic eutectic phases. The number of brittle eutectic phases strongly influenced crack propagation. According to Zhou et al. [37] and Wang et al. [38], the thermal stress caused by a high temperature gradient and different coefficients of thermal expansion between the ceramic particle and matrix cause cracking of WC-W_2C particles (Figure 6b,d). Often, cracks in WC-W_2C particles and matrix continued to spread, merging to form macrocracks, as shown in Figure 11b. Xu et al. [39] also observed cracks of WC particles in the composite layer on the matrix of a nickel alloy padded with the LMD method. Researchers proposed that the critical value of residual stress was determined mainly by microstructural and thermal stresses and was the main cause of MMC fracture. Such microcracks, under the influence of dynamic loading, can initiate brittle cracks. Surface topography tests on the impact fractures of all composite layers and the base material (Figure 11c) showed a brittle fracture mechanism.

It was also found that WC-W_2C melted at a much higher temperature during surfacing than the Ni-Cr-B-Si matrix material (C2, C3, C4 filler materials), as shown in Figure 6b–d. Depending on the amount of heat supplied to the material and the shape and size of the tungsten carbide particles, a small number of particles of the hard matrix strengthening phase dissolved during surfacing, which was also observed by Badish and Kirchgaßner [13]. A similar situation occurred in the case of synthetic polycrystalline diamond particles covered by a tungsten coating (PD-W), where, after partial melting of the coating, the tungsten was transferred to the matrix solution (Figure 8). The same observation was made by Telasang et al. [40], who investigated the microstructure and mechanical properties of laser-powder deposited layers using the direct metal deposition (DMD) technique on an AISI H13 tool steel substrate. As a consequence, during solidification, the tungsten and carbon pads entered the matrix solution. In general, the dissolution of tungsten and carbon strengthened the matrix and increased its abrasion resistance [9,10,41], but also reduced the fracture toughness and overall strength of the surfacing layers. According to Yan et al. [42], particle fracture is usually related to the highest principal stress, which increases upon increasing the particle size. As a result, small ceramic particles whose greatest principal stress was negligible were more difficult to fracture. In addition, the ceramic inclusions tended to break more easily at large aggregates where there was a high stress concentration; however, no cracks were observed in the particles of the polycrystalline synthetic diamond PD-W. The formation of an intermediate layer (~25 µm) between the polycrystalline synthetic diamond particle and the cobalt or nickel matrix (Figure 8), by melting the tungsten coating improved the stress transfer from the matrix to particles.

4.2. Influence of Density and Porosity on the Cracking of Surfaced Composite Layers under an Impact Load

Fracture toughness is a complex function of not only the physical properties of the matrix, or the type, proportion, and size of the hard phase reinforcing phase particles, but also the state of inter-grain and interfacial boundaries, as well as the stress in non-metallic inclusions and their surroundings. According to Bućko et al. [43], another important parameter is the density of a composite—often determined by its porosity. Figure 13 shows the joint effect of the matrix and ceramic inclusions, expressed by the specific density of the composite surfaced layer on the K_{Ic} value.

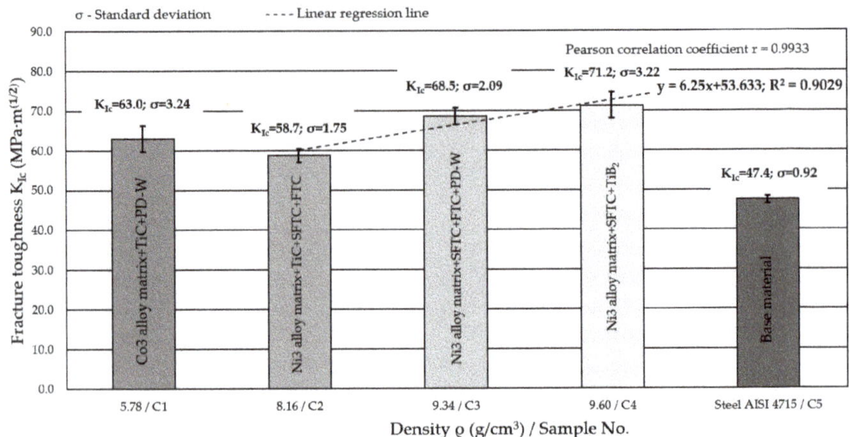

Figure 13. Influence of composite layer density on the critical value of the stress intensity factor K_{Ic}.

The lowest fracture toughness was obtained by the base material (AISI 4715 steel) without a surfacing layer, for which the calculated average K_{Ic} value did not exceed 48 MPa·m$^{\frac{1}{2}}$. In the case of composite layers surfaced by plasma with powders on a Ni3 alloy matrix, the highest value of the stress intensity factor K_{Ic} = 72.1 MPa·m$^{\frac{1}{2}}$ was recorded for the layer containing spherical particles of fused WC-W$_2$C and fine particles of TiB$_2$ (C4 filler material). Moreover, in the case of nickel-based filler materials (C2, C3, C4 filler materials), a linear relationship was observed between the density and fracture toughness of the surfaced composite layers. The coefficient of determination R^2 = 0.9029 indicated that the presented regression equation is very useful for predicting the value of K_{Ic} using the specific density ρ. The high Pearson's linear correlation coefficient r = 0.9933 suggests its significant influence on cracking. A large standard deviation σ for the calculated K_{Ic} values obtained at higher densities and decreasing upon decreasing the density, indicates that this material property is an important parameter that controls the fracture toughness of composites. The decrease in density decreased the fracture toughness. This has been confirmed by Rabin et al. [44] and Grabowy et al. [45]. It should be remembered that the graph shown in Figure 13 is merely an illustration of the qualitative tendency and not the quantitative dependence of the presented values. This is because the data come from measurements made on samples differing in both the type and size of the strengthening phase. On the other hand, the composite layer made of a powder on the matrix of Co3 alloy (C1 filler material) showed an average fracture toughness (K_{Ic} = 63 MPa·m$^{\frac{1}{2}}$) with a comparatively low density of the composite material (ρ = 5.78 g/cm³). The relatively good fracture toughness of this layer was attributed to the strong interfacial bonding of the polycrystalline synthetic diamond (PD-W) with the cobalt alloy matrix (Figure 7a). The density of the padded layer depended on the type and volume fraction of the hard matrix reinforcing phase and also on the total porosity of the composite layer. According to Bober et al. [12], when hard phase particles with a high melting point are well-wetted by the liquid metal matrix, correct surfacing welds are formed, and the usage degree of the hard reinforcing phase is high. On the other hand, when the wettability is insufficient, the ceramic particles are displaced from the liquid weld pool, and the degree of recovery of the strengthening particles is low. This usually leads to a series of welding imperfections, especially gas bubbles, external pores, and cracks in the padding layer. In addition to ensuring good wettability of the composite components, the density difference between the hard reinforcing phase and the matrix greatly influences the porosity of the surfacing welds. A significant difference in the specific mass of the matrix metal and the strengthening phase can lead to the uneven distribution or agglomeration of the strengthening particles in the surfacing weld metal. According to Gawdzińska et al. [46], local densification of ceramic particles may result in pores caused by insufficient saturation of capillary spaces between the reinforcement phase and liquid matrix metal. The test results of composite surfacing welds made of powder on a Ni3 alloy matrix do not unequivocally confirm the thesis that a reduction in fracture toughness is tantamount to an increase in the porosity of the layer (Figure 14).

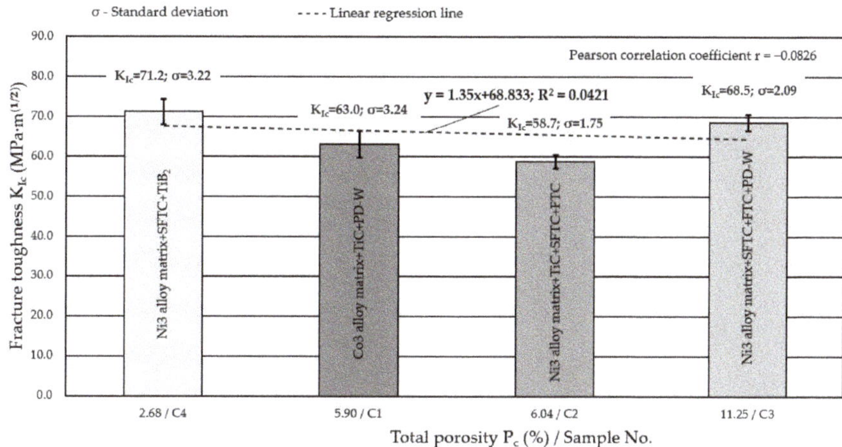

Figure 14. Influence of porosity of composite layers on the critical value of the stress intensity factor K_{Ic}.

Even when the total porosity P_c rises above 6%, other factors were responsible for the fracture toughness; therefore, it is possible to obtain composites that differ significantly in their K_{Ic} with a comparatively high total porosity ($P_c > 6$%). For the same reasons, an increase in K_{Ic} was observed for filler material C3. The analysis of Figure 14 does not allow us to conclude that with the same chemical composition of the matrix, the greatest impact on the critical value of K_{Ic} can be expected in the case of a high porosity P_c. The low value of the correlation coefficient (-0.0826) does not show a rectilinear relationship between K_{Ic} and P_c, which suggests that other factors have a more significant influence on the cracking mechanism, e.g., the type and volume fraction of hard-phase particles reinforcing the composite. The composite layer padded with powder on the matrix of the Co3 alloy (C1 filler material) with a total porosity $P_c = 5.9$% showed a crack resistance (K_{Ic}) of 63 MPa·m$^{\frac{1}{2}}$.

5. Conclusions

Based on experimental tests and analytical calculations of the critical values determining the fracture toughness of the innovative metal matrix composite layers, it can be concluded that:

(1) In the case of composite layers based on a nickel alloy matrix, the critical value of the stress intensity factor K_{Ic} decreased linearly upon increasing the powder particle size and increased linearly upon increasing the density of the composite layer.

(2) Taking into account the comparable size of the powder particles as well as the density and porosity of the composite material, it was shown that the composite layer based on the cobalt alloy matrix (C1 filler material) had a slightly higher fracture toughness than the composite layer based on the nickel alloy matrix (C2 filler material). The critical value of the stress intensity factor K_{Ic} for the composite layer on the Co3 alloy matrix was 4.3 MPa·m$^{\frac{1}{2}}$ higher.

(3) The nucleation of cracks in composite layers most often occurred inside the particles of the hard strengthening phases (TiC and WC-W$_2$C). They propagated in the matrix along the very brittle dendritic and interdendritic eutectic phases of the composite microstructure, which were susceptible to cracking. Spherical particles made of polycrystalline synthetic diamond PD-W covered with a tungsten coating showed high thermal stability and no susceptibility to cracking during surfacing using the PPTAW method.

(4) Thermal stresses occurring during plasma surfacing and the different thermal expansion coefficients of the ceramic reinforcement particles (TIC and WC-W$_2$C) and the metallic composite matrix may have been the cause of crack initiation and propagation in the composite layers.

6. Patents

The procedure for granting a patent (No. P435997) was initiated before the Patent Office of the Republic of Poland.

Author Contributions: Conceptualization, A.C.; methodology, A.C.; software, M.Ż.; validation, A.C. and M.Ż.; formal analysis, A.C; investigation, A.C.; resources, A.C. and M.Ż.; data curation, A.C.; writing—original draft preparation, A.C.; writing—review and editing, M.Ż.; visualization, A.C.; supervision, A.C.; project administration, A.C.; funding acquisition, A.C. and M.Ż. All authors have read and agreed to the published version of the manuscript.

Funding: The research was founded by Silesian University of Technology Rector's habilitation grant 10/050/RGH_20/1006 and pro-quality grant 10/050/BKM21/1018.

Institutional Review Board Statement: Not applicable.

Informed Consent Statement: Not applicable.

Data Availability Statement: The data are not publicly available due to the initiation of a patent procedure (No. P435997).

Conflicts of Interest: The authors declare no conflict of interest. The funders had no role in the design of the study; in the collection, analyses, or interpretation of data; in the writing of the manuscript, or in the decision to publish the results.

References

1. Zhang, G.; Xiong, W.; Yao, Z.; Chen, S.; Chen, X.; Yang, Q. Synthesis of ultrafine (Ti, W, Nb)C solid solution powders by microwave-assisted carbothermal reduction. *Ceram. Int.* **2014**, *40*, 7621–7625. [CrossRef]
2. Brezinová, J.; Viňáš, J.; Guzanová, A.; Živčák, J.; Brezina, J.; Sailer, H.; Vojtko, M.; Džupon, M.; Volkov, A.; Kolařík, L.; et al. Selected Properties of Hardfacing Layers Created by PTA Technology. *Metals* **2021**, *11*, 134. [CrossRef]
3. Bober, M. Composite coatings deposited by the plasma transferred arc–characterization and coating formation. *Weld. Technol. Rev.* **2011**, *83*, 43–47. [CrossRef]
4. Sue, A.; Sreshta, H.; Qiu, B.H. Improved Hardfacing for Drill Bits and Drilling Tools. *J. Therm. Spray Technol.* **2011**, *20*, 372–377. [CrossRef]
5. Vallauri, D.; Atias Adrian, I.C.; Chrysanthou, A. TiC-TiB$_2$ composites: A review of phase relationships, processing and properties. *J. Eur. Ceram. Soc.* **2008**, *28*, 1697–1713. [CrossRef]
6. Wen, G.; Li, S.; Zhang, B.; Guo, Z. Reaction synthesis of TiB2-TiC composites with enhanced toughness. *Acta Mater.* **2001**, *49*, 1463–1470. [CrossRef]
7. Baoshuai, D.; Zengda, Z.; Xinhong, W.; Shiyao, Q. Laser cladding of in situ TiB$_2$/Fe composite coating on steel. *Appl. Surf. Sci.* **2008**, *254*, 6489–6494. [CrossRef]
8. Tijo, D.; Masanta, M.; Das, A.K. In-situ TiC-TiB2 coating on Ti-6Al-4V alloy by tungsten inert gas (TIG) cladding method: Part-I. Microstructure evolution. *Surf. Coat. Technol.* **2018**, *344*, 541–552. [CrossRef]
9. Czupryński, A. Microstructure and Abrasive Wear Resistance of Metal Matrix Composite Coatings Deposited on Steel Grade AISI 4715 by Powder Plasma Transferred Arc Welding Part 2. Mechanical and Structural Properties of a Nickel-Based Alloy Surface Layer Reinforced with Particles of Tungsten Carbide and Synthetic Metal-Diamond Composite. *Materials* **2021**, *14*, 2805. [CrossRef]
10. Czupryński, A. Microstructure and Abrasive Wear Resistance of Metal Matrix Composite Coatings Deposited on Steel Grade AISI 4715 by Powder Plasma Transferred Arc Welding Part 1. Mechanical and Structural Properties of a Cobalt-Based Alloy

Surface Layer Reinforced with Particles of Titanium Carbide and Synthetic Metal–Diamond Composite. *Materials* **2021**, *14*, 2382. [CrossRef]

11. Cherepova, T.; Dmitrieva, G.; Tisov, G.; Dukhota, O.; Kindrachuk, M. Research on the properties of Co-TiC and Ni-TiC HIP-sintered alloys. *Acta Mech. Autom.* **2019**, *13*, 57–67. [CrossRef]

12. Bober, M.; Senkara, J. Study of the structure of composite coatings Ni-WC deposited by plasma transferred arc. *Weld. Technol. Rev.* **2016**, *88*, 67–70. [CrossRef]

13. Badish, E.; Kirchgaßner, M. Influence of Welding Parameters on Microstructure and Wear Behavior of a Typical NiCrBSi Hardfacing Alloy Reinforced with Tungsten Carbide. *Surf. Coat. Technol.* **2008**, *202*, 6016–6022. [CrossRef]

14. Milanti, A.; Koivuluoto, H.; Vuoristo, P.; Bolelli, G.; Bozza, F.; Lusvarghi, L. Microstructural Characteristics and Tribological Behavior of HVOF-Sprayed Novel Fe-Based Alloy Coatings. *Coatings* **2014**, *4*, 98–120. [CrossRef]

15. Czupryński, A. Comparison of Properties of Hardfaced Layers Made by a Metal-Core-Covered Tubular Electrode with a Special Chemical Composition. *Materials* **2020**, *13*, 5445. [CrossRef]

16. Poloczek, T.; Czupryński, A.; Żuk, M.; Chruściel, M. Structure and tribological properties of wear-resistant layers produced in process of plasma powder surfacing. *Weld. Technol. Rev.* **2019**, *91*, 35–41. [CrossRef]

17. Durejko, T.; Łazińska, M.; Dworecka-Wójcik, J.; Lipiński, S.; Varin, R.A.; Czujko, T. The Tribaloy T-800 Coatings Deposited by Laser Engineered Net Shaping (LENSTM). *Materials* **2019**, *12*, 1366. [CrossRef] [PubMed]

18. Liu, L.; Minasyan, T.; Ivanov, R.; Aydinyan, S.; Hussainova, I. Selective laser melting of TiB_2-Ti composite with high content of ceramic phase. *Ceram. Int.* **2020**, *46*, 21128–21135. [CrossRef]

19. EN 14700. *Welding Consumables. Welding Consumables for Hard-Facing*; CEN: Brussels, Belgium, 2014.

20. ISO 12737. *Metallic Materials—Determination of Plane-Strain Fracture Toughness*; ISO: Geneva, Switzerland, 2010.

21. ISO 15653. *Metallic Materials—Method of Test for the Determination of Quasistatic Fracture Toughness of Welds*; ISO: Geneva, Switzerland, 2018.

22. Wang, L.; Liu, H.; Huang, C.; Liu, X.; Zou, B.; Zhao, B. Microstructure and mechanical properties of $TiC-TiB_2$ composite cermet tool materials at ambient and elevated temperature. *Ceram. Int.* **2016**, *42*, 2717–2723. [CrossRef]

23. Evans, A.G.; Wilshaw, T.R. Quasi-static solid particle damage in brittle solids—I. Observations analysis and implications. *Acta Metall.* **1976**, *24*, 939–956. [CrossRef]

24. Bai, L.L.; Li, J.; Chen, J.L.; Song, R.; Shao, J.Z.; Qu, C.C. Effect of the content of B_4C on microstructural evolution and wear behaviors of the laser-clad coatings fabricated on Ti6Al4V. *Opt. Laser Technol.* **2015**, *76*, 33–45. [CrossRef]

25. Wang, C.; Miranda, J.; Yang, Y.; Chung, Y. Investigation of hardness and fracture toughness properties of Fe/VC multilayer coatings with coherent interfaces. *Surf. Coat. Technol.* **2016**, *288*, 179–184. [CrossRef]

26. Michalak, M.; Łatka, L.; Sokołowski, P. Comparison of mechanical properties of the plasma sprayed coatings by powder and suspension. *Weld. Technol. Rev.* **2017**, *89*, 56–60. [CrossRef]

27. ISO 14175. *Welding Consumables—Gases and Gas Mixtures for Fusion Welding and Allied Processes*; ISO: Geneva, Switzerland, 2008.

28. ISO 148. *Metallic Materials—Charpy Pendulum Impact Test—Part 1: Test Method*; ISO: Geneva, Switzerland, 2016.

29. ASTM D792-00. *Standard Test Methods for Density and Specific Gravity (Relative Density) of Plastics by Displacement*; ASTM International: West Conshohocken, PA, USA, 2000.

30. ISO 17639. *Destructive Tests on Welds in Metallic Materials—Macroscopic and Microscopic Examination of Welds*; ISO: Geneva, Switzerland, 2013.

31. Werner, K.; Rozumek, D.; Wojsyk, K.; Zajączkowski, J. Determination of the critical values of toughness on crack of steels and welded joints. *Weld. Technol. Rev.* **2018**, *90*, 56–61. [CrossRef]

32. Werner, K.; Wojsyk, K. The analysis of the possibility of fragile cracking steel units of welded constructions. *Weld. Technol. Rev.* **2015**, *87*, 93–95.

33. Liu, D.; Zhang, S.Q.; Li, A.; Wang, H.M. Microstructure and tensile properties of laser melting deposited TiC/TA15 titanium matrix composites. *J. Alloys Compd.* **2009**, *485*, 156–162. [CrossRef]

34. Grum, J.; Sturm, R. A new experimental technique for measuring string and residual stresses during a laser remelting process. *J. Mater. Process. Technol.* **2004**, *147*, 351–358. [CrossRef]

35. Madadi, F.; Shamanian, M.; Ashrafizadeh, F. Effect of pulse current on microstructure and wear resistance of Stellite6/tungsten carbide claddings produced by tungsten inert gas process. *Surf. Coat. Technol.* **2011**, *205*, 4320–43288. [CrossRef]

36. Steinbrech, R.W. R-Curve Behavior of Ceramics. In *Fracture Mechanics of Ceramics. Fracture Mechanics of Ceramics (Composites, R-Curve Behavior, and Fatigue)*; Bradt, R.C., Hasselman, D.P.H., Munz, D., Sakai, M., Shevchenko, V.Y., Eds.; Springer: Boston, MA, USA, 1992; Volume 9. [CrossRef]

37. Zhou, S.; Zeng, X.; Hu, Q.; Huang, Y. Analysis of crack behavior for Ni-based WC composite coatings by laser cladding and crack-free realization. *Appl. Surf. Sci.* **2008**, *255*, 1646–1653. [CrossRef]

38. Wang, J.; Li, L.; Tao, W. Crack initiation and propagation behavior of WC particles reinforced Fe-based metal matrix composite produced by laser melting deposition. *Opt. Laser Technol.* **2016**, *82*, 170–182. [CrossRef]

39. Xu, J.S.; Zhang, X.C.; Xuan, F.Z.; Tian, F.Q.; Wang, Z.D.; Tu, S.T. Tensile properties and fracture behavior of laser cladded WC/Ni composite coatings with different contents of WC particle studied by in-situ tensile testing. *Mater. Sci. Eng. A-Struct.* **2013**, *560*, 744–751. [CrossRef]

40. Telasang, G.; Majumdar, J.D.; Padmanabhan, G.; Manna, I. Wear and Corrosion Behavior of Laser Surface Engineered AISI H13 Tool Steel. *Surf. Coat. Technol.* **2015**, *261*, 69–78. [CrossRef]
41. D'Oliveira, A.S.C.M.; Tigrinho, J.J.; Takeyama, R.R. Coatings Enrichment by Carbide Dissolution. *Surf. Coat. Technol.* **2008**, *202*, 4660–4665. [CrossRef]
42. Yan, Y.W.; Geng, L.; Li, A.B.; Fan, G.H. Finite Element Analysis about Effects of Particle Size on Deformation Behavior of Particle Reinforced Metal Matrix Composites. *Key Eng. Mater.* **2006**, *353–358*, 1263–1266.
43. Bućko, M.M.; Pyda, W. Microstructural aspects of toughening of cubic zirconia-alumina composites. *Composites* **2003**, *3*, 39–46.
44. Rabin, B.H.; German, R.M. Microstructure effects on tensile properties of tungsten-Nickel-Iron composites. *Metall. Mater. Trans. A* **1988**, *19*, 1523–1532. [CrossRef]
45. Grabowy, M.; Maciewicz, K.; Łuszcz, M.; Pędzich, Z.; Bućko, M.M. Mechanical properties of ATZ-type composites obtained in the sintering process of zirconia powders with different chemical composition. *Ceram. Mater.* **2019**, *71*, 286–294.
46. Gawdzińska, K.; Jackowski, J.; Szweycer, M. Porosity variations of castings made of saturated metal composites. *Composites* **2001**, *1*, 68–71.

[MDPI logo]

Article

Influence of Preheating Temperature on Structural and Mechanical Properties of a Laser-Welded MMC Cobalt Based Coating Reinforced by TiC and PCD Particles

Artur Czupryński [1],[*] and Mirosława Pawlyta [2]

[1] Department of Welding Engineering, Faculty of Mechanical Engineering, Silesian University of Technology, Konarskiego 18A, 44-100 Gliwice, Poland

[2] Materials Research Laboratory, Silesian University of Technology, Konarskiego 18A, 44-100 Gliwice, Poland; miroslawa.pawlyta@polsl.pl

[*] Correspondence: artur.czuprynski@polsl.pl; Tel.: +48-322371443

Abstract: This article presents research on the structural and mechanical properties of an innovative metal matrix composite (MMC) coating designed for use in conditions of intense metal-mineral abrasive wear. The layer, which is intended to protect the working surface of drilling tools used in the oil and natural gas extraction sector, was padded using the multi-run technique on a sheet made of AISI 4715 low-alloy structural steel by Laser Direct Metal Deposition (LDMD) using a high-power fiber laser (FL). An innovative cobalt alloy matrix powder with a ceramic reinforcement of crushed titanium carbide (TiC) and tungsten-coated synthetic polycrystalline diamond (PCD) was used as the surfacing material. The influence of the preheating temperature of the base material on the susceptibility to cracking and abrasive wear of the composite coating was assessed. The structural properties of the coating were characterized by using methods such as optical microscopy, scanning electron microscopy (SEM), energy dispersion spectroscopy (EDS), transmission electron microscopy (TEM) and X-ray diffraction analysis (XRD). The mechanical properties of the hardfaced coating were assessed on the basis of the results of a metal-mineral abrasive wear resistance test, hardness measurement, and the observation of the abrasion area with a scanning laser microscope. The results of laboratory tests showed a slight dissolution of the tungsten coating protecting the synthetic PCD particles and the transfer of its components into the metallic matrix of the composite. Moreover, it was proved that an increase in the preheating temperature of the base material prior to welding has a positive effect on reducing the susceptibility of the coating to cracking, reducing the porosity of the metal deposit and increasing the resistance to abrasive wear.

Keywords: LDMD; cladding; deposition; titanium carbide; synthetic polycrystalline diamond

check for **updates**

Citation: Czupryński, A.; Pawlyta, M. Influence of Preheating Temperature on Structural and Mechanical Properties of a Laser-Welded MMC Cobalt Based Coating Reinforced by TiC and PCD Particles. *Materials* **2022**, *15*, 1400. https://doi.org/10.3390/ma15041400

Academic Editor: Christopher C. Berndt

Received: 31 December 2021
Accepted: 3 February 2022
Published: 14 February 2022

Publisher's Note: MDPI stays neutral with regard to jurisdictional claims in published maps and institutional affiliations.

1. Introduction

Laser Direct Metal Deposition (LDMD) hardfacing—with direct feeding of metallic powder to the weld pool—is a very modern and advanced welding technology. The idea behind the process is to slightly melt the surface of the material with a heat source, a laser beam generated by a high-power laser, and simultaneously deliver a stream of metal powder to the weld pool through a nozzle. The additional material supplied from outside then becomes the main component of the produced coating, and as a result of the high temperature of the process, it melts with the substrate. The coatings produced using this method exhibit excellent metallurgical bonding to the substrate, low dilution (i.e., low mixing between the clad material and the substrate material, high density, little or no cracking and good mechanical properties. The LDMD technique and other types of laser surfacing can produce coatings from almost any metal material, both on new and regenerated elements, which are used in the following industries: aerospace, energy, petrochemical, automotive, and medical [1,2].

Recently, there has been a great deal of interest in the use of the laser cladding method to produce composite surface layers on the surface of components made of ferrous and nonferrous alloys [3,4]. Metal matrix composites (MMCs), due to their properties such as high hardness, excellent wear and corrosion resistance, are currently the most interesting—in terms of research and being the most desirable materials in industrial applications—group of additional materials for laser surfacing [5,6]. The combinations of metal matrix and reinforcing particles are very diverse, and the most popular are composites based on iron, nickel, cobalt, magnesium and aluminum alloys in combination with reinforcing particles such as tungsten carbide (WC) [7,8], silicon carbide (SiC) [9], boron carbide (B_4C) [10] and titanium carbide (TiC) [11], the latter being used in this study.

Recently, many publications have referred to studies in which cobalt and its alloys were used as a matrix; however, such materials are seldom described as matrix materials in composite layers [12,13], and more often as homogeneous layers [14–16]. Cobalt–chromium alloys, called Stellites, which are characterized by high hardness, resistance to corrosive environments and, above all, high wear resistance (including metal–metal, metal–mineral and metal–liquid resistance) constitute a particularly important group of hardfacing materials used for the production of laser-clad surfaces on elements designed to work in conditions of high abrasive wear and extreme temperatures (up to 1050 °C), often combined with an aggressive working environment. In Stellites, the addition of chromium plays an important role—it is the main carbide-forming factor, and, as an alloying element in the matrix, it increases the strength and the resistance to corrosion, oxidation, and the effect of sulphur compounds. The very hard carbides present in the structure of Stellites create microscopic unevenness between sliding surfaces, which ensures high resistance to abrasive wear [17]. Despite this, efforts are still being made to improve surface properties and reduce the thickness of the overlay coating, resulting in an increase in the reliability and service life of machine and device parts and thus contributing to minimizing production losses [18–20]. For this reason, more innovative additional materials for surfacing and methods of their application are constantly being sought.

Composites based on cobalt alloys offer extremely broad prospects for research in this area. TiC particles are often used as reinforcements for composite layers in cobalt alloy matrices produced by laser surfacing processes [21–23]. They are used as both volumetric and surface reinforcement. Shasha et al. [13] investigated the mechanical properties of a coating based on a cobalt alloy reinforced in situ with TiC, TiB_2, Cr_5Si_3, WB, SiC, Co3Ti and NiC particles produced on the surface of a TA15 titanium alloy using a cross-flow CO_2 laser. The results showed that the diversified and numerous interstitial phases dispersed between fine dendritic structures in a matrix mainly composed of γ-Co, α-Ti solid solution improved by more than threefold the hardness of the coating compared to the substrate. Compared to the titanium alloy, the abrasion resistance of the coating was significantly improved, and the wear rate of the coating was about 1/12 of the titanium alloy. The abrasive wear mechanism of the hardfaced coating was mixed.

TiC-Co LMD (Laser Metal Deposition) composite coatings produced on a ductile iron substrate were investigated by Tong et al. [21]. The results showed that an increase in the size of the primary or secondary dendrite can be inhibited by the separation of TiC after its dissolution in the molten metal pool. The hardness on the surface of the coating gradually increased (to 1246.6 HV0.2) with a decrease in laser power or an increase in the scanning rate. Attention was paid to the importance of the size and shape of the hard phase particles and the type of matrix material. Unfortunately, the paper did not present any tribological tests. Zahang [22] produces a composite TiC-Co layer laser welded onto a substrate made of 2Cr13 steel. The layer was intended for preventive protection of the working surface of machine parts and power devices exposed to abrasive wear and impact load. The structure of the shell was highly diversified and consisted of several sublayers. As a result of the impact of the laser beam, the surface material became harder, while fusion and a transition zone occurred further within the coating. The coating structure consisted of supersaturated cobalt dendrites with dispersed TiC particles. The test results showed that, for steel, it was possible to strengthen the welded material by simply hardening it. However, in the case of superalloys, as a result of mutual diffusion and the mixing of the alloy and substrate components, the chemical composition and structure of the top coat changed, which was undesirable. Moreover, it was found that a higher amount of the hard TiC phase promoted the growth of the metal matrix grain. The abrasive wear test showed relatively good metal-to-metal wear resistance.

Many more pieces of experimental data have been published on laser-welded composite layers employing a matrix of cobalt alloys reinforced with WC [24–27]. Most of the research concerns the dissolution of WC in the matrix and its influence on the final properties of the coatings. According to Zanzarin et al. [25], the dissolution of WC during the laser surfacing process increased with an increase in the amount of heat supplied, an increase in the temperature of the preheating of the base material or a decrease in the tungsten and carbon content in the matrix. Dissolving WC can increase the susceptibility to cracking and reduce the resistance to abrasive wear. An interesting alternative to WC, due to its lower density and much higher values of Young's modulus, compressive strength, flexural strength, fracture toughness and hardness, is a synthetic polycrystalline diamond (PCD). However, the decomposition temperature of pure synthetic PCD (1450 °C) is much lower than the decomposition temperature of tungsten carbide (2870 °C). There are no data in the scientific literature on the use of synthetic PCD as a matrix reinforcement of MMC composite layers produced by LMD.

The aim of the study was to determine the effect of the preheating temperature of the base material on the structure, susceptibility to cracking, and abrasion resistance of a laser-welded coating consisting of composite powder in a cobalt alloy matrix. A novelty in the presented results is the development of an innovative composition of the matrix reinforcement consisting of a composite containing super-hard phases in the form of ceramic particles from finely crushed TiC and spherical particles made of synthetic PCD, the latter being protected against thermal decomposition with a tungsten coating. Our results showed that there is a possibility of producing a composite laser-welded coating using the ceramic reinforcement—the cobalt matrix phase system, the microstructure and abrasive properties of which are appropriate to protect the working surface three-cone toothed bits [28]. Currently, laser surfacing of the working surface of drilling cutter teeth is not used under production conditions. Typically, these elements are protected against wear by hardfacing using the oxyacetylene welding (OAW) method with a composite stick (tubular electrode) with a powder core, Figure 1b. The implementation of the LMD surfacing technology in place of the gas surfacing technique used up to now for this industrial application may contribute to an improvement in the quality of surface layers by minimizing noncompliance, stresses, and welding deformations, as well as the share of base metal dilution in the weld, even to levels below 3%. In addition, it will allow for obtaining the required chemical composition of the padding weld to be already obtained in the first layer, contributing to greater production efficiency resulting from the automation and robotization of the padding process, and allowing for the finishing time of the padded surface to be shortened.

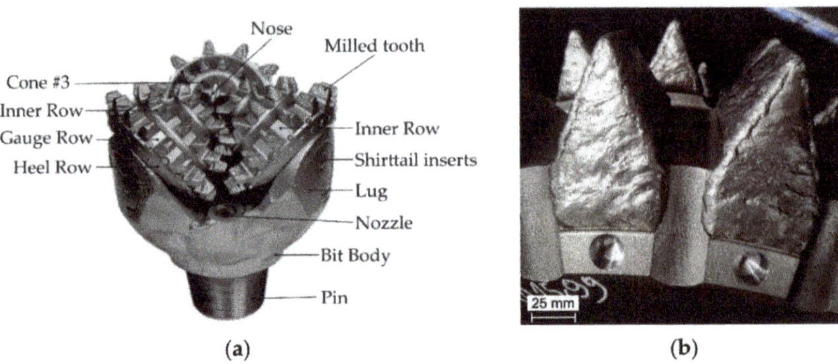

Figure 1. Three-cone bit with milled teeth: (**a**) general view of the drill bit with milled teeth, (**b**) working surface of the teeth after oxyacetylene hardfacing.

2. Materials and Methods

2.1. Materials

The material used for surfacing was an innovative metallic powder of the Co3 alloy group (Höganäs AB, Höganäs, Sweden) according to EN 147000 [29], characterized by an increased weight fraction of carbon (up to 3%) and tungsten (up to 14%). The metal powder that constituted the composite matrix material was mechanically mixed in a Turbula T2F laboratory turbulent mixer (Glen Mills Inc., Clifton, NJ, USA). The powder consisted of a hard reinforcing phase composed of crushed sharp-edged titanium carbide TiC and spherical synthetic polycrystalline diamond PCD (Harmony Industry Diamond, Zhengzhou, China) with a tungsten coating. The proportion of the reinforcing phase components to the metal matrix components was 60% to 40% by weight with the hard phase content composed of 90% TiC with a particle size ranging from 140 to 250 μm and 10% PCD with a particle size of 180 μm, Figure 2. The chemical composition of the composite powder, claimed by patent P.435997, is presented in Table 1. The base material was a low alloy structural steel of AISI 4715 (Table 2).

Table 1. Chemical composition of AISI 4715 low-alloy structural steel according to manufacturer data (TimkenSteel Ltd., Canton, OH, USA).

Element, wt.(%)									
Fe	C	Mn	Cr	Mo	Ni	Si	P	S	CEV [1]
Bal.	0.12–0.18	0.65–0.95	0.40–0.70	0.45–0.60	0.65–1.00	0.15–0.35	≤0.015	≤0.015	0.66

[1] CEV—Carbon Equivalent Value.

Table 2. Chemical composition of powder.

Chemical Composition of the Matrix, wt.(%)									Ceramic Reinforcement of the Matrix, wt.(%)	
Co	Cr	W	Ni	Fe	Mn	Mo	C	Si	TiC	PCD-W
Bal.	24–28	12–14	≤3	<5	≤2	≤1	2.5–3	≤1	90	10

Notes: Carbide to matrix ratio: 60/40 wt.(%).

2.2. Laser Processing

The laser powder surfacing process was carried out using the LDMD technique—with direct feeding of the composite powder to the weld pool—on a robotic stand equipped with a modern surfacing system using a YLS-4000 ytterbium fibere laser system (IPG Photonics Corporation, Oxford, MA, USA) with a wavelength of λ = 1070 nm and a maximum laser beam power of 4000 W mounted on a REIS RV30-26 six-axis robot (Reis Robotics, Obernburg am Main, Germany), Figure 3.

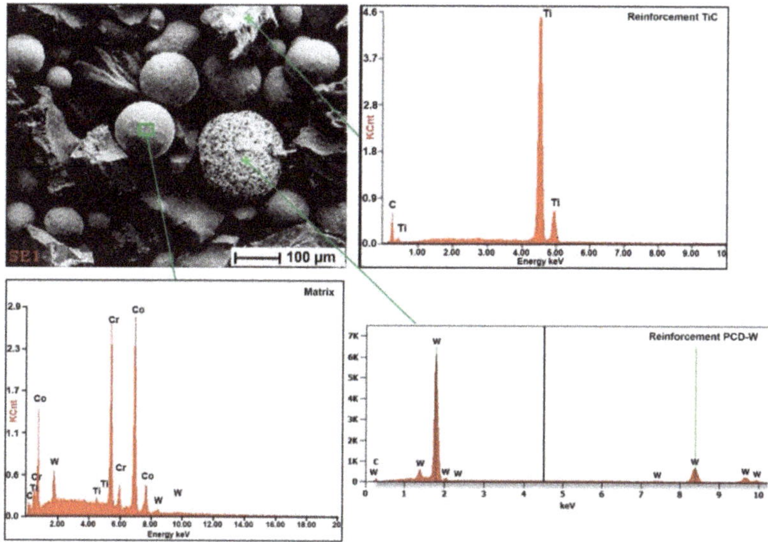

Figure 2. SEM image of the composite powder particle morphology and EDS spectra obtained for selected grains.

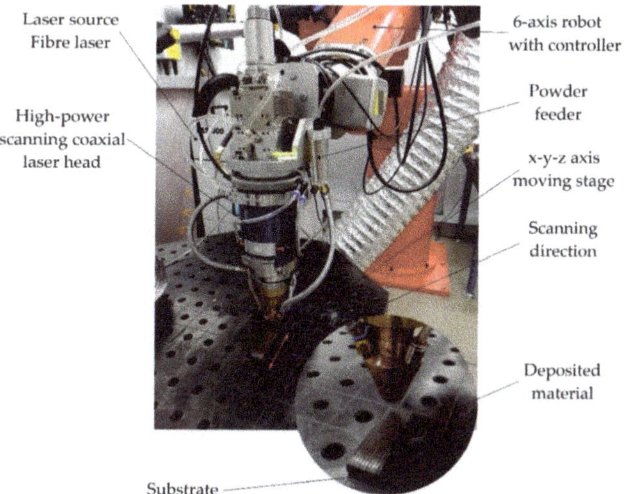

Figure 3. View of the test stand used for robotic surfacing with the LDMD.

The diameter of the laser spot at the focal plane, 20 mm from the nozzle tip, measured with a UFF100 Laserscope (Prometec GmbH, Aachen, Germany), was 5 mm, with energy distribution in the TEM01* beam. The powder cladding set-up consisted of a computer-controlled powder feeding system and a coaxial cladding nozzle integrated with a Computerized Numerical Control (CNC) five-axis gantry. Coaxial powder injection was realized by using nozzle gas, carrier gas and shielding gas—Argon 5.0 (99.999%). Cladding was performed at the following gas flow rates: nozzle gas (Ar) = 15 L/min, carrier gas (Ar) = 2.8 L/min and shielding gas (Ar) = 12 L/min. In order to determine the range of optimum surfacing parameters, a series of single-run padding welds was carried out with laser power values of 1400, 1600, 1800, 2000 and 2200 W with a surface speed from 4 to 12 mm/s and a powder feed rate from 15 to 30 g/min. The optimum parameters for the surfacing of the composite coating (Table 3) were defined as the parameters that ensured a uniform distribution of the powder across the entire surface of the liquid metal in the weld pool, the correct depth of fusion g < 2 mm, a layer height h < 5 mm and a percentage of dilution of the base metal in the surface layer D < 5% (Figure 4).

Table 3. Optimum parameters for laser surfacing with Co3+TiC+PCD powder on AISI 4715 steel.

Process Parameters	Value of the Parameter
Laser Power (W)	1800
Scanning Speed (mm/s)	8
Laser Spot Size, (mm)	5
Powder Feed Rate (g/min)	24
Overlap Ratio (%)	33
Heat Input [1] (J/mm)	225

[1] defined as the laser power divided by the scanning speed.

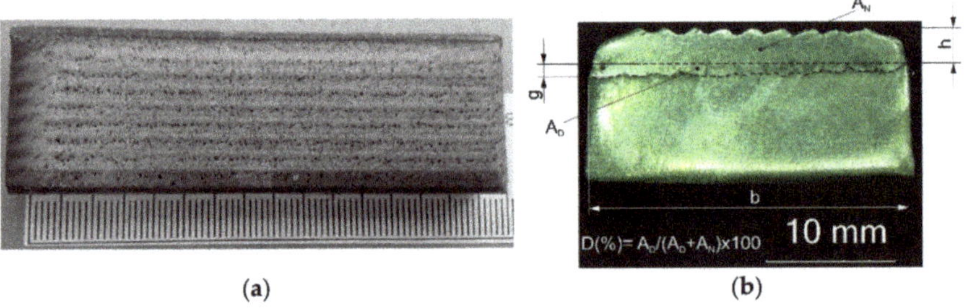

Figure 4. LDMD coating with composite powder: (**a**) a view from the face of the padding welds, (**b**) a cross section of the coating.

Test samples were produced with dimensions of 75 × 25 × 10 mm—in a series of three, welded either without or with preheating of the substrate using an oxyacetylene torch to temperatures of 100, 200 or 300 °C, and the samples were marked with the symbols T0 (no heating), T100, T200 or T300, respectively.

2.3. Methodology of Research

In order to assess the quality of the surfacing welds and to determine the number of cracks in the surfacing layers at different preheating temperatures of the substrate, non-destructive testing i.e., visual tests (VT) and penetration tests (PT) were carried out. The structural and mechanical properties were determined on the basis of the analysis of the results of macro- and microscopic metallographic tests, chemical composition, X-ray diffraction, hardness, and porosity measurements as well as metal-mineral abrasive wear resistance tests.

2.3.1. NDT Tests

Visual tests (VT) and penetrant tests were carried out in accordance with the guidelines contained in the relevant standards, i.e., ISO 17637 [30] and ISO 3452-2 [31]. VT consisted of determining, using the naked eye and a digital pen microscope, the location and assessment of the surface quality features of the coating, such as cracks, porosity, spikes, undercut, imperfect shapes, and dimension. Before starting the actual test, the tested surface was prepared by thoroughly cleaning it of all impurities and drying it. The system of color preparations (system design type II, sensitivity 2) PT ISO 3452-2 II Cd-2 and EN 571-1 was used for penetration tests.

2.3.2. Metallographic Examination and X-ray Diffraction Analysis

Microscopic examinations were performed on metallographic specimens prepared by standard methods. The etching reagent was a mixture of concentrated hydrochloric acid and nitric acid in a volume ratio of 3:1, the so-called 'aqua regia', and the digestion time of the sample was experimentally selected. The observation and saving of macro- and microstructure images were performed using an Olympus SZX9 stereoscopic microscope (Olympus Corporation, Tokyo, Japan) equipped with a Moticam 5.0+ digital camera (Motic (Xiamen) Electric Group Co Ltd., Xiamen, China) and Motic Images Plus 3.0 software (Motic (Xiamen) Electric Group Co Ltd., Xiamen, China). SEM studies were performed on a Zeiss Supra 35 microscope (Carl Zeiss SMT, Oberkochen, Germany) equipped with an EDS spectrometer for chemical composition analysis. For transmission electron microscopy (TEM) observations, samples were prepared employing the Focused Ion Beam (FIB) technique using an SEM/Ga-FIB Helios NanoLabTM 600i microscope (FEI Company, Hillsboro, OR, USA). TEM investigations were performed using an S/TEM Titan 80-300 microscope (FEI Company, Hillsboro, OR, USA) equipped with a Cetcor Cs probe corrector (CEOS, Heidelberg, Germany) and an EDS and EELS spectrometer for chemical composition analysis. Crystal Maker (version 10.4.1) and Single Crystal software (CrystalMaker Software Limited, Oxfordshire, UK) were used to simulate the crystal structure and diffraction patterns. The phase composition of the hardfaced coating was determined by X-ray diffraction tests performed on at Panalytical X'Pert Pro MPD diffractometer (Malvern Panalytical Ltd., Malvern, UK), using filtered radiation (Kβ Fe filter) from cobalt anode lamps ($\lambda K\alpha = 0.179$ nm). The diffractogram was recorded in Bragg—Brentano geometry, using a PIXcel 3D detector on the axis of the diffracted beam over an angle range of 20–110 [2θ] (step = 0.05°, count time per step = 100 s). The diffractograms obtained were analyzed using dedicated Panalytical High Score Plus software (Malvern Panalytical Ltd., Malvern, UK) together with the PAN-ICSD structural database. Quantitative phase analysis of the X-rays was performed using the Rietveld method.

2.3.3. Density Measurement and Examination of the Porosity of the Coating

The density of the top coat was measured according to ASTM D792 [32] using a Radwag AS 220.R2 analytical laboratory balance (Radwag, Warsaw, Poland) with the Archimedes method of density measurement.

The porosity of the surface coating was assessed using a μCT Nanotom 180N micro-tomography device (Ge Sensing & Inspection Technologies GmbH, Wunstorf, Germany) equipped with an X-ray tube with a maximum voltage of 180 kV. Tomographic images were recorded using a Hamamastu 2300 × 2300 pixel decoder (Hamamatsu Photonics K.K., Hamamatsu, Japan). The virtual reconstruction of the shell structure was mapped using proprietary GE datosX ver.2.1.00 software (GE Sensing & Inspection Technologies GmbH, Hürth, German). All tomographic images were taken at a source voltage of 140 kV and 200 μA with a 360° rotation of the element in 2400 steps. The exposure time was set at 500 ms, the frame averaging at 3 and the image skipping at 1, while the scanning time of a single element t = 80 min. Porosity analysis was performed using the MyVGL program (Volume Graphics GmbH, Heidelberg, Germany).

2.3.4. Hardness Measurements

The hardness on the surface of all coatings (samples symbols T0, T100, T200 and T300) was measured using the Rockwell C method, according to the procedure described in ISO 6508-1 [33]. The tests were carried out using a Sunpoc Super Rockwell SHRS-450M hardness tester (Guizhou Sunpoc Tech Industry Co., Ltd., Guizhou, China) on the minimally sanded outer surface of the hardfacing coating, at five measuring points located on the axis of the sample with a 10 mm space between them. Hardness measurements in the cross section of the overlay coating were performed using the Vickers method, according to the guidelines contained in the ISO 6507-1 standard [34]. The measurement was carried out on metallographic specimens (T0, T100, T200 and T300) at 10 measurement points, separately for the hard phase particles and the metal matrix, and the hardness was determined on the HV 0.05 scale. The tests were performed using a hardness tester with an automatic measuring track and a Future-Tech FM-ARS 9000 image analysis system (Future-Tech Corp., Kawasaki, Japan).

2.3.5. Abrasive Wear Test

The metal-mineral wear resistance of the surface layer was tested according to the ASTM G 65-00 standard, Procedure A [35]. The abrasive wear process was tested with the use of a 'rubber wheel' type device (Figure 5).

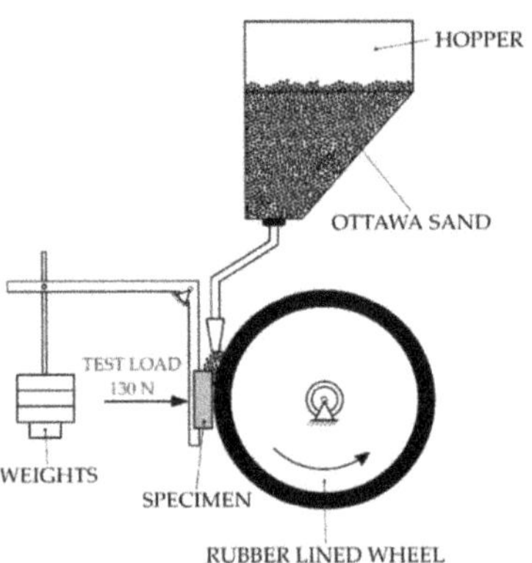

Figure 5. Schematic diagram of the stand for testing resistance to metal-mineral abrasive wear according to the ASTM G65 standard.

The surface coating abrasion test, without heating and with preheating of the base material to temperatures of 100, 200 and 300 °C, was carried out for four series of two samples for each initial temperature of the base material. During the test, which lasted about 30 min, the friction wheel made 6000 revolutions against the test material with a force of 130 N and a flow rate of the abrasive material (A. F. S. Testing Sand 50–70 mesh) of 335 g/min.

Before and after the abrasion test, the samples were weighed on a laboratory balance with an accuracy of 0.0001 g. The average coating density was determined using a laboratory balance, on the basis of three density measurements of samples taken from the tested materials and weighed at room temperature in air and liquid. Using the measured average density of the hardfaced coating and the average mass loss of the sample after the abrasion test, the volumetric mass loss was calculated according to Equation (1):

$$\text{Volume loss}\ \left[\text{mm}^3\right] = \frac{\text{mass loss [g]}}{\text{density}\ \left[\frac{\text{g}}{\text{cm}^3}\right]} \times 1000 \tag{1}$$

The surface abrasion area of the coating after the ASTM G65 test was observed using a 3D LEXT™ OLS5100 laser scanning microscope (Olympus Corporation, Tokyo, Japan), allowing a very precise measurement of the shape of the abrasion profile and surface roughness.

Furthermore, based on planimetric tests carried out on macroscopic samples (Figure 4b), the share of base metal dilution substrate material in the composite layer was estimated according to Equation (2):

$$D = \frac{A_N}{A_N + A_D} \times 100\% \tag{2}$$

where A_N is the area of the fusion of the layer into the base material, and A_D is the rim area of the weld overlay.

3. Results and Discussion

3.1. Non-Destructive Testing Results

Non-destructive testing of laser-welded composite layers of cobalt Co3 powder containing of TiC and PCD ceramic particles on the substrate of low-alloy AISI 4715 structural steel allowed the type, location and size of surface imperfections to be determined. The results of penetration tests of surface coatings without heating or with preheating of the base material to temperatures of 100, 200 or 300 °C are shown in Figure 6.

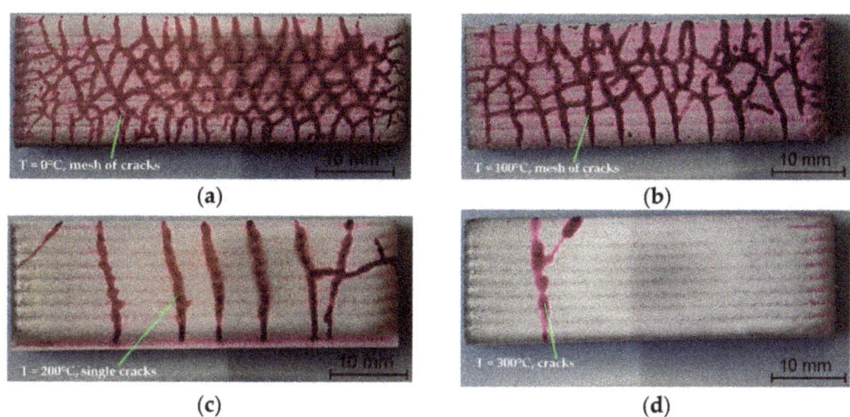

(a)　　　　　(b)

(c)　　　　　(d)

Figure 6. View of the coating laser-welded with composite powder after penetration tests (PT): (a) surfacing without heating, surfacing with preheating of the base material to temperatures of (b) 100 °C, (c) 200 °C, (d) 300 °C.

In terms of the quality of workmanship, the coatings were characterized by a high regularity of the outer plane and the symmetry of the successive overlapping, weld beads. VT and PT of the surface of the layers showed only the presence of radial cracks (1051) and transverse cracks (1021) [36]. The lack of preheating of the substrate before surfacing resulted in the appearance of a dense network of cracks, Figure 6a. Increasing the preheating temperature to 100 °C caused the crack mesh density to decrease, and in the temperature range from 200 to 300 °C, a significant reduction was observed in the number of cracks in the layer. Only single cracks were detected spreading throughout the coating width in a line perpendicular to the welding direction, Figure 6c,d. A higher content of matrix alloying additives and a higher proportion of hard phase ceramic particles reinforcing the composite contributed to an increase in abrasion resistance, but at the same time increased the tendency for surface cracks to form in the padding weld. Cracks appeared during cooling due to the difference in the thermal expansion coefficients of the base material and the padding weld material [37]. The cracks most often penetrated deep into the base material and in no way weakened the adhesion of the surface coating.

According to [38], in wear resistant MMC layers, cracks are acceptable if they appear across the seams and the distance between them is not less than 20–30 mm. In the case of the composite tested, this condition was met for the preheating temperature of 300 °C. To obtain a crack-free surfaced coating, it was necessary to select the preheating temperature and control the inter-pass temperature. Preheating the material before surfacing reduces shrinkage stress [39] and the hardness in the heat-affected zone [40], while also reducing the risk of hydrogen cracks [41]. In the analyzed case, the factors that influenced the necessity to use preheating were the high carbon equivalent value, CEV = 0.66% and the high-power density of the heat source, which translated into low linear welding energy. For a given configuration of the base material and filer metal, the preheating temperature should exceed 300 °C and should be maintained throughout the welding cycle. The size and thickness of the welded material are also important in this case. Elements with larger dimensions, such as toothed cones, should be heated more (even up to 500–700 °C), bearing in mind that too high a preheating temperature can, however, contribute to the thermal decomposition of the carbide phase [42]. The effect of the preheating temperature of the base material before surfacing on the number of cracks in the composite layer is shown in Figure 7.

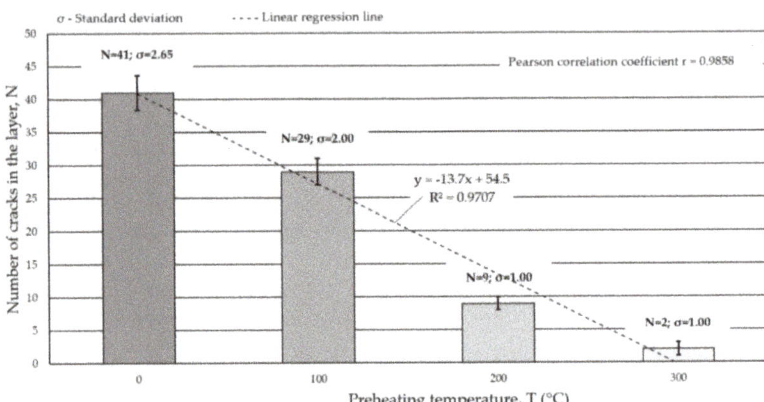

Figure 7. Influence of the preheating temperature of the base material before surfacing on the number of cracks in the composite layer.

3.2. Metallographic Test Results and Results of the XRD Analysis

The results of microscopic metallographic observations of the composite coating allowed the matrix structure as well as the type, distribution and dimensions of the ceramic reinforcement to be determined. SEM observations were carried out at 80, 500 and 1500 times magnification. Secondary Electron (SE) as well as Back Scattered Electron (BSE) detectors were used for image acquisition, the latter to more clearly show the chemical contrast, ensuring the highest image quality. For detailed structural analysis of the areas containing carbide reinforcement particles, a transmission electron microscope (TEM) was used to determine the grain size, structural defects and cracks in the TiC and PCD particles. The results of the microscopic observations are shown in Figures 8 and 9.

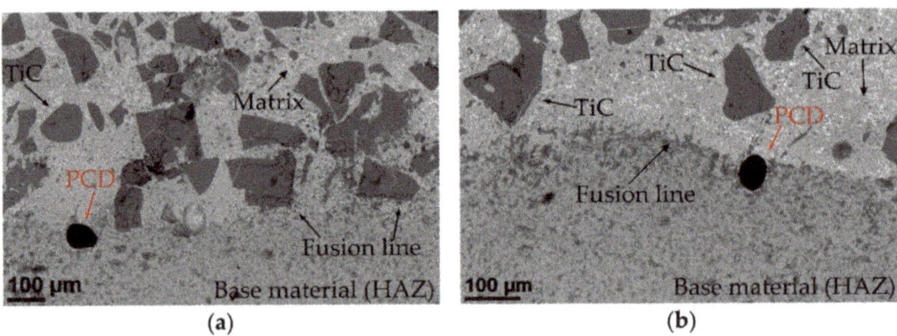

Figure 8. Area of the composite shell fusion line. SEM images (BSE detector): (**a**) left side of the coating section and (**b**) right side of the coating section.

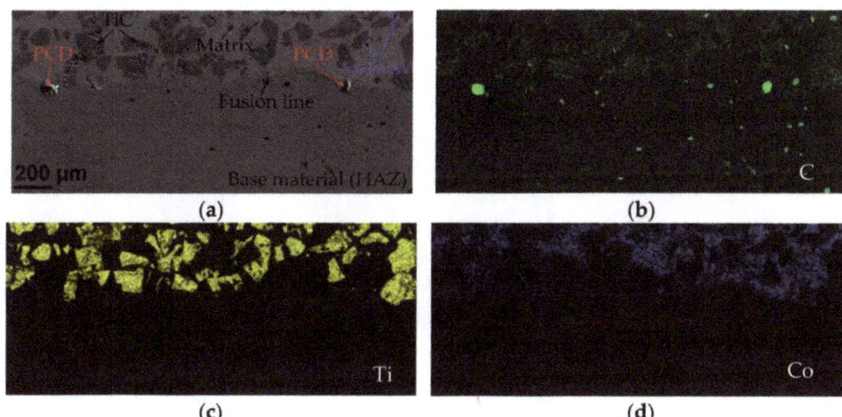

Figure 9. Area of the fusion line of the composite shell. (**a**) SEM image (SE detector). EDS maps of of element distribution, (**b**) carbon—green, (**c**) titanium—yellow, (**d**) cobalt—blue.

Scanning microscopy studies showed that the metal matrix of the composite coating was densely and evenly filled with a large number of inclusions of ceramic TiC particles and single particles of synthetic PCD. Figure 8 shows the fusion zone of the coating deposited against the base material. Two areas separated by a fusion line are visible, differing in morphology and chemical composition. In the surfacing layer, geometrically irregular TiC particles with grain sizes often exceeding 100 µm can be observed. TiC is known to have good mechanical properties and is a frequently used ceramic additive for composite binders (MMCs) designed to protect elements exposed to high abrasive wear in combination with impact stress [43]. However, like most ceramics, titanium carbide is brittle and, due to the reduction of plasticity by strong bonds, has a tendency to undergo catastrophic fracture [44,45]. According to Sun et al. [46], cracking occurs during plasma surfacing (PPTAW), in the first phase of cooling of the liquid metal in the weld pool, due to the concentration of tensile stresses in carbide defects. The tests conducted showed that, during the LDMD method, the tendency of the TiC particles to fracture clearly decreased with an increasing preheating temperature of the base material.

Figure 10 shows a single grain of titanium carbide with a size of approximately 200 µm. In the matrix surrounding it, much finer grains (of a size of a few micrometers) enriched with tungsten and titanium are noticeable.

In BSE images (Figure 8), they are visible as brighter spots. To characterize them in detail, TEM studies were performed. Figure 11a shows a smaller area (approximately 4 µm × 4 µm) taken from the surroundings of the TiC.

It included two fragments of grains containing carbon, titanium and tungsten, and a cobalt matrix in which the presence of Cr, C, Ni, Ti and W (Figure 11b) was confirmed. Due to the use of the EELS spectrometer, it is possible to clearly confirm the presence of C in the analyzed grains containing Ti and W. The EDS analysis (Figure 11d) showed that the atomic share of Ti and W was 80% and 20%, respectively (ESD does not allow for a reliable quantitative analysis of light elements; therefore, the C content is not included here). The presence of Cu in the spectrum resulted from its presence in the structure of the microscope and can be ignored. Based on Selected Area Electron Diffraction (SAED) (Figure 11c), the analyzed grains were identified as the C1Ti0.8W0.2 phase (98-007-7553, cubic structure, space group Fm-3m).

In addition, two large grains of synthetic PCD were observed in the fusion line, Figure 9. The particles, which were partially fused into the base material, had an oval shape and sizes of up to several dozen micrometers. Single grains of sintered diamond were also observed in the central part of the cross section of the surface coating, as well as in the vicinity of the padding weld. The results of the analysis of the chemical composition of the synthetic PCD grain embedded in the substrate material and the matrix fragment surrounding it are presented in Figure 12.

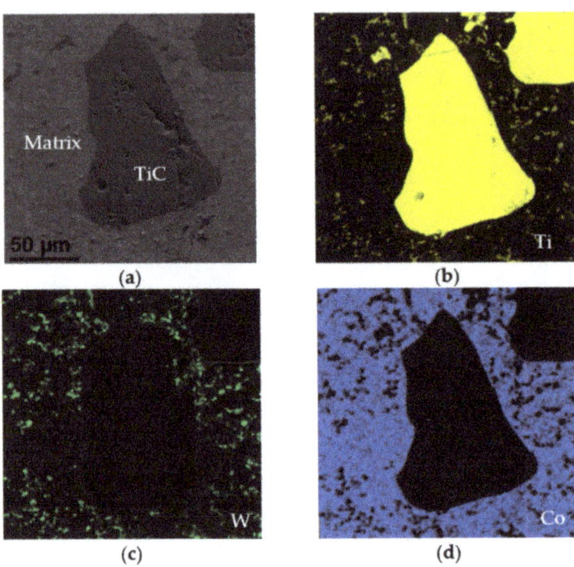

Figure 10. Titanium carbide particle: (**a**) SEM image (SE detector) and EDS maps of element distribution, (**b**) titanium—yellow, (**c**) tungsten—green, (**d**) cobalt—blue.

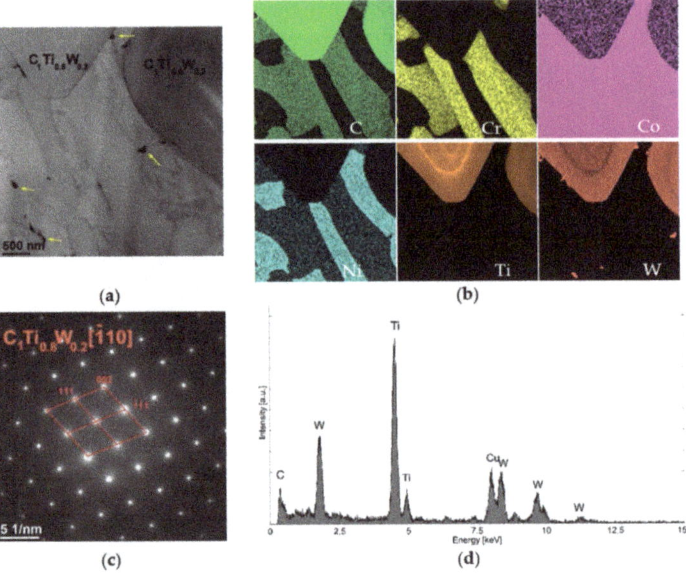

Figure 11. Structure of the area containing C1Ti0.8W0.2 carbides: (**a**) TEM image, (**b**) maps of C, Cr, Co, Ni, Ti and W distribution (C and Ti determined by EELS, other by EDS), (**c**) SAED electron diffraction of C1Ti0.8W0.2 in the [−110] direction and (**d**) EDS spectrum of C1Ti0.8W0.2 carbide.

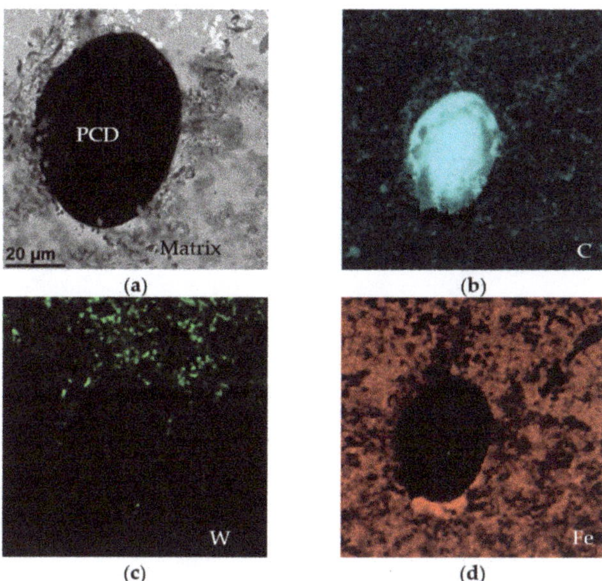

Figure 12. PCD particle embedded in the substrate material: (**a**) SEM image (BSE detector) and element distribution maps: (**b**) carbon—blue, (**c**) tungsten—green, (**d**) iron—red.

In order to fully characterize the grain structure of the PCD particles, TEM observations were performed. It is worth noting that, because of the large differences in the hardness of the components of the tested material (diamond-matrix), making lamellas using the FIB technique turned out to be extremely difficult. The obtained lamellas had a heterogeneous thickness, contained significant amounts of Pt deposited for protective purposes and were subject to strong deformation as a result of stresses. Figure 13 shows a TEM image of a lamella with a fragment of a PCD particle.

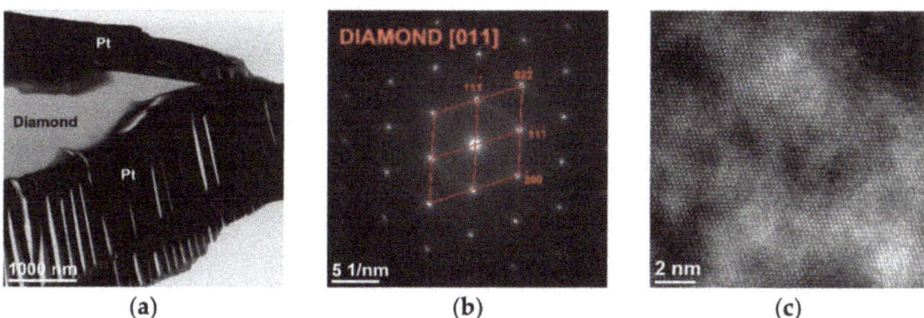

Figure 13. Synthetic PCD: (**a**) TEM image—BF, (**b**) SAED diffraction of the PCD particle towards [011], (**c**) HRTEM image.

Electron diffraction has a selected area aperture (Figure 13b) of a diamond structure (96-901-1576 cubic structure, Fd-3m space group), while high resolution HRTEM imaging of its crystal showed a non-defective structure (Figure 13c).

It was noted that some of the tungsten from the protective coating that covered the PCD particles went into solution, Figure 12a,c. The rest of the element chemically reacted with the carbon contained in the diamond to form a thin WC coating. This coating had good thermal stability, ensured the cohesion of the particle, increased its strength, and also improved the thermal conductivity of the alloy, which could indirectly increase its resistance to abrasive wear.

In Figure 11a, there are several small irregular particles that contain tungsten. The size of the particles did not exceed several dozen nanometers. Precipitates of this type were only located in the area of the fusion line, where the iron content in the matrix was high. An example of such a precipitate located on the C1Ti0.8W0.2 carbide boundary (marked A) is shown in Figure 14a.

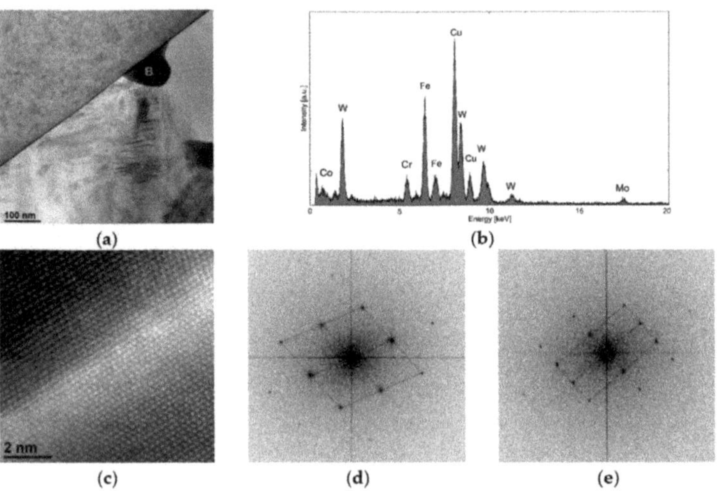

Figure 14. Precipitate of Cr0.4Fe0.475W0.125 on the boundary of C1Ti0.8W0.2 carbide: (**a**) TEM image, (**b**) EDS analysis of the precipitate of Cr0.4Fe0.475W0.125 from the area marked as B, (**c**) STEM– HAADF image of the boundary between the C1Ti0.8W0.2 carbide and the precipitate of Cr0.4Fe0.475W0.125, (**d**) Fourier transformation of the area denoted as A (identified as C1Ti0.8W0.2 in the [–110] direction), (**e**) Fourier transformation of the area denoted as B (identified as Cr0.4Fe0.475W0.125 in the [–110] direction).

The precipitate (marked B) contained 46% Fe, 22% W, 12% Cr, 16% Co and 4% Mo (Figure 14b). Based on the STEM-HAADF image obtained and the Fourier transformation (Figure 14c–e), the analyzed precipitate was identified as the Cr0.4Fe0.475W0.125 phase (98-062-6001, cubic structure, space group I-43m).

The properties of cobalt alloys are largely due to their crystallographic structures. In the developed powder, alloying elements such as Fe, Ni and C stabilize the cubic structure A1 of cobalt, which below a temperature of 417 °C is transformed into crystals of the densely packed hexagonal A3 lattice stabilized by Cr, W and Mo. At ambient temperature, a metastable regular flat-centered phase is often present in cobalt alloys instead of the hexagonal phase. This phase, defined as cobalt (alloy) austenite, is a solid solution of Cr, Ni, Fe, W, Mo or Mn in cobalt [42].

The chemical composition of the coating, especially in the area of the fusion line, differed from the chemical composition of the filler metal used for surfacing due to the proportion of the substrate in the layer. The powder used for the surfacing contained up to 5% Fe, while in the layer an increased presence of this element was found. The matrix microstructure of the composite was chemically heterogeneous. The dendritic area was cobalt austenite strengthened by a solution of elements such as chromium, tungsten or molybdenum. Interdendritic eutectics were rich in chromium, tungsten, silicon and carbides. The Cr alloy was designed to provide corrosion resistance and strengthen the solid solution by creating M_7C_3 and $M_{23}C_6$ carbides. Tungsten reached a content of 22%

in some areas of the coating, contributing to the strengthening of the solid solution and favoring the formation of MC and M6C carbides and intermetallic phases [47].

X-ray diffraction phase and quantification analyses were performed to identify the phases present in the layer. The diffractogram and the result of the X-ray qualitative phase analysis are presented in Figure 15. The analysis showed less than 35% of γ-Co (cobalt austenite) and over 51% of regular carbide with the TiC structure, as well as 14% of hexagonal carbide with a WC structure. The indication of WC probably comes from carbide phases formed on the surface of the synthetic PCD.

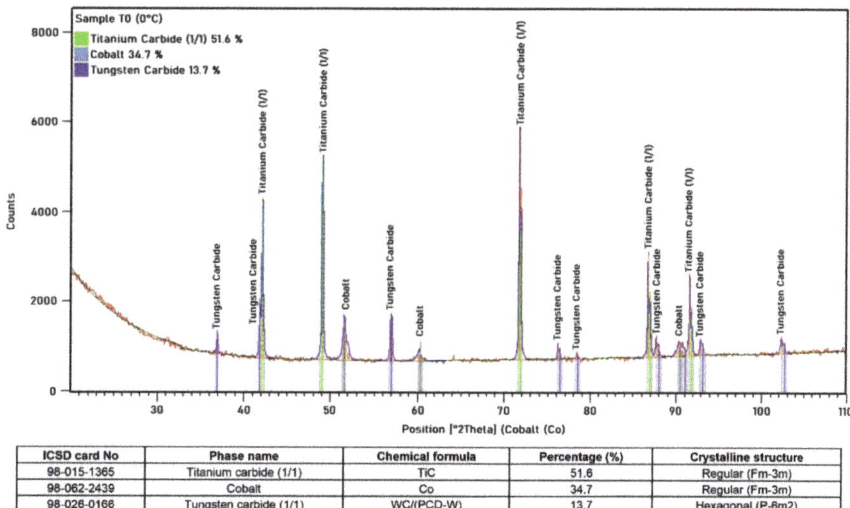

ICSD card No	Phase name	Chemical formula	Percentage (%)	Crystalline structure
98-015-1365	Titanium carbide (1/1)	TiC	51.6	Regular (Fm-3m)
98-062-2439	Cobalt	Co	34.7	Regular (Fm-3m)
98-026-0166	Tungsten carbide (1/1)	WC/(PCD-W)	13.7	Hexagonal (P-6m2)

Figure 15. X-ray diffraction pattern of the composite coating welded by LDMD with marked reference lines of the identified crystalline phases and the results of X-ray qualitative phase analysis.

3.3. Hardness Measurements' Test Results

Measurements of the hardness of the outer surface and the cross-section of the coating are presented in Table 4 and Figure 16.

Table 4. Results of Rockwell C hardness measurement on the outer surface of the coating laser welded with Co3+TiC+PCD composite powder on AISI 4715 low-alloy structural steel.

Specimen Number	Measurement Point Number 1	2	3	4	5	Average Hardness of the Tested Samples	Standard Deviation	Dilution Ratio, (%)
T 0	59.8	61.8	60.9	61.7	60.3	60.9	0.9	2.6
T 100	60.4	59.5	61.2	59.2	60.0	60.1	0.8	3.3
T 200	60.2	57.9	59.4	58.7	59.9	59.2	0.9	6.5
T 300	57.4	58.5	59.7	59.0	58.8	58.7	0.8	8.2

It was assumed that, as the amount of heat supplied to the substrate material increased, the dissolution of the PCD particles would increase, which would affect the hardness of the coating in two ways. First, it was expected that, as PCD particle dissolution increased, the hardness would decrease as the volume fraction of the ceramic phase decreased. At the same time, it was expected that the average microhardness of the alloy matrix would increase as a result of the dissolution of tungsten and carbon. Furthermore, it was taken into account that the hardness of the coating would also be influenced by the dilution ratio of the substrate due to the dissolution of iron and carbon from the substrate. The results of the tests presented in Table 4 show that, even in the case of a large share of the base material in the padding weld, the hardness measurements on the surface of the layer show only a slight downward trend. The share of base material did not affect the final microhardness of the alloy matrix (Figure 16), and thus did not affect the overall hardness of the coating.

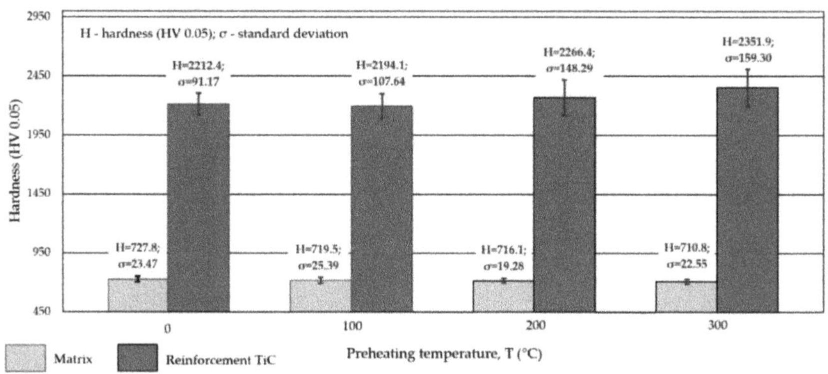

Figure 16. HV hardness measurement results on the cross-section of the coating laser-padded with Co3+TiC+PCD-W composite powder on AISI 4715 low-alloy structural steel.

In the Co-Cr-W-Ni-Fe-Mn-Mo-C alloy, there was no intensive dissolution of the tungsten coating protecting the PCD particle, and tungsten and carbon were already present in the matrix. The transfer of these two elements to the weld metal slightly enriched the alloy matrix with tungsten and carbon and had no significant effect on the microhardness of the cross section of the composite coating.

The hardness of the surface layer tested on the outer surface, depending on the preheating temperature of the substrate material, varied in a range from 58.7 to 60.9 HRC, Table 4. An increase in the preheating temperature of the substrate material by 300 °C in relation to the material not heated before surfacing resulted in a slight decrease in hardness, which was slightly more than 2 HRC. However, the measurements carried out on the cross section of the top coat showed that, depending on the preheating temperature, the average microhardness of the areas between the dendritic areas of the alloy on the cobalt alloy matrix ranged from slightly more than 710 HV0.05 (60.5 HRC) to almost 728 HV0.05 (61.3 HRC).

With an increase in the preheating temperature of the base material, a slight decrease in the average hardness of the composite matrix was observed. The reason for this may be the partial dissolution of the PCD particles and the transfer of tungsten and carbon into the metallic matrix. This explanation is confirmed by the research conducted by Zanzarin et al. [25] and Janicki [27]. Moreover, a slight increase in the microhardness value was found as the measuring point moved away from the fusion line toward the coating surface. This trend may have been influenced by the dissolution of the iron and carbon composite from the substrate into the matrix. The microstructure of the coating in the area adjacent to the heat affected zone differed from that of the padding weld subsurface coating because of the method of crystallization of the liquid metal. Higher hardness may have also resulted from a large number of hard phase particles at the surface of the padding weld. The average hardness of the TiC particles that made up the reinforcement of the matrix was about 2256 HV0.05, and it should be noted that the hardness of the synthetic PCD particles could not be measured.

3.4. The Results of Density Measurement and Testing the Coating Porosity

On the basis of the measurements of the mass of the samples taken from the composite layers, calculations were carried out regarding the variation of the specific density of the composite and the degree of its porosity with regard to the preheating temperature of the base material. The results of the measurements and calculations for three samples in each series are presented in Table 5. Examples of images of the structure of the composite coating obtained as a result of the μCT analysis are shown in Figure 17.

Table 5. The results of the density measurements together with the calculations of the degree of porosity of the top coat laser-welded with the Co3+TiC+PCD-W composite powder on the AISI 4715 low-alloy structural steel.

Physical Quantity	Average Value of the Measured Quantity for Samples			
	T0	T100	T200	T300
Density ρ (g/cm^3)	5.7785	6.0176	6.3805	6.6169
Standard Deviation σ_ρ	0.4092	0.2908	0.2254	0.1978
Open Porosity P_o (%)	6.6467	6.0041	2.7647	1.8161
Closed Porosity P_c (%)	3.5316	2.0263	0.8925	1.0648
Apparent Density ρ_a (g/cm^3)	5.1903	5.5344	6.1472	6.4263
Total Porosity P_c (%)	10.1783	8.0304	3.6572	2.8809

(a)

(b)

(c)

Figure 17. View of the μCT structure of the composite coating welded using the LDMD method, sample T300: (**a**) view of the longitudinal section from the side of the padding weld face, (**b**) view of the coating cross-section surface, (**c**) view of the longitudinal section surface from the base material side.

The composite layer tested was not free of internal porosity. The overall porosity of the coating decreased from more than 10% to less than 3% as the substrate preheat temperature increased. The lower amount of heat supplied to the welded material and the narrow fusion zone—although advantageous for technological and aesthetic reasons—may have impeded the escape of metal vapors from the steam channel and thus promoted the formation of bubbles and porosity in the padding weld. Preheating the base material before padding reduces the rate of crystallization of liquid metal in the weld pool and promotes the desorption of accumulated gases [26,48].

3.5. Abrasive Wear Test Results

The metal-mineral abrasion resistance of the composite coating was determined by calculating its average volume loss after the ASTM G65 test, Table 6. The results obtained were related to the average share of the base metal dilution in the surface layer, Table 4. The character of the abrasive wear of the top coat was assessed on the basis of visual tests (Figure 18) and observations using a scanning laser microscope (Figure 19), determining the average height of the abrasion area profile depending on the preheating temperature of the base material, Figure 20.

Table 6. The results of the metal-mineral abrasive wear resistance test of the composite coating welded with the LDMD method according to ASTM G65.

Specimen Designation	Mass before Test, (g)	Mass after Test, (g)	Mass Loss, (g)	Average Mass Loss, (g)	Clad Layer Density, (g/cm^3)	Average Volume Loss, (mm^3)
No preheating						
T 0_1	149.8935	149.8203	0.0732	0.0806	5.7785	13.9482
T 0_2	149.8704	149.7824	0.0880			
Preheating temperature, T = 100 °C						
T 100_1	149.4675	149.3991	0.0684	0.0613	6.0176	10.1867
T 100_2	150.2985	150.2443	0.0542			
Preheating temperature, T = 200 °C						
T 200_1	150.8568	150.8377	0.0191	0.0184	6.3805	2.8837
T 200_2	149.7372	149.7195	0.0177			
Preheating temperature, T = 300 °C						
T 300_1	149.9748	149.9639	0.0109	0.0101	6.6169	1.5263
T 300_2	149.8394	149.8301	0.0093			

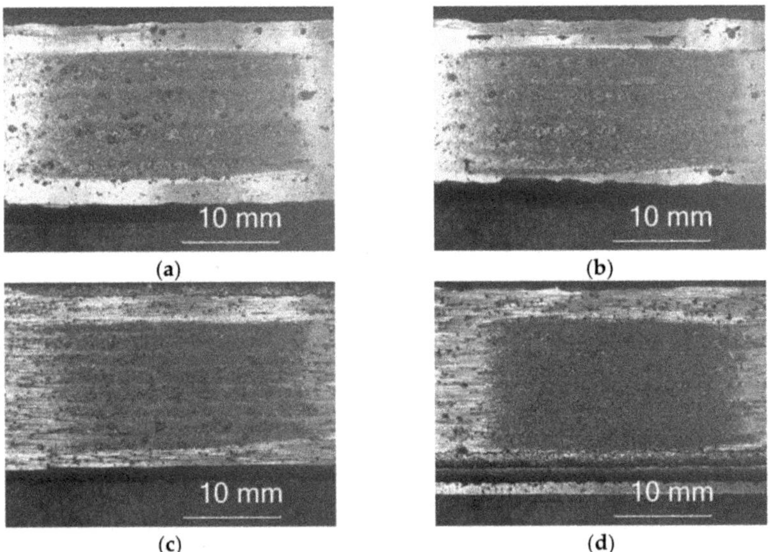

Figure 18. View of the surface of the abrasion area after the metal-mineral abrasion test of a composite coating welded by LDMD: (**a**) without heating the substrate and with heating to a temperature of: (**b**) T = 100 °C, (**c**) T = 200 °C, (**d**) T = 300 °C.

The metal-mineral abrasion test according to ASTM G65, Procedure A, was carried out under medium stress because the sand grains after their interaction with the surface of the test sample were only partially crushed. The mass loss of the composite coating after the abrasive test decreased with increasing preheating temperature of the base material and density of the padding weld. The maximum mass loss of the composite coating, which exceeded 0.08 g, was recorded for the sample made without preheating the base material before padding, and the smallest for the padded coating with preheating the base material to 300 °C. The increase in mass loss is related to the decrease in the density of the composite coating caused by its external and internal porosity, Table 5. The high stress in the layer is also important due to rapid heat dissipation from the unheated substrate, which causes cracks on the surface of the coating and promotes the fragmentation of TiC particles as result of brittle fracture. Analysis of surface condition after the abrasion test showed an abrasive wear mechanism. The strongly embedded and evenly distributed hard phase in the form of TiC and PCD particles in the matrix of the cobalt alloy constituted a natural and effective barrier for the abrasive medium.

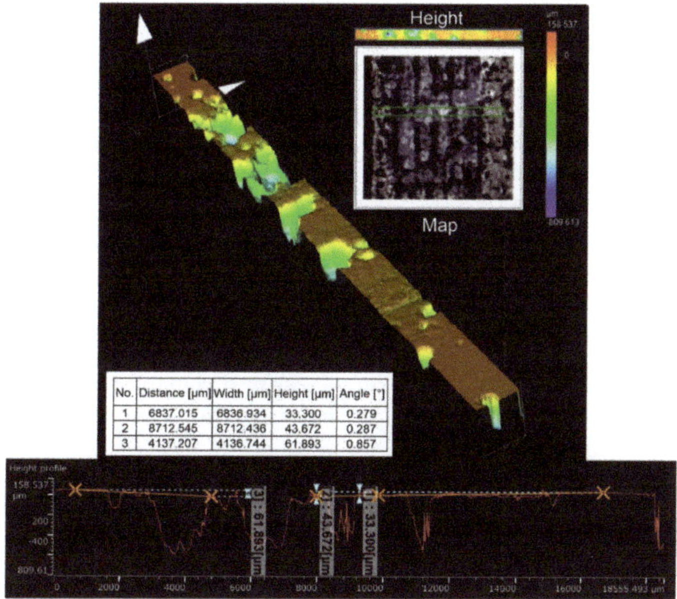

Figure 19. Image and surface profile after the metal-mineral abrasion test of the composite coating (T300 sample) obtained with a scanning laser microscope.

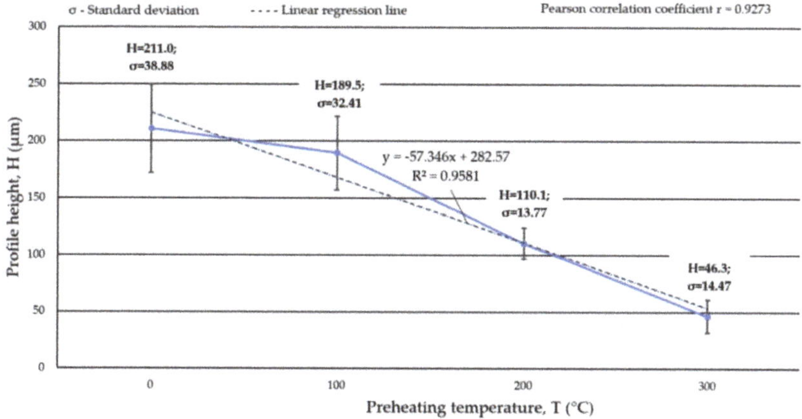

Figure 20. Dependence of the height of the profile of the abrasion area of the composite coating surface on the preheating temperature of the substrate material.

The main wear mechanism of the hardfaced coating was micro-cutting in the form of continuous scratches parallel to the weld axis and, to a much lesser extent, grooving of the surface. The course of the cracks deviated from the rectilinear direction in some places, which indicated the effectiveness of the strengthening of the base material and confirmed the presence of hard phases in the structure. The sand formed scratches on the surface of the composite coating with a profile height ranging from 33 to about 240 μm. The width of the crack profile ranged from 400 as much as over 2000 μm, with the deepest and widest dimensions of the wear profile observed at the overlapping point of successive runs of padding welds. In the area of surface abrasion, single spherical craters were found, remaining after the synthetic PCD particles peeled from the matrix. The wear mechanism noted for microcutting a coating welded by LDMD with a Co-Cr-W-Mo alloy binder was confirmed by Lin et al. [49]. In relation to previously published data [43] and our own research [28] on the wear resistance of layers padded with composite binders in the cobalt alloy matrix with the addition of a hard phase in the form of TiC particles, a beneficial effect on abrasive wear was observed from the addition of synthetic PCD particles. In the case of abrasion-resistant composite layers on a cobalt matrix, the abrasive wear resistance increases with an increase in the volume fraction of the ceramic reinforcement. In the case of the designed alloy, the high resistance to abrasive wear was the result of the type, size and shape in the particles of the hard matrix strengthening phase and the test conditions specified in ASTM G65. In the working conditions of the drilling tool, the influence of natural factors is necessary, which include mechanical load, soil structure, hydrogeological considerations and humidity conditions. These factors were not included in the abrasion resistance assessment of the tested composite coating. Currently, LDMD surfacing technology from the teeth of the prototype three-cone bite (Figure 21) using a designed alloy is being refined. The finished tool will be tested under field conditions. The surfacing layer will be subjected to an assessment of resistance to abrasive, impact and thermal wear.

Figure 21. View of the tooth surface of the prototype three-bite auger filled with LDMD with a designed composite filler.

4. Conclusions

The purpose of this research was to assess the effect of the preheating temperature of the base material—low-alloy structural steel of grade AISI 4715—on the susceptibility to cracking, resistance to metal-mineral abrasive wear and metallographic structure LDMD coating, which was innovative in terms of its chemical composition and the type of hard matrix reinforcement phase. In the case analyzed, the substrate material was a low-alloy structural steel AISI 4715 for which the chemical equivalent of carbon CEV = 0.65%. It is assumed that, with CEV values higher than 0.60%, steel is considered difficult to weld, and therefore requires the use of additional treatments regardless of the size and weight of the element being welded. Then, precautions such as preheating the material and maintaining this temperature throughout the surfacing process should be applied. After surfacing, it is also recommended to cool the element very slowly using heating mats, as well as often additional heat treatment. Preheating and meeting the temperature regime during surfacing, especially large components such as a drill bit with milled teeth, are critical factors in terms of surfacing efficiency and weld quality. This treatment allows for an increase in the speed of surfacing, a better melting and flow of the alloy, and a reduction in the likelihood of thermal decomposition of the carbide phase. The results allow us to conclude that the selected chemical composition of the metallic phases as well as the type, amount and size of the hard phase particles of the developed composite powder allow for high accuracy, repeatability of dosing and near perfect melting with the use of laser powder deposition systems. Preheating the base material above 300 °C significantly reduces the susceptibility to cracking and porosity of the metal deposit, reduces internal stresses in the TiC (preventing particle brittleness) and significantly increases resistance to metal-mineral abrasive wear. The process of LDMD with direct feeding of the powder to the weld pool helps to maintain the structural and thermal stability of the synthetic PCD particles. The tungsten from the protective coating of the PCD particles slightly enriches the matrix of the composite and does not significantly increase the hardness measured on the outer surface and the cross section of the surface coating.

5. Patents

The procedure for granting a patent (No. P435997) was initiated before the Patent Office of the Republic of Poland.

Author Contributions: Conceptualization, A.C.; methodology, A.C.; software, M.P.; validation, A.C. and M.P.; formal analysis, A.C. and M.P.; investigation, A.C.; resources, A.C.; data curation, A.C.; writing—original draft preparation, A.C.; writing—review and editing, A.C.; visualization, A.C.; supervision, A.C.; project administration, A.C.; funding acquisition, A.C. All authors have read and agreed to the published version of the manuscript.

Funding: The research was founded by the Silesian University of Technology Rector's habilitation grant 10/050/RGH_20/1006.

Data Availability Statement: The data are not publicly available due to the initiation of a patent procedure (No. P435997).

Conflicts of Interest: The authors declare no conflict of interest. The funders had no role in the design of the study; in the collection, analyses, or interpretation of data; in the writing of the manuscript, or in the decision to publish the results.

References

1. Bhattacharya, S.; Dinda, G.P.; Dasgupta, A.K.; Mazumder, J. Microstructural evolution of AISI 4340 steel during Direct Metal Deposition process. *Mater. Sci. Eng. A* **2011**, *528*, 2309–2318. [CrossRef]
2. Lisiecki, A. Study of Optical Properties of Surface Layers Produced by Laser Surface Melting and Laser Surface Nitriding of Titanium Alloy. *Materials* **2019**, *12*, 3112. [CrossRef] [PubMed]
3. Yang, S.; Liu, W.; Zhong, M.; Wang, Z. TiC reinforced composite coating produced by powder feeding laser cladding. *Mater. Lett.* **2004**, *58*, 24, 2958–2962. [CrossRef]
4. Zhang, P.; Pang, Y.; Yu, M. Effects of WC Particle Types on the Microstructures and Properties of WC-Reinforced Ni60 Composite Coatings Produced by Laser Cladding. *Metals* **2019**, *9*, 583. [CrossRef]
5. Popov, V.V.; Pismenny, A.; Larianovsky, N.; Lapteva, A.; Safranchik, D. Corrosion Resistance of Al–CNT Metal Matrix Composites. *Materials* **2021**, *14*, 3530. [CrossRef] [PubMed]
6. Senthil Kumar, K.; Karthikeyan, S.; Gokul Rahesh, R. Experimental investigation of wear characteristics of aluminium metal matrix composites, Mater. *Today Proc.* **2020**, *33*, 139–3142. [CrossRef]
7. Czupryński, A. Comparison of Properties of Hardfaced Layers Made by a Metal-Core-Covered Tubular Electrode with a Special Chemical Composition. *Materials* **2020**, *13*, 5445. [CrossRef]
8. Yang, J.; Liu, F.; Miao, X.; Yang, F. Influence of laser cladding process on the magnetic properties of WC–FeNiCr metal–matrix composite coatings. *J. Mater. Process. Technol.* **2012**, *212*, 1862–1868. [CrossRef]
9. Li, Q.; Song, G.M.; Zhang, Y.Z.; Lei, T.C.; Chen, W.Z. Microstructure and dry sliding wear behavior of laser clad Ni-based alloy coating with the addition of SiC. *Wear* **2003**, *254*, 222–229. [CrossRef]
10. Niu, X.; Chao, M.J.; Zhou, X.W.; Wang, D.S.; Yuan, B. Research on in-situ synthesis of B4C particulate reinforced Ni-based composite coatings by laser cladding. *Chin. J. Lasers* **2005**, *32*, 1583–1588.

11. Nurminen, J.; Näkki, J.; Vuoristo, P. Microstructure and properties of hard and wear resistant MMC coatings deposited by laser cladding. *Int. J. Refract. Hard Meter.* **2009**, *27*, 472–478. [CrossRef]

12. Kotarska, A.; Poloczek, T.; Janicki, D. Characterization of the Structure, Mechanical Properties and Erosive Resistance of the Laser Cladded Inconel 625-Based Coatings Reinforced by TiC Particles. *Materials* **2021**, *14*, 2225. [CrossRef] [PubMed]

13. Shasha, L.; Yuhang, W.; Weiping, Z. Microstructure and Wear Resistance of Laser Clad Cobalt-based Composite Coating on TA15 Surface, Rare Met. *Mater. Eng.* **2014**, *43*, 1041–1046. [CrossRef]

14. Kusmoko, A.; Dunne, D.; Li, H.; Nolan, D. Laser cladding of stainless steel substrates with Stellite 6. *Mater. Sci. Forum.* **2013**, *773–774*, 573–589. [CrossRef]

15. Díaz, E.; Amado, J.M.; Montero, J.; Tobar, M.J.; Yáñez, A. Comparative study of Co-based alloys in repairing low Cr-Mo steel components by laser cladding. *Phys. Procedia.* **2012**, *39*, 368–375. [CrossRef]

16. Kusmoko, A.; Dunne, D.; Li, H.J. Wear behaviour of Stellite 6 coatings produced on an austenitic stainless steel substrate by laser cladding using two different heat inputs. *Appl. Mech. Mater.* **2014**, *619*, 13–17. [CrossRef]

17. Sousa, V.F.C.; Silva, F.J.G. Recent Advances on Coated Milling Tool Technology—A Comprehensive Review. *Coatings* **2020**, *10*, 235. [CrossRef]

18. Lisiecki, A. Tribology and surface engineering. *Coatings* **2019**, *9*, 663. [CrossRef]

19. Mele, C.; Bozzini, B. Localised corrosion processes of austenitic stainless steel bipolar plates for polymer electrolyte membrane fuel cells. *J. Power Sources* **2010**, *195*, 3590–3596. [CrossRef]

20. Tomków, J.; Czupryński, A.; Fydrych, D. The abrasive wear resistance of coatings manufactured on high-strength low-alloy (HSLA) offshore steel in wet welding conditions. *Coatings* **2020**, *10*, 219. [CrossRef]

21. Tong, W.; Zhang, X.; Li, W.; Liu, Y.; Li, Y.; Guo, X. Effect of Laser Process Parameters on the Microstructure and Properties of TiC Reinforced Co-Based Alloy Laser Cladding Layer. *Acta Metall. Sin.* **2020**, *56*, 1265–1274. [CrossRef]

22. Zhang, W. Research on Microstructure and Property of TiC-Co Composite Material Made by Laser Cladding. *Phys. Procedia* **2012**, *25*, 205–208. [CrossRef]

23. Tong, W.; Zhao, Z.; Zhang, X.; Wang, J.; Guo, X.; Duan, X.; Liu, Y. Microstructure and properties of TiC/Co-based alloy by laser cladding on the surface of nodular graphite cast iron. *Acta Metall. Sin.* **2017**, *53*, 472–478. [CrossRef]

24. Bartkowski, D.; Kinal, G. Microstructure and wear resistance of Stellite-6/WC MMC coatings produced by laser cladding using Yb:YAG disk laser. *Int. J. Refract. Met. Hard Mater.* **2016**, *58*, 157–164. [CrossRef]

25. Zanzarin, S.; Bengtsson, S.; Molinari, A. Study of carbide dissolution into the matrix during laser cladding of carbon steel plate with tungsten carbides-stellite powders. *J. Laser Appl.* **2015**, *27*, S29209. [CrossRef]

26. Paul, C.P.; Alemohammad, E.; Toyserkani, E.; Khajepour, A.; Corbin, S. Cladding of WC-12 Co on low carbon steel using a pulsed Nd:YAG laser. *Mater. Sci. Eng. A* **2007**, *464*, 170–176. [CrossRef]

27. Janicki, D. High power direct diode laser cladding of Stellite 6+WC coatings. In Proceedings of the IX International Congress-Machines, Technologies, Materials, Varna, Bulgaria, 8–11 September 2012; pp. 27–30.

28. Czupryński, A. Microstructure and Abrasive Wear Resistance of Metal Matrix Composite Coatings Deposited on Steel Grade AISI 4715 by Powder Plasma Transferred Arc Welding Part 1.Mechanical and Structural Properties of a Cobalt-Based Alloy Surface Layer Reinforced with Particles of Titanium Carbide and Synthetic Metal–Diamond Composite. *Materials* **2021**, *14*, 2382. [CrossRef]

29. *EN 14700*; Welding Consumables. Welding Consumables for Hardfacing. CEN: Brussels, Belgium, 2014.

30. *ISO 17637*; Non-Destructive Testing of Welds—Visual Testing of Fusion-Welded Joints. ISO: Geneva, Switzerland, 2016.

31. *ISO 3452*; Non-Destructive Testing—Penetrant Testing—Part 2: Testing of Penetrant Materials. ISO: Geneva, Switzerland, 2013.

32. *ASTM D792*; Standard Test Methods for Density and Specific Gravity (Relative Density) of Plastics by Displacement. American Society for Testing and Materials: West Conshohocken, PA, USA, 2020.

33. *ISO 6508*; Metallic Materials—Rockwell Hardness Test—Part 1: Test Method. ISO: Geneva, Switzerland, 2016.

34. *ISO 6507*; Metallic Materials—Vickers Hardness Test—Part 1: Test Method. ISO: Geneva, Switzerland, 2018.

35. *ASTM G65-00*; Standard Test Method for Measuring Abrasion Using the Dry Sand/Rubber Wheel Apparatus. American Society for Testing and Materials: West Conshohocken, PA, USA, 2015.

36. *ISO 6520*; Welding and Allied Processes—Classification of Geometric Imperfections in Metallic Materials—Part 1: Fusion Welding. ISO: Geneva, Switzerland, 2009.

37. Czupryński, A.; Żuk, M. Matrix Composite Coatings Deposited on AISI 4715 Steel by Powder Plasma-Transferred Arc Welding. Part 3. Comparison of the Brittle Fracture Resistance of Wear-Resistant Composite Layers Surfaced Using the PPTAW Method. *Materials* **2021**, *14*, 6066. [CrossRef]

38. Gates, J. Wear plate and materials selection for sliding abrasion. *Aust. J. Min.* **2003**, *16*, 28–32.

39. Kik, T.; Moravec, J.; Švec, M. Experiments and Numerical Simulations of the Annealing Temperature Influence on the Residual Stresses Level in S700MC Steel Welded Elements. *Materials* **2020**, *13*, 5289. [CrossRef]

40. Kik, T.; Górka, J.; Kotarska, A.; Poloczek, T. Numerical Verification of Tests on the Influence of the Imposed Thermal Cycles on the Structure and Properties of the S700MC Heat-Affected Zone. *Metals* **2020**, *10*, 974. [CrossRef]

41. Schaupp, T.; Schroeder, N.; Schroepfer, D.; Kannengiesser, T. Hydrogen-Assisted Cracking in GMA Welding of High-Strength Structural Steel—A New Look into This Issue at Narrow Groove. *Metals* **2021**, *11*, 904. [CrossRef]

42. Bober, M.; Senkara, J.; Li, H. Comparative Analysis of the Phase Interaction in Plasma Surfaced NiBSi Overlays with IVB and VIB Transition Metal Carbides. *Materials* **2021**, *14*, 6617. [CrossRef] [PubMed]
43. Cherepova, T.; Dmitrieva, G.; Tisov, G.; Dukhota, O.; Kindrachuk, M. Research on the properties of Co-TiC and Ni-TiC HIP-sintered alloys. *Acta Mech. Et Autom.* **2019**, *13*, 57–67. [CrossRef]
44. Grum, J.; Šturm, R. A new experimental technique for measuring strain and residual stresses during a laser remelting process. *J. Mater. Process. Technol.* **2004**, *147*, 351–358. [CrossRef]
45. Gireń, B.G.; Szkodo, M.; Steller, J. The influence of residual stresses on cavitation resistance of metals—An analysis based on investigations of metals remelted by laser beam and optical discharge plasma. *Wear* **1999**, *233–235*, 86–92. [CrossRef]
46. Sun, B.; Zhang, W.Y.; Lu, J.B.; Wang, Z.X. Microstructure and Hardness of TiC/Ni-Based Coating by Plasma Cladding. *Adv. Mater. Res.* **2012**, *510*, 734–737. [CrossRef]
47. Balagna, C.; Spriano, S.; Faga, M.G. Characterization of Co-Cr-Mo alloys after a thermal treatment for high wear resistance. *Mater. Sci. Eng. C* **2012**, *32*, 1868–1877. [CrossRef]
48. Poloczek, T.; Kotarska, A. Effect of laser cladding parameters on structure properties of cobalt-based coatings. *IOP Conf. Ser. Mater. Sci. Eng.* **2020**, *916*, 012085. [CrossRef]
49. Lin, W.C.; Chen, C. Characteristics of thin surface layers of cobalt-based alloys deposited by laser cladding. *Surf. Coat. Technol.* **2006**, *200*, 4557–4563. [CrossRef]

Article

Characterization of the Structure, Mechanical Properties and Erosive Resistance of the Laser Cladded Inconel 625-Based Coatings Reinforced by TiC Particles

Aleksandra Kotarska *, Tomasz Poloczek and Damian Janicki

Welding Department, Faculty of Mechanical Engineering, Silesian University of Technology, Konarskiego Street 18A, 44-100 Gliwice, Poland; Tomasz.Poloczek@polsl.pl (T.P.); Damian.Janicki@polsl.pl (D.J.)
* Correspondence: Aleksandra.Kotarska@polsl.pl

Abstract: The article presents research in the field of laser cladding of metal-matrix composite (MMC) coatings. Nickel-based superalloys show attractive properties including high tensile strength, fatigue resistance, high-temperature corrosion resistance and toughness, which makes them widely used in the industry. Due to the insufficient wear resistance of nickel-based superalloys, many scientists are investigating the possibility of producing nickel-based superalloys matrix composites. For this study, the powder mixtures of Inconel 625 superalloy with 10, 20 and 40 vol.% of TiC particles were used to produce MMC coatings by laser cladding. The titanium carbides were chosen as reinforcing material due to high thermal stability and hardness. The multi-run coatings were tested using penetrant testing, macroscopic and microscopic observations, microhardness measurements and solid particle erosive test according to ASTM G76-04 standard. The TiC particles partially dissolved in the structure during the laser cladding process, which resulted in titanium and carbon enrichment of the matrix and the occurrence of precipitates formation in the structure. The process parameters and coatings chemical composition variation had an influence on coatings average hardness and erosion rates.

Keywords: Inconel 625; metal matrix composite; laser cladding; erosive wear

Citation: Kotarska, A.; Poloczek, T.; Janicki, D. Characterization of the Structure, Mechanical Properties and Erosive Resistance of the Laser Cladded Inconel 625-Based Coatings Reinforced by TiC Particles. *Materials* **2021**, *14*, 2225. https://doi.org/10.3390/ma14092225

Academic Editors: Claudio Mele and Artur Czupryński

Received: 19 March 2021
Accepted: 21 April 2021
Published: 26 April 2021

Publisher's Note: MDPI stays neutral with regard to jurisdictional claims in published maps and institutional affiliations.

1. Introduction

Metal-matrix composite (MMC) coatings have been constantly developed in recent times due to the increasing demand and requirements of the industry in the field of surface wear resistance. Production of MMC coatings on machine parts can significantly extend their service life and at the same time reduce costs of the regeneration or replacement of worn parts [1–5]. Nickel-based superalloys show attractive properties including high tensile strength, fatigue resistance, high-temperature corrosion and oxidation resistance in aggressive environments, together with high-temperature toughness and ductility. The combination of these properties makes these alloys widely used in many industries including aerospace, chemical and energy industry [6,7]. In addition, nickel-based superalloys are also used as coatings on machine parts to improve their corrosion resistance. The studies conducted by Abioye et al. [8] and Nemecek et al. [9] show the positive effect of laser cladding of Inconel 625 coatings on the corrosion resistance of the S355 and AISI 304 steel surfaces. The production of composite coatings based on nickel-based superalloys has a high application potential due to the combination of the unique properties of these alloys and the increased wear resistance of the surface. The previously conducted studies [10–13] show that the production of nickel-based superalloys coatings reinforced with WC, Cr_3C_2, VC, TiC, TiB_2 particles can improve the hardness and wear resistance of the coatings. Taking into account the advantageous properties of the nickel-based superalloys, it is reasonable to select the reinforcing material with high thermal stability in order to ensure high wear resistance in both low and high temperatures. Titanium carbide (TiC) is not only characterized by high hardness (2859–3200 HV) and strength (240–390 MPa), low density (4.92 g/cm^3), but also has a high melting point (3180 °C) and high thermal stability, which makes it highly attractive for producing nickel-based superalloys matrix composites [14–17]. The laser cladding process is commonly used for the production of MMC coatings. This technology is characterized by many advantages, including high heating and cooling speeds, which leads to the creation of unique structures, the negligible influence of the base material on the chemical composition and properties of the produced coatings, high precision and efficiency [18,19]. Moreover, laser cladding can be used for producing both in-situ and ex-situ MMC coatings [20–23].

Research on the production of TiC reinforced nickel-based superalloys coatings using laser cladding technology was previously conducted by Gopinath et al. [24]. The authors used Inconel 718 as coating matrix material and their research was focused on the effect of changing the thermal cycle in the molten pool on the final MMC coating structure. Jiang et al. [25] produced Inconel 625 based coatings reinforced with nano-TiC particles by micro particles insertion and partial dissolution in the structure. As a result, the hardness and modulus of the produced coating increased in comparison to the Inconel 625 coating. Lian et al. [26] tested the dependence of laser cladding of Ni-based coatings reinforced by TiC particles parameters on coatings hardness and wear resistance. Bakkar et al. [27] investigated the high volume percent TiC reinforced (25–70%) Inconel 625 composites. Cao and Gu [28] investigated the microstructure and properties of Inconel 625 matrix composite coatings with 2.5 wt.% TiC nano-particles. The previously conducted researches in this field [24–28] show that nickel-based superalloys composite coatings reinforced by 2.5–50% TiC particles are characterized by 10–45% higher hardness and up to 6% higher wear resistance than metallic nickel-based superalloys coatings. The results [27] also show that the increased TiC particles content of 50–70% may cause defects in the microstructure due to lack of matrix penetration.

The aim of the following research was the production of Inconel 625-based MMC coatings reinforced with 10, 20 and 40 vol.% of TiC using the laser cladding process. The tests were performed to determine the influence of laser cladding process parameters on the structure, hardness and solid particle erosive resistance of the produced MMC coatings in comparison to Inconel 625 metallic coatings.

2. Materials and Methods

For the study, the flat surface of the base material (10 mm thick, as-received S355JR low-alloy steel, Cognor, Stalowa Wola, Poland) was prepared by grinding, cleaning and degreasing with ethyl alcohol (Stanlab, Lublin, Poland). For the coating production, Inconel 625 (Metcoclad 625, Oerlikon, Westbury, NY, USA, gas atomized spheroidal powder) and TiC (Goodfellow, Huntington, UK, 50–150 µm, purity 99.8%) powders were used. The chemical composition of the base material and Metcoclad 625 powder are presented in Table 1. Powder mixtures of Metcoclad 625 with the addition of 10, 20 and 40 vol.% TiC were prepared, mixed and dried for 1 h at 50 °C. The laser cladding process was carried out without preheating.

Table 1. Chemical composition of S355JR base material and Metcoclad 625 powder.

Material Designation	C	Mn	Si	P	S	Cr	Ni	Mo	Nb	Al	Cu	Fe
						(wt.%)						
S355JR	0.2	1.5	0.2–0.5	max 0.04	max 0.04	max 0.3	max 0.3	-	-	max 0.02	max 0.03	balance
Oerlikon Metcoclad 625	-	-	-	-	-	20.0–23.0	58.0–63.0	8.0–10.0	3.0–5.0	-	-	max 5.0

The laser cladding process was perfomed on the stand equipped with a disc laser TRUMPF Trudisc 3302 (TRUMPF, Ditzingen, Germany) (Table 2), a numerically controlled system for positioning the processed material in relation to the laser head and gravitational powder feeder system. For the laser cladding process, the laser beam focus (diameter of 200 μm) was set 30 mm above the base material surface. For the laser cladding process, argon was used as shielding gas (10 L/min) and powder transporting gas (3 L/min). The powder during the laser cladding process was injected directly into the molten pool. To determine the optimal parameters of laser cladding, single-pass coatings were produced with a laser power range of 1400–2300 W, cladding speed of 0.1–0.25 m/min, powder feed rate of 0.03–0.05 g/mm and heat input of 500–560 J/mm. The parameters' range was chosen based on the previous experience [11,19]. The proceeding analysis of single-pass coating geometry, dilution and TiC particles distribution throughout the volume of the coatings allowed determination of the optimal parameters for producing multi-run coatings (Table 3). The multi-run coatings were produced with a 40% overlap. For each set of parameters, one multi-run coating was prepared.

Table 2. Technical specifications of TRUMPH Trudisc 3302 laser.

Property	Value
Wavelength (μm)	1.3
Maximum output power (W)	3300
Laser beam divergence (mm·rad)	<8.0
Fibre core diameter (μm)	200
Collimator focal length (mm)	200
Focusing lens focal length (mm)	200
Beam spot diameter (μm)	200
Fiber length (m)	20

Table 3. Laser cladding parameters.

Designation	Powder TiC Content (vol.%)	Laser Power (W)	Speed (m/min)	Powder Feed Rate (g/mm)	Coating Thickness (mm)	Coating TiC Content (vol.%)	Dilution (%)
I-01	-	2100	0.25	0.04	1.6	-	3.3
I-02	-	2100	0.25	0.05	2.1	-	2.1
M-01	10	2100	0.25	0.04	1.7	8.8	25.5
M-02	10	2100	0.25	0.05	2.1	9.8	12.6
M-03	20	2100	0.25	0.04	1.8	18.3	17.5
M-04	20	2100	0.25	0.05	2.2	19.6	9.8
M-05	40	2100	0.25	0.04	1.9	38.6	14.6
M-06	40	2100	0.25	0.05	2.3	39.7	7.5

The research included the penetrant tests for the verification of the presence of the cracks (colour contrast technique, penetrant MR 68 NF, developer MR 70, cleaner MR 79, MR Chemie, Unna, Germany) in the multi-run coatings surfaces, the macrostructure observations of fabricated coatings, the microstructure observations and energy dispersive spectroscopy (EDS) analysis using Scanning Electron Microscope (SEM) Phenom World PRO (Thermo Fisher Scientific, Waltham, MA, USA). The coating's general chemical composition was estimated based on the 4 cross-sectional surface EDS analysis results (magnification 1000×, accelerating voltage 15 kV) for each coating. For the etching, the mixture of HNO_3 (Chempur, Piekary Śląskie, Poland), HCl (Chempur, Piekary Śląskie, Poland), acetic acid (Stanlab, Lublin, Poland) and glycerol (Poch, Gliwice, Poland) (etchant 89 according to ASTM E 407-99) was used [29]. The coating's dilution rate was measured using Equation, where F_{BM} is the melted cross-sectional area of the substrate and RA is the cross-sectional area of reinforcement of the clad. The cross-sectional areas were obtained using AutoCad 2018 software (Autodesk, CA, USA).

$$U = \frac{F_{BM}}{F_{BM} + RA} \times 100 \; [\%]$$

X-ray diffraction (XRD) analysis was proceeded using a PANalytical X'Pert PRO diffraction system (Malvern Panalitycal, Malvern, UK) with filtered radiation from the lamp with a cobalt anode. The X-ray diffraction patterns were recorded from the ground coatings surfaces. The diffraction profiles were obtained in the 2θ range between $25°$ and $130°$ in continuous scan mode with a step size of $0.1444°$. The counting time per step was 22.695 s. To assess the produced MMC coatings properties, the Vickers microhardness measurements were performed using Wilson 401MVD Vickers microindentation tester (Wilson Instruments, Instron Company, Norwood, MA, USA) and the solid particle erosive tests (device manufactured in Welding Department, Silesian University of Technology, Gliwice, Poland) were carried out according to the ASTM G76-04 standard [30]. The microhardness measurements were performed with a 200 g load and dwell time of 12 s in three lines across the beads at a distance of 0.7, 1.0 and 1.3 mm from the surface (Figure 1a). The distance between consecutive measuring points was 0.5 mm. Additionally, the microhardness measurements were completed in three lines through the coatings from the surface to the base material with the distance between consecutive measuring points of 0.1 mm (Figure 1b). For the test solid particle erosive test, the Al_2O_3, 50 μm diameter, abrasive particles in dry air were used as erodent. The velocity of abrasive particles was 70 m/s and its feed rate was 2 g/min. The test lasted for 10 min. The tested sample surface was located at a distance of 10 mm to the nozzle. The test was carried out for each sample with an impingement angle of $90°$ and $30°$. For each angle, three tests were performed. As a result of the solid particle erosive test, mass loss was obtained using a laboratory scale with an accuracy of 0.0001 g. The erosion rate was counted for each sample according to ASTM G76–04 standard [30]. After the erosive tests, the received craters were observed on Scanning Electron Microscope ZEISS SUPRA 35 (ZEISS, Jena, Germany).

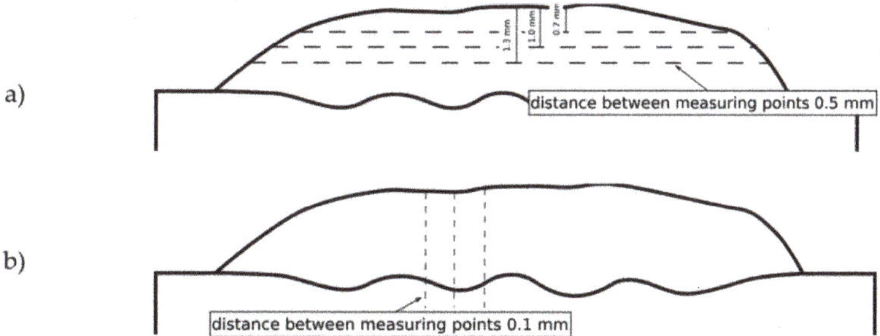

Figure 1. The Vickers microhardness measuring lines scheme, (**a**) measurements across the beads, (**b**) measurements from the coating surface to the base material.

3. Results and Discussion

The macrograph of representative single-pass composite coating is presented in Figure 2. The observations allowed to find that the laser cladded single-pass coating is metallurgically bounded with the base material surface. On the basis of the macroscopic observations of the single-pass coatings, it has been found that the optimal range of laser cladding parameters is very narrow. In the case of the coatings produced with the lowest laser beam power of 1400 W and speed of 0.15 m/min, the insufficient penetration and the TiC particles accumulation on the coatings surface, causing a lack of proper distribution, were observed. The increase in laser beam power to 1850 W, with constant heat input, caused slightly better TiC particles distribution, while in the case of the highest laser beam power 2300 W and speed 0.25 m/min, the penetration and coatings dilution was too high (maximum of 57.2%). The optimal penetration, dilution and reinforcing particles distribution was received for the coatings fabricated by 2100 W power laser beam, with the speed of 0.25 m/min and powder feed rate of 0.04 and 0.05 g/mm.

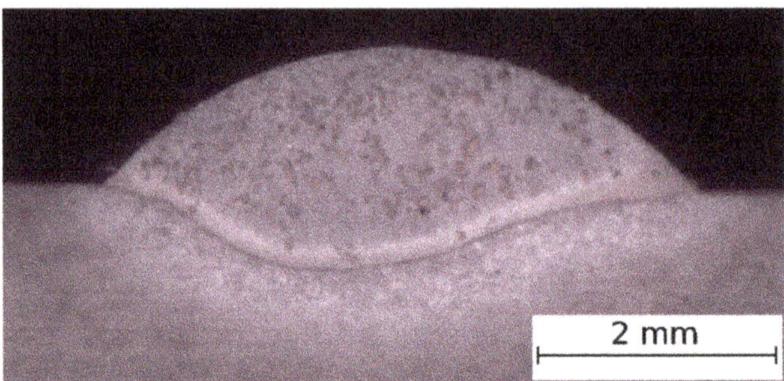

Figure 2. The macrograph of single-pass coating (40% TiC, laser beam power 2100 W, speed 0.25 m/min, powder feed rate 0.05 g/mm).

The surface views after penetrant testing of the produced multi-run coatings are presented in Figure 3. The indications visible around coatings and in the beginning and end areas of the beads resulted from the surface roughness. On the M-02 (Figure 3d) coating can be seen two non-linear indications, which, because of the surface roughness, can be qualified as false indications. The largest linear indications caused by cracks can be observed on the surfaces of coatings M-05 (Figure 3g) and M-06 (Figure 3h), which were produced using a powder mixture with the highest TiC contribution (40 vol.%). The results of this study indicate the negative impact of increased volume fraction of TiC in the Inconel 625 matrix on coating cracking during the laser cladding process with the same parameters. This is due to the increased brittleness of the MMC coating along with the increase in the proportion of the reinforcing phase. As previously investigated [31], the presence of cracks on the surface may deteriorate the erosion resistance.

The macrographs of produced coatings are presented in Figure 4. The thicknesses, dilutions and measured TiC contents of multi-run laser cladded coatings are summarized in Table 3. The average chemical compositions of coatings cross-sectional regions received from EDS are presented in Table 4. These observations and results allowed to assess the impact of powder feed rate and volume fraction of titanium carbide in the powder mixture on coating thickness, penetration, dilution and uniformity of TiC dispersion in the structure. The results show that together with the increase in powder feed rate, the thickness of the coatings increase. The increased TiC particles content in the powder mixture also caused the increase in the thickness of the coatings fabricated with the same parameters. The measured TiC particles content was in each coating lower than the carbides content in the powder mixture used in the laser cladding process. This phenomenon is directly associated with the fusion and dilution of the coating with the base material, which is causing the volume increase of the coating material. The coatings characterized by higher dilution in each laser cladding parameters set (M-01, M-03, M-05) show a higher decrease in measured volume TiC content (Table 3). At a constant laser beam power of 2100 W and a cladding speed of 0.25 m/min, the use of a lower powder feed rate (0.04 g/mm) resulted in the formation of a higher penetration and dilution of coatings with the same chemical composition. However, the higher powder feed rate (0.05 g/mm) resulted in defects near the fusion line in coatings M-04 (Figure 4f) and M-06 (Figure 4h). In the case of increasing the volume fraction of titanium carbide with the use of constant laser cladding parameters, the penetration and dilution of each composite coating decreased. However, the lowest dilution and penetration were measured for metallic Inconel 625 coatings. The higher penetration can be attributed to the increase in laser radiation absorption level by the presence of TiC particles in the powder mixture, which results in increased heat generation.

As a result, the temperature gradient in the molten pool is higher and the mechanism of convective mixing of liquid metal occurs more intensively, leading to higher penetration. On the other hand, the increase of TiC content in the powder mixture leads to Marangoni convection inhibition [32]. As a result, a decrease in penetration can be observed. The coating's dilution influences the average iron composition, which for composite coatings varies from about 3.3 to 17.98 wt.%. Based on the macroscopic observations, it can be also observed that the powder feed rate change in the tested range does not have a significant effect on the uniformity of titanium carbide dispersion in the structure. Coatings with the lowest volume fraction of titanium carbides show the lowest uniformity of its dispersion in the structure. Titanium carbides accumulated in clusters mainly in the upper part of the coatings with a 10 and 20% volume fraction. It is related to the density of this carbide, which is lower than that of the matrix material. Along with increasing the proportion of titanium carbide to 40%, the homogeneity of the coatings improved.

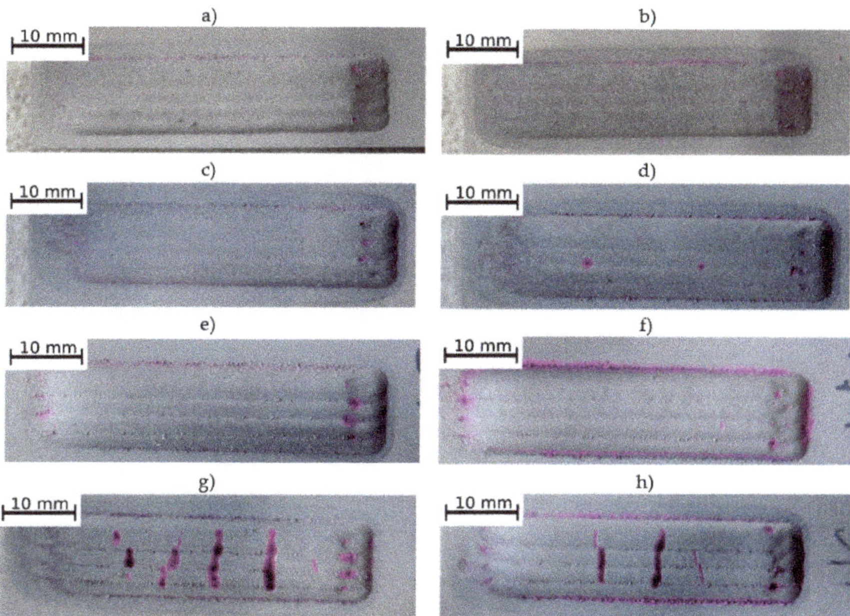

Figure 3. The penetrant test results of laser cladded Inconel 625/TiC coatings (**a**) I-01, (**b**) I-02, (**c**) M-01, (**d**) M-02, (**e**) M-03, (**f**) M-04, (**g**) M-05, (**h**) M-06 (coatings designation according to Table 3).

Table 4. The produced coatings average chemical composition received from EDS surface analysis.

Desigantion	Ni	Cr	Mo	Nb	Fe	Ti
			(wt.%)			
I-01	60.74 ± 1.56	19.82 ± 0.49	10.16 ± 0.79	4.61 ± 0.12	4.67 ± 1.14	-
I-02	63.64 ± 0.63	20.71 ± 0.28	9.47 ± 0.53	4.4 ± 0.58	1.78 ± 0.25	-
M-01	49.13 ± 1.07	16.2 ± 0.41	8.1 ± 0.58	4.7 ± 0.36	17.98 ± 0.95	2.65 ± 0.49
M-02	54.07 ± 4.72	17.59 ± 1.43	9.27 ± 1.09	4.17 ± 0.91	7.89 ± 2.55	3.51 ± 1.05
M-03	50.31 ± 1.28	16.64 ± 0.19	9.73 ± 1.41	3.89 ± 0.34	14.38 ± 0.81	5.37 ± 2.08
M-04	55.13 ± 1.9	18.12 ± 0.5	10.14 ± 1.32	5.1 ± 0.17	5.11 ± 1.58	6.41 ± 1.26
M-05	44.08 ± 2.93	14.11 ± 2.16	7.43 ± 0.87	4.33 ± 0.51	14.48 ± 3.91	14.55 ± 3.75
M-06	47.04 ± 4.94	16.02 ± 1.78	8.59 ± 1.35	4.82 ± 0.76	3.3 ± 1.92	19.73 ± 8.48

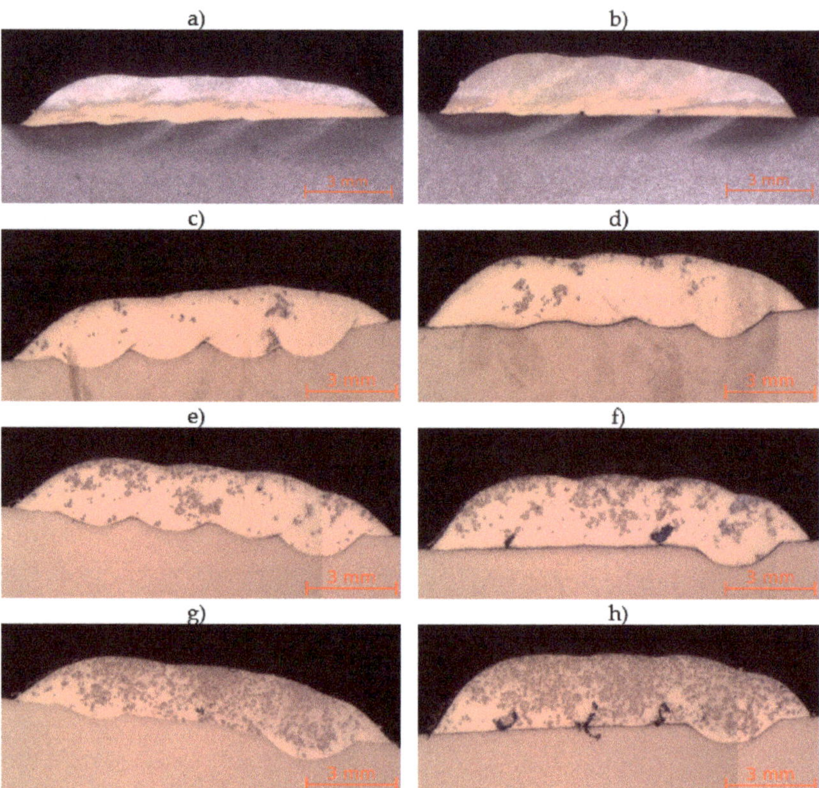

Figure 4. The laser cladded Inconel 625/TiC coatings macrostructures (**a**) I-01, (**b**) I-02, (**c**) M-01, (**d**) M-02, (**e**) M-03, (**f**) M-04, (**g**) M-05, (**h**) M-06 (coatings designation according to Table 3).

The microstructure of metallic Inconel 625 coatings is presented on Figure 5. The produced composite coating's microstructure (Figure 6) consists of Inconel 625 matrix and TiC reinforcing particles (RPs). The matrix microstructure consists of austenite dendrites, confirmed by XRD analysis (Figure 7) and minor secondary phases. The austenite dendrites according to EDS analysis consists (Figure 8) mainly of nickel, chromium and iron, while secondary phases are rich in carbon, niobium, molybdenum and titanium. Since carbon and titanium are not present in the Metcoclad 625 powder and are present in the composite coating's matrix (secondary phases), it can be assumed that the presence of these elements in the matrix is caused by the partial dissolution of titanium carbide RPs. The phenomenon of TiC particles dissolution in Inconel 718 matrix composite coatings depending on molten pool lifetime was investigated in more detail by Gopinath et al. [23]. This is also confirmed by the analysis of the microstructure of metallic Inconel 625 coatings (Figure 5), in which no such precipitates were observed. The metallic Inconel 625 coatings show typical dendritic microstructure with minor constituents in the interdendritic regions, which were previously investigated and reported by Cieslak et al. [33,34]. The columnar dendrites' growth is caused by temperature gradient and occurs in an opposite to the heat transfer direction.

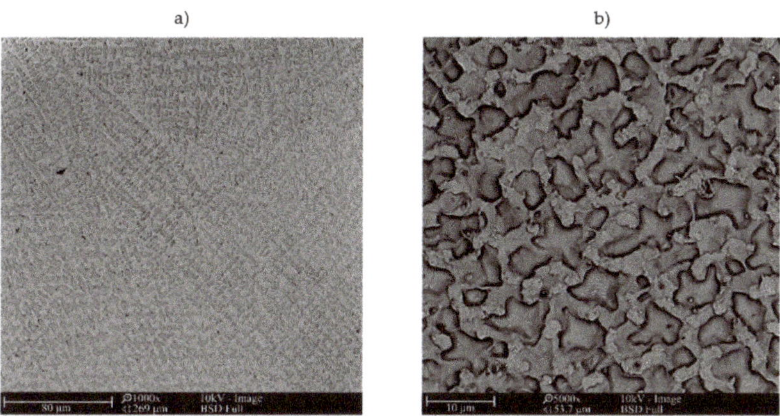

Figure 5. The SEM microstructure of the central bead area of the laser cladded Inconel 625 coating I-01, magnification (**a**) 1000×, (**b**) 5000× (coating designation according to Table 3).

Figure 6. The SEM microstructures of central beads area of laser cladded Inconel 625/TiC coatings; (**a**) M-01 coating, (**b**) M-04 coating, (**c**) M-04 coating (coatings designation according to Table 3).

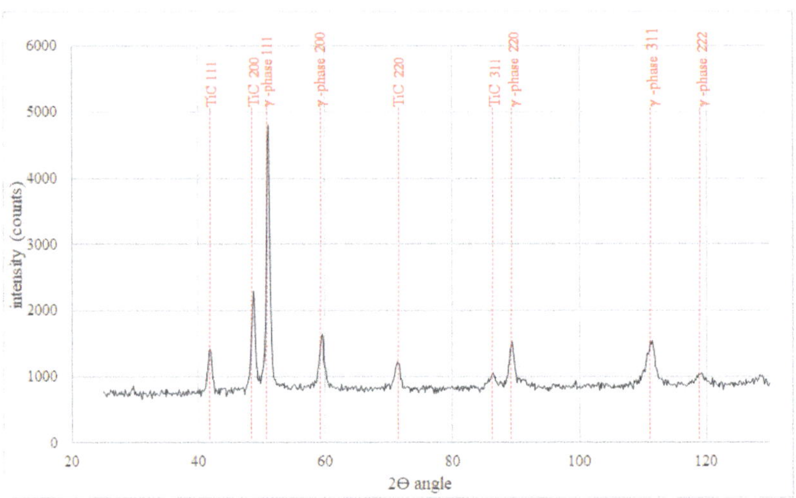

Figure 7. The XRD results of M-05 coating according to Table 3.

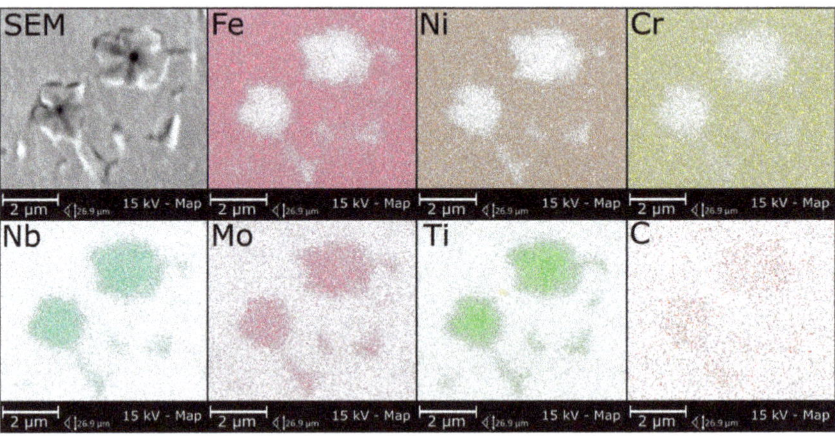

Figure 8. The EDS analysis results of the secondary phases in the matrix, representative coating.

The secondary phases (Figure 6c) were formed in the structure of the composite coatings as a result of enrichment of the matrix with carbon and titanium and are characterized by blocky and dendritic morphology. As can be observed in Figures 6 and 8, the secondary phases show a gradient distribution of chemical composition. The EDS analysis (Figure 8) revealed that the inner, darker part of these precipitates is rich in titanium. It also allowed to find that Mo and Nb atoms dissolved in the secondary phase's crystal lattice. On the basis of the XRD and EDS analysis, it can be assumed that the secondary phases are titanium carbides in the crystal lattice, of which the niobium and molybdenum atoms have been dissolved. In the matrix, the microstructure can also be observed minor eutectic precipitates (Figure 6c) formed on secondary phases. This means that the secondary phases formed first during crystallization, and during further cooling, they behave as crystal nucleus for eutectic precipitates between austenite dendrites.

In the overlap area microstructure (Figure 9), as in the case of the central beads area, the austenite dendrites and secondary phases together with minor eutectic precipitates formed on them can be observed. In addition, due to the higher dissolution of RPs in this area, large dendritic precipitates occur. The higher RPs dissolution resulted in liquid metal enrichment in titanium and carbon in this area. Moreover, as can be observed in Figure 9, TiC RPs are more rounded than in the central bead area and the lighter shell formed around RPs. By comparing the overlap area microstructure with the results obtained by Gopinath et al. [24], it can be assumed that the molten pool lifetime was extended in the overlap area in comparison to the central bead area. The composition of shell around RPs and dendritic precipitates formed in overlap area was tested using EDS analysis (Figure 10). The EDS analysis revealed that both shells around RPs and dendritic precipitates are rich in C, Nb, Mo and Ti. The morphology of formed in overlap area dendritic precipitates is characteristic for TiC particles formed in situ [23]. Therefore, during crystallization in the overlap area, dendritic titanium carbides were formed which dissolved niobium and molybdenum atoms in their crystal lattice.

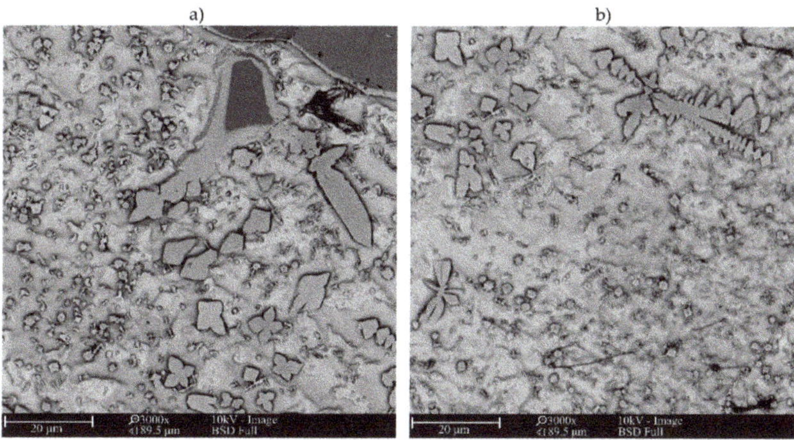

Figure 9. The SEM microstructures of overlap areas of laser cladded Inconel 625/TiC (**a**) M-03 coating, (**b**) M-03 coating (coating designation according to Table 3).

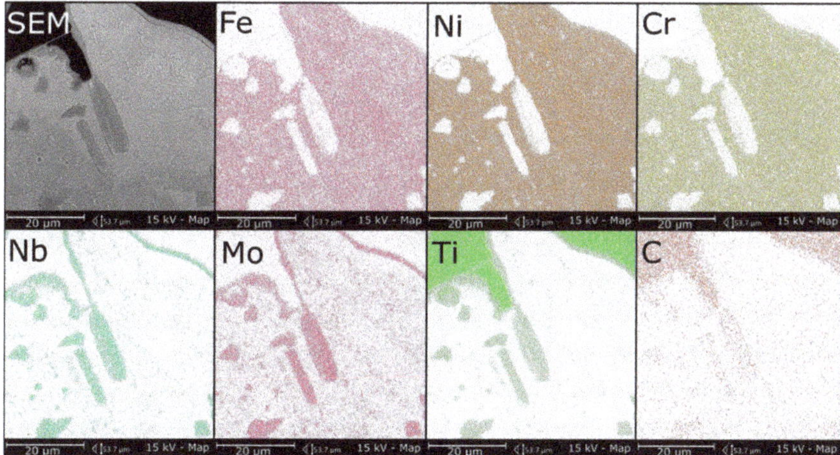

Figure 10. The EDS analysis results of the overlap area, representative coating.

The average Vickers microhardness and erosion rates of produced coatings are presented in Table 5. Figure 11 shows the microhardness distribution of the coatings. The average microhardness of fabricated composite coatings varies from 258 to 342 µHV 0.2. In comparison to metallic Inconel 625 coatings produced with the same laser cladding parameters, the addition of 10 ÷ 40 vol.% of TiC particles to powder mixture led to the average microhardness increase of 5 ÷ 50%. These results are consistent with previous research [25,27]. For composite coatings, the highest average microhardness was measured for M-06 coating with 40 vol.% TiC content, while the lowest average microhardness was measured for M-01 coating with 10 vol.% TiC content. With the increase of TiC content in the structure of coatings fabricated with constant parameters, the average microhardness increased. In the case of coatings with the same chemical composition of the used powder mixture, an increase in the powder feed rate parameter (with a constant power of the laser beam and a constant cladding speed) resulted in an increase in the average microhardness of the coatings, which is associated with lower coating dilution and higher measured TiC particles content. Along with the increase in coatings dilution, the average microhardness of the coatings decreased due to mixing with the base material. The highest values of the standard deviation of microhardness measurements were reported in the case of coatings with the highest content of high hardness titanium carbide. In the case of any of the tested coatings, no significant and repeated changes in hardness were observed in the area of overlapping subsequent beads. The slight decrease in average microhardness towards the end of the measuring lines, which can be observed in Figure 11a, is caused by higher dilution of the first bead, which can be observed on the macrographs. The microhardness distribution from the coatings surface to the base material (Figure 11b) show the highest hardness near the surface and a slight decrease towards the base material. This phenomenon is related to a higher proportion of carbides in the upper area of the coating, due to their lower density than the matrix material, and flowing upwards in the molten metal pool.

Table 5. The average Vickers microhardness and erosion rates of Inconel 625/TiC laser cladded coatings (coatings designation according to Table 3).

Designation	Average Microhardness (μHV 0.2)	Average Erosion Rate, (mg/min)	
		30°	90°
I-01	245.5 ± 11.7	0.65 ± 0.07	0.35 ± 0.01
I-02	235.9 ± 9.5	0.68 ± 0.04	0.37 ± 0.04
M-01	257.7 ± 17.6	0.54 ± 0.09	0.24 ± 0.02
M-02	299.1 ± 21.4	0.51 ± 0.07	0.26 ± 0.04
M-03	274.4 ± 18.5	0.49 ± 0.03	0.27 ± 0.03
M-04	313.5 ± 24.7	0.49 ± 0.05	0.30 ± 0.02
M-05	307.1 ± 38.8	0.45 ± 0.10	0.27 ± 0.01
M-06	342.0 ± 38.3	0.43 ± 0.08	0.29 ± 0.02

The erosion tests showed that for both tested impingement angles, the erosion rates are higher for Inconel 625 metallic coatings than for TiC reinforced composite coatings. Thus, the addition of TiC particles to the powder mixture in 10, 20 and 40 vol.% caused the increase in erosive wear resistance of the Inconel 625 laser cladded coatings. The average erosion rates of all tested coatings with the impingement angle of 30° are higher than the average erosion rates received after the tests carried out with the impingement angle 90°. The dependence of increased erosion wear at a smaller impingement angle (20–30°) compared to the angle of 90° is characteristic for plastic materials [35]. The test results achieved for the 30° impingement angle show that in comparison to the Inconel 625 metallic coating laser cladded with the same parameters, the erosion rate of the TiC reinforced composite coatings decreased by 17 ÷ 37%. In the case of the tests carried out with the impingement angle of 90°, the composite coatings showed 19 ÷ 31% lower erosion rates in comparison to metallic coatings laser cladded with the same parameters. Together with the increase in TiC particle volume content in the powder mixture used for laser cladding of composite coatings, the erosion rate achieved during tests with 30° impingement decreased, while for the tests with 90° impingement angle, the lowest erosion rates were achieved for the powder mixture with 10 vol.% of TiC particles (Figure 12). In this case, with the increase of RPs volume content, the erosion rates slightly increased. It is directly attributed to the increase in the fraction of the TiC phase that is the fraction of brittle material in the coating. Brittle materials are characterized by low erosion resistance at an impingement angle of 90° [35]. Thus, an increase in erosion rate under these conditions with the increase in the TiC fractions is associated with a higher extent of the brittle mechanism of material loss from the eroded surface (Figure 13).

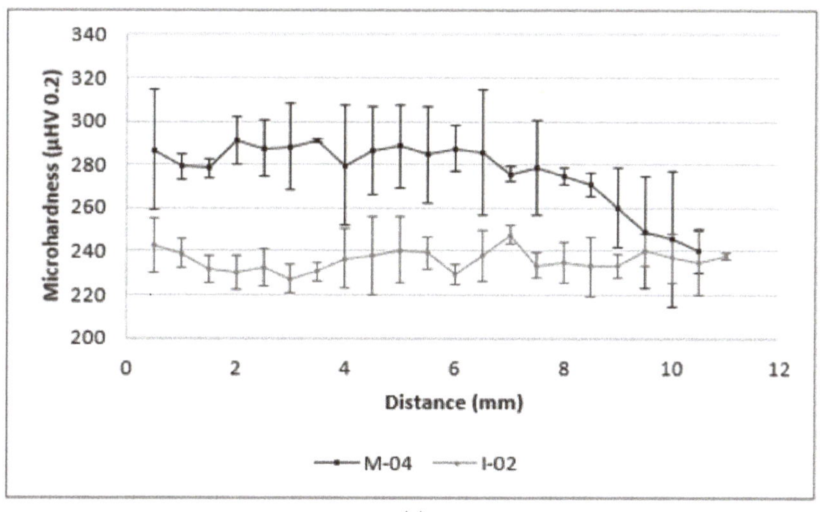

(a)

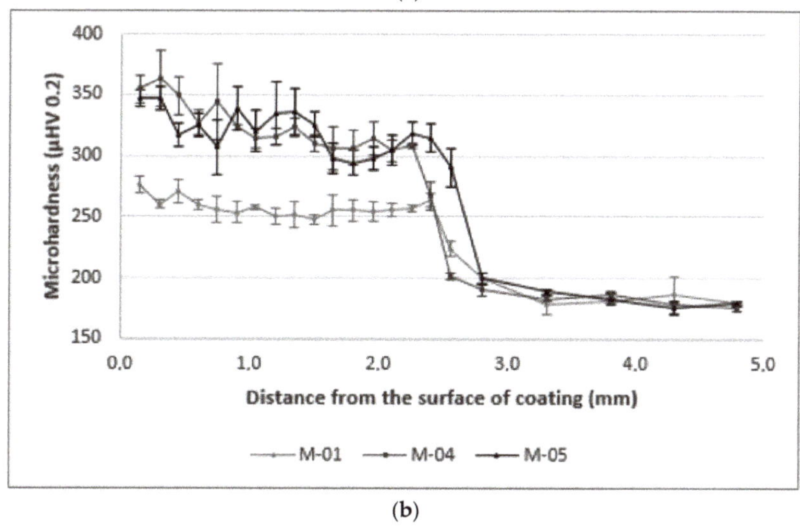

(b)

Figure 11. The Vickers microhardness distribution of the coatings; (**a**) across the subsequent beads according to Figure 1a, (**b**) from the surface to the base material according to Figure 1b (coatings designation according to Table 3).

SEM observations of craters (Figures 13 and 14) after solid particle erosive tests allowed specification of the erosion mechanism of tested coatings. During proceeded erosion, the plastic deformation occurred on the coating's matrix surface. On the analyzed micrographs, the coatings matrix and TiC particles can be observed. In the case of surface tested with 30° impingement angle (Figure 14) in the matrix, scars and narrow grooves occur, which proves the plastic deformation and micro-cutting of the material as a result of the interaction with erosive particles. Observations of the surface of titanium carbides show that the mechanism of their erosive destruction is different. On the surface of the TiC particles, sharp edges are visible, which are the result of brittle destruction and detachment of a part of the material during interaction with accelerated erodent particles. Figure 14b also shows a titanium carbide crack. Due to the different properties of TiC, the mechanism of its erosive destruction is brittle.

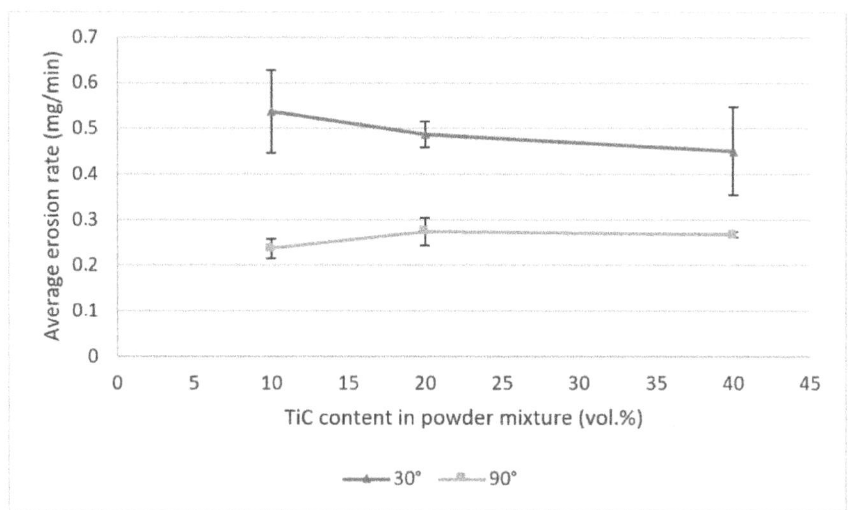

Figure 12. The effect of TiC content in the powder mixture on the average erosion rate of laser cladded composite coatings (laser beam power 2100 W, cladding speed 0.25 m/min, powder feed rate 0.04 g/mm).

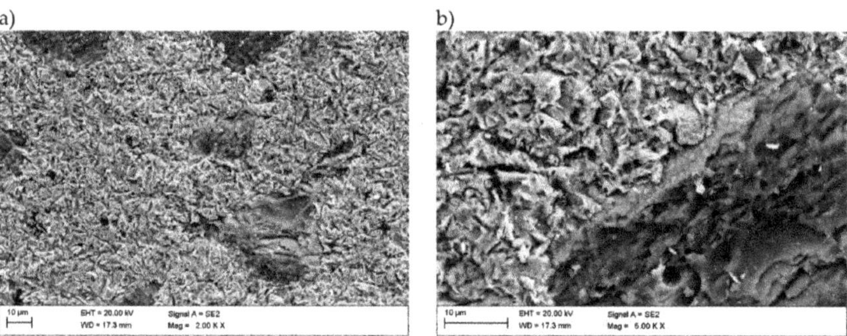

Figure 13. SEM micrographs of craters received after erosion test with impingement angle 90°, magnification (**a**) 2000×, (**b**) 5000×.

The SEM micrographs of craters after erosion test with 90° impingement angle are presented in Figure 13. In this case, the matrix of the coatings was also plastically deformed, but the visible grooves are shorter due to a different trajectory of erosive particles. Similarly, the titanium carbides showed a brittle erosive destruction mechanism. In this case, large smooth areas were observed on the surface of titanium carbides, resulting from the fatigue and brittle detachment of a part of the material due to the interaction with erodent particles.

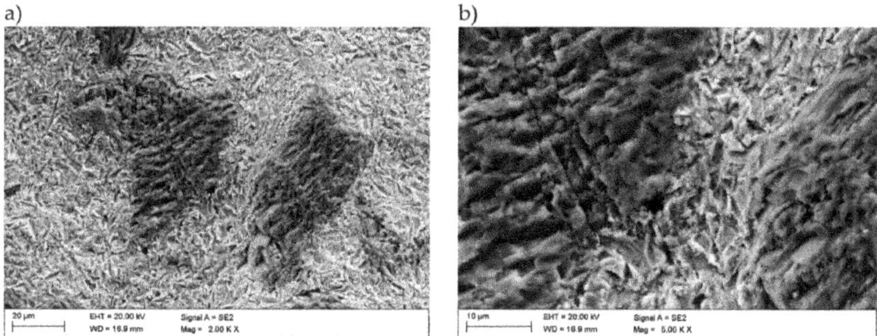

Figure 14. SEM micrographs of craters received after erosion test with impingement angle 30°, magnification (**a**) 2000×, (**b**) 5000×.

4. Conclusions

The research on the production of Inconel 625-based MMC laser cladded coatings reinforced by TiC particles allowed the following conclusions to be drawn:

- The laser cladding process can be used for the production of homogeneous Inconel 625-based MMC coatings reinforced by TiC particles. With a constant laser beam power and cladding speed, along with the increase in powder feed rate, the penetration of base material decreased together with coatings dilution. The powder feed rate has no significant influence on the homogeneity of the coating. The highest homogeneity of the coatings was received using a powder mixture with the highest TiC content (40 vol.%).
- The TiC particles partially dissolved in the structure during the laser cladding process. The enrichment of the matrix in carbon and titanium had an impact on its structure, in which, besides the austenite dendrites, the blocky secondary phases rich in niobium, molybdenum, titanium and carbon appeared. In the overlap area of composite coatings, the increased dissolution of TiC particles occurred, which resulted in the additional formation of dendritic precipitates rich in niobium, molybdenum, titanium and carbon in this area.
- The average microhardness of the composite coatings produced for this research is higher than for the metallic Inconel 625 coatings and varies from 258 to 342 µHV 0.2. The average microhardness increased with the increase in the TiC content in MMC coatings. On the other hand, the increased dilution of the coatings produced with a lower powder feed rate resulted in the average microhardness decrease. No significant and repeated changes in hardness were observed in the area of overlapping subsequent beads.
- In comparison to metallic Inconel 625 laser cladded coatings, the addition of TiC reinforcing particles caused the erosion rates decrease, for both 30° and 90° impingement angles, by 17 ÷ 37% and 19 ÷ 31% respectively. Together with the increase in TiC particle content in MMC coatings, the erosion rate for 30° impingement angle decreased, while for 90° it slightly increased due to the higher content of the brittle phase in the structure. The study allowed definition of the erosive wear mechanism of the coatings. The matrix has been plastically deformed during interaction with the erosive particles, while the TiC particles showed a brittle and fatigue mechanism of erosive wear.

Author Contributions: Conceptualization, A.K. and D.J.; Data curation, T.P.; Formal analysis, A.K., T.P. and D.J.; Funding acquisition, A.K.; Investigation, T.P.; Methodology, A.K. and D.J.; Writing—original draft, A.K. and T.P.; Writing—review and editing, A.K. and D.J. All authors have read and agreed to the published version of the manuscript.

Funding: Publication supported by Own Scholarship Fund of the Silesian University of Technology in the year 2019/2020.

Institutional Review Board Statement: Not applicable.

Informed Consent Statement: Not applicable.

Data Availability Statement: Data sharing not applicable.

Conflicts of Interest: The authors declare no conflict of interest.

References

1. Peat, T.; Galloway, A.; Toumpis, A.; McNutt, P.; Iqbal, N. The erosion performance of particle reinforced metal matrix composite coatings produced by co-deposition cold gas dynamic spraying. *Appl. Surf. Sci.* **2017**, *396*, 1623–1634. [CrossRef]
2. Janicki, D. High power diode laser cladding of wear resistant metal matrix composite coatings. *Solid State Phenom.* **2013**, *199*, 587–592. [CrossRef]
3. Lisiecki, A.; Ślizak, D.; Kukofka, A. Robotized fiber laser cladding of steel substrate by metal matrix composite powder at cryogenic conditions. *Mater. Perform. Charact.* **2019**, *8*, 1214–1225. [CrossRef]
4. Czupryński, A. Flame Spraying of Aluminum Coatings Reinforced with Particles of Carbonaceous Materials as and Alternative for Laser Cladding Technologies. *Materials* **2019**, *12*, 3467. [CrossRef]
5. Czupryński, A. Properties of Al_2O_3/TiO_2 and ZrO_2/CaO Flame-Sprayed Coatings. *Mater. Technol.* **2017**, *51*, 205–212.
6. Feng, K.; Chen, Y.; Deng, P.; Li, Y.; Zhao, H.; Lu, F.; Li, R.; Huang, J.; Li, Z. Improved high-temperature hardness and wear resistance of Inconel625 coatings fabricated by laser cladding. *J. Mater. Process. Technol.* **2017**, *243*, 82–91. [CrossRef]
7. Hu, H.X.; Zheng, Y.G.; Qin, C.P. Comparison of Inconel 625 and Inconel 600 in resistance to cavitation erosion and jet impingement erosion. *Nucl. Eng. Des.* **2010**, *240*, 2721–2730. [CrossRef]
8. Abioye, T.E.; McCartney, D.G.; Clare, A.T. Lacer cladding of Inconel 625 wire for corrosion protection. *J. Mater. Process. Technol.* **2015**, *217*, 232–240. [CrossRef]
9. Nemecek, S.; Fidler, L.; Fiserova, P. Corrosion resistance of laser clads of Inconel 625 and Metco 41C. *Phys. Procedia* **2014**, *56*, 294–300. [CrossRef]
10. Janicki, D. Laser cladding of Inconel 625-based composite coatings reinforced by porous chromium carbide particles. *Opt. Laser Technol.* **2017**, *94*, 6–14. [CrossRef]
11. Janicki, D.; Musztyfaga-Staszuk, M. Direct diode laser cladding of Inconel 625/WC composite coatings. *Stroj. Vestn. J. Mech. E* **2016**, *62*, 363–372. [CrossRef]
12. Nurminen, J.; Nakki, J.; Vuoristo, P. Microstructure and properties of hard and wear resistant MMC coatings deposited by laser cladding. *Int. J. Refract. Met. Hard Mater.* **2009**, *27*, 472–478. [CrossRef]
13. Tang, B.; Tan, Y.; Zhang, Z.; Xu, T.; Sun, Z.; Li, X. Effects of process parameters on geometrical characteristics, microstructure and tribological properties of TiB_2 reinforced Inconel 718 alloy composite coatings by laser cladding. *Coatings* **2020**, *10*, 76. [CrossRef]
14. Cui, C.; Guo, Z.; Wang, H.; Hu, J. In situ TiC particles reinforced grey cast iron composite fabricated by laser cladding of Ni-Ti-C system. *J. Mater. Process. Technol.* **2007**, *183*, 380–385. [CrossRef]
15. Galevsky, G.V.; Rudneva, V.V.; Garbuzova, A.K.; Valuev, D.V. Titanium carbide: Nanotechnolgy, properties, application. *IOP Conf. Ser. Mater. Sci. Eng.* **2015**, *91*, 012017. [CrossRef]
16. Dong, B.-X.; Qiu, F.; Li, Q.; Shu, S.-L.; Yang, H.-Y.; Jiang, Q.-C. The synthesis, structure, morphology characteristics and evolution mechanisms of nanosized titanium carbides and their further applications. *Nanomaterials* **2019**, *9*, 1152. [CrossRef]
17. Mhadhbi, M.; Driss, M. Titanium Carbide: Synthesis, Properties and Applications. *J. Brill. Eng.* **2021**, *2*, 1–11.
18. Lisiecki, A.; Ślizak, D.; Kukofka, A. Laser cladding of Co-based metallic powder at cryogenic conditions. *J. Achiev. Mater. Manuf. Eng.* **2019**, *95*, 20–31. [CrossRef]
19. Poloczek, T.; Kotarska, A. Effect of laser cladding parameters on structure properties of cobalt-based coatings. *IOP Conf. Ser. Mater. Sci. Eng.* **2020**, *916*, 012085. [CrossRef]
20. Man, H.C.; Zhang, S.; Cheng, F.T.; Guo, X. In situ formation of a TiN/Ti metal matrix composite gradient coating on NiTi by laser cladding and nitriding. *Surf. Coat. Technol.* **2006**, *200*, 4961–4966. [CrossRef]
21. Van Nghia, T.; Sen, Y.; Tuan Anh, P. Microstructure and properties of Cu/TiB_2 wear resistance composite coating on H13 steel prepared by in-situ laser cladding. *Opt. Laser Technol.* **2018**, *108*, 480–486.
22. Kotarska, A. Laser surface alloying of ductile cast iron with Ti + 5% W mixture. *Weld. Tech. Rev.* **2019**, *91*, 5. [CrossRef]
23. Janicki, D. Microstructure and sliding wear behaviour of in-situ TiC-reinforced composite surface layers fabricated on ductile cast iron by laser alloying. *Materials* **2018**, *11*, 75. [CrossRef] [PubMed]

24. Gopinath, M.; Mulick, S.; Nath, A.K. Development of process maps based on molten pool thermal history during laser cladding of Inconel 718/TiC metal matrix composite coatings. *Surf. Coat. Technol.* **2020**, *399*, 126100.
25. Jiang, D.; Hong, C.; Zhong, M.; Alkhayat, M.; Weisheit, A.; Gasser, A.; Zhang, H.; Kelbassa, I.; Poprawe, R. Fabrication of nano-TiCp reinforced Inconel 625 composite coatings by partial dissolution of micro-TiCp through laser cladding energy input control. *Surf. Coat. Technol.* **2014**, *249*, 125–131. [CrossRef]
26. Lian, G.; Zhang, H.; Zhang, Y.; Yao, M.; Huang, X.; Chen, C. Computational and Experimental Investigation of Micro-Hardness and Wear Resistance of Ni-Based Alloy and TiC Composite Coating Obtained by Laser Cladding. *Materials* **2019**, *12*, 793. [CrossRef]
27. Bakkar, A.; Ahmed, M.M.Z.; Alsaleh, N.A.; El-Sayed Seleman, M.M.; Ataya, S. Microstructure, wear, and corrosion characterization of high TiC content Inconel 625 matrix composites. *J. Mater. Res.* **2019**, *8*, 1102–1110. [CrossRef]
28. Cao, S.; Gu, D. Laser metal deposition additive manufacturing of TiC/Inconel 625 nanocomposites: Relation of densification, microstructures and performance. *J. Mater. Res.* **2015**, *30*, 3616–3628. [CrossRef]
29. *ASTM E407-99: Standard Practice for Microetching Metals and Alloys*; ASTM International: West Conshohocken, PA, USA, 1999.
30. *ASTM G76-04: Standard Test Method for Conducting Erosion Tests by Solid Particle Impingement Using Gas Jets*; ASTM International: West Conshohocken, PA, USA, 2004.
31. Kotarska, A.; Janicki, D.; Górka, J.; Poloczek, T. Solid particle erosion of laser surface melted ductile cast iron. *Arch. Foundry Eng.* **2020**, *20*, 105–111.
32. Janicki, D. *Shaping the Structure and Properties of Surface Layers of Ductile Cast Iron by Laser Alloying*; Wydawnictwo Politechniki Śląskiej: Gliwice, Poland, 2018; p. 50.
33. Cieslak, M.J.; Headley, T.J.; Romig, A.D.; Kollie, T. A melting and solidification study of alloy 625. *Metall. Trans. A* **1988**, *19A*, 2319–2331. [CrossRef]
34. Cieslak, M.J. The welding and solidification metallurgy of Alloy 625. *Weld. J.* **1991**, *70*, 49–56.
35. Stachowiak, G.W.; Batchelor, A.W. *Engineering Tribology*, 4th ed.; Butterworth-Heinemann: Oxford, UK, 2014; pp. 551–556.

 *materials*

Article

Comparative Analysis of the Phase Interaction in Plasma Surfaced NiBSi Overlays with IVB and VIB Transition Metal Carbides

Mariusz Bober [1],* , Jacek Senkara [1] and Hong Li [2],*

1 Faculty of Mechanical and Industrial Engineering, Warsaw University of Technology, Narbutta 85, 02-524 Warsaw, Poland; jacek.senkara@pw.edu.pl
2 Faculty of Materials and Manufacturing, Beijing University of Technology, Beijing 100124, China
* Correspondence: mariusz.bober@pw.edu.pl (M.B.); lih_bjut@163.com (H.L.)

Abstract: Important applications of transition metal carbides (TMCs) are as wear resistant composite layers deposited by plasma transferred arc welding (PTAW) and laser methods. Growing interest in them has also been observed in additive manufacturing and in HEA technology (bulk composite materials and layers), and in the area of energy conversion and storage. This paper presents the results of comparative studies on interfacial interactions in the NiBSi−TMCs system for two border IVB and VIB TM groups of the periodic table. Model (wettability and spreadability) and application experiments (testing of the PTAW-obtained carbide particle−matrix boundaries) were performed. Fe from partially melted steel substrates is active in the liquid NiBSi−TMCs system. It was revealed that the interaction of TMCs with the liquid NiBSi matrix tends to increase with the group number, and from the top to bottom inside individual groups. Particles of IVB TMCs are decomposed by penetration of the liquid along the grain boundaries, whereas those of VIB are decomposed by solubility in the matrix and secondary crystallization. No transition zones formed at the interfacial boundaries of the matrix−IVB group TMCs, unlike in the case of the VIB group. The experimental results are discussed using the data on the TMC electronic structure and the physicochemical properties.

Keywords: transition metal carbides; NiBSi alloy; phase interaction; PTAW; composite layers

check for
updates

Citation: Bober, M.; Senkara, J.; Li, H. Comparative Analysis of the Phase Interaction in Plasma Surfaced NiBSi Overlays with IVB and VIB Transition Metal Carbides. *Materials* **2021**, *14*, 6617. https://doi.org/10.3390/ma14216617

Academic Editors: Claudio Mele and Artur Czupryński

Received: 27 September 2021
Accepted: 29 October 2021
Published: 3 November 2021

Publisher's Note: MDPI stays neutral with regard to jurisdictional claims in published maps and institutional affiliations.

1. Introduction

Surface composite layers (SCLs) consisting of a metal matrix strengthened with hard particles of high-melting phases (simple carbides, borides, nitrides, or more complex phases) combine the properties of an abrasion-resistant and relatively plastic matrix and hard ceramics, often with the effect of synergy. Such coatings can be deposited onto large surfaces, on selective areas (flat or curved), or on the edges. They meet the industrial needs wherever high wear resistance is required at parallel dynamic loads, often in corrosive environments and in elevated temperatures. They have been known and used for a relatively long time. Despite the fact that almost all welding methods can be applied to obtain such SCLs, they are usually obtained from powders by padding, re-melting, or thermal spraying techniques. Today, the implementation of laser and plasma beams is mainly used for this purpose (see [1–5], for instance). In general, variants of all of the following methods are utilized in the production of SCLs [6–10]:

- injection of refractory particles to the previously initiated liquid weld pool,
- simultaneous introduction of metal−ceramic powder mixtures to the beam or arc,
- melting pre-placed powder mixtures at the surface,
- strengthening of the matrix by in-situ formed particles.

During the preparation of SCLs, regardless of the method, an interaction occurs between the solid refractory particles and the liquid matrix material. As a result, interfacial boundaries are formed. This is the case for all types of upper layer formations in the presence of a liquid phase, namely alloying, buttering, cladding, overlaying, and hardfacing. Phenomena at the interfaces are crucial from the point of view of layer formation and its subsequent exploitation [11–13]. If the strengthening phase particles are wetted well with the liquid matrix, a good adhesion and low interfacial energy are achieved, and they are kept in the liquid pool. Their distribution in the matrix also depends on wettability. The particle—matrix interface type depends on the chemical affinity of the contacting phases and can be formed as a result of an adhesive (in inert systems) or more complex interaction in the diffusion or in the reaction-controlled systems [14]. The type of resulting interface—with or without an intermediate zone—is important for the sake of the load transfer to the strengthening phase and the level of residual stresses, and hence for the behavior of the whole SCL under the external load. Moreover, in the case of service at elevated temperatures, the transition zone may enlarge, change, or even degrade due to the volume or reactive diffusion.

Recently, there has been increased interest in SCLs with transition metal carbides (TMCs). An impetus was brought about by the new applications in other advanced technologies. Apart from being used as only surface layers of machine parts, they have found applications in additive manufacturing technologies for the production of massive 3D printed wear resistance parts using laser or electron beams [15,16]. Their new capabilities and increasing interest are also related to the applications of TMCs in the areas of energy storage and conversion in batteries and fuel cells [17,18], as well as in catalysis [19,20]. Another developing direction is a new class of high-entropy materials of composite structure containing TMCs [21–25]. This causes a need for more in-depth research on composites with TMCs. While many publications in the field have focused on the methods of producing SCLs, their structure and tribological properties, computer modeling, and process simulation, much less information has emerged about the interfacial interactions.

The aim of this work is to analyze the interaction between the matrix and individual carbides in the plasma transferred arc welding (PTAW) process so as to find the regularities resulting from the position of the transition metal forming the carbide in the periodic table of elements. This paper presents the results of comparative studies on interfacial interactions in the system of NiBSi-TMCs for two border TM groups (IVB and VIB). Basic and application experiments were performed for this purpose. The first relied on the investigation of the wetting and spreading of the liquid alloy droplets on a flat carbide surface at a macroscopic scale under controlled temperature conditions. In the framework of application research, composite overlays were made from powders using the PTAW method with subsequent testing of the carbide particles—matrix boundaries. During the technological process, a dynamic interaction occurred between the liquid phase and the curved surfaces of the small strengthening particles.

The NiBSi alloy was selected as the matrix material for the SCLs because it is produced commercially in the form of a powder for the manufacture of overlays that exhibit excellent corrosion, abrasion, and wear-resistance at up to 600 °C [26]. The alloy comes from the Ni-rich corner of the Ni-B-Si system, but with a relatively low melting range due to the presence of B and Si additives that act as a melting point depressants [27].

2. Materials and Methods

2.1. Materials

Two forms of materials were applied in the research:

- Specially prepared massive solids for the model wettability and spreadability tests;
- Commercial powders for plasma surfacing.

2.1.1. Composite Matrix Material

The nominal chemical composition of the NiBSi alloy of the purchased powder according to the manufacturer certificate is presented in Table 1, along with the composition determined by the atomic absorption spectrophotometry (AAS) method. The granularity of this powder was in the range of 45 ÷ 150 μm. Its morphology is presented in Figure 1a.

Table 1. The chemical compositions of NiBSi matrix materials used.

Material		Elements, Wt%				
		C	Si	B	Fe	Ni
Manufacturer's certificate	Commercial powder	0.03	2.40	1.40	0.40	Bal.
AAS measured	Commercial powder	-	2.18	1.21	0.09	Bal.
	Cast alloy φ 3 mm	-	2.45	1.17	0.20	Bal.

Figure 1. The morphology of powders for the production of composite layers using the plasma surfacing method: (**a**) NiBSi, (**b**) TiC, (**c**) ZrC, (**d**) Cr$_3$C$_2$, (**e**) Mo$_2$C, and (**f**) WC.

The solid alloy of the corresponding composition was used in the wettability and spreadability tests. It was obtained by melting the pure elements in a vacuum induction furnace. In order to avoid segregation of the components during the ingot crystallization and following troublesome plastic forming, the liquid melt was sucked off by pure Ar gas directly from the crucible into quartz tubes with an internal diameter of 3 mm. The rods formed in this way were cut into pieces with equal lengths of 3 mm. The compatibility of the chemical composition of the material produced in this way with the commercial powder was confirmed by the AAS analysis (Table 1). The solidus and liquidus temperatures of the obtained alloy, determined by the differential thermal analysis (DTA) method, were 1075.5 and 1086 °C, respectively.

2.1.2. Strengthening Carbide Phase

The TMCs powders of the IVB and VIB group of the periodic table were used for the strengthening phase. Commercial powders of carbides of titanium (TiC), zirconium (ZrC), chromium (Cr$_3$C$_2$), molybdenum (Mo$_2$C), and tungsten (WC), of a similar grain size 80–200 µm, were used for the preparation of the composite layers. The morphology of the particular powders is shown in Figure 1b–f. Carbide powders were mixed with the matrix NiBSi powder before hardfacing.

The carbide substrates for the wetting and spreading tests were made in the form of continuous layers over molybdenum plates using the magnetron sputtering method. In this way, possible errors due to the harmful impact of activators and residual oxide inclusions on wettability were avoided, if commercial sintered carbide materials were used instead. Mo plates were selected due to their high melting point (2883 K) and low thermal expansion coefficient ($\lambda = 4.9 \times 10^{-6}$ K^{-1}) [28] so as to prevent a high stress level at the boundary with the carbide layers. Then, 18 × 18 mm plates were cut from a 1 mm thick molybdenum sheet and polished. These prepared substrates were covered with coatings of five different carbides (i.e., titanium, zirconium, chromium, molybdenum, and tungsten carbide) in the process of magnetron sputtering of suitable metal targets in the reactive gas atmosphere (mixture of argon and acetylene). The details of the whole procedure are described in [29]. As a result, continuous coatings adhesively bonded to the molybdenum plates were obtained. The thicknesses of these coatings are presented in Table 2. Measured surface roughness was small and comparable reflecting the roughness of the Mo plates (R_a parameter < 0.8 μm for all cases). X-ray diffraction (XRD) phase analysis confirmed the composition of the five planned carbides (Figure 2).

Table 2. Thicknesses of the deposited carbide coatings (average of five measurements).

Type of Coating	TiC	ZrC	Cr_3C_2	Mo_2C	WC
Coating thickness [μm]	4.7	3.0	2.4	7.8	1.3

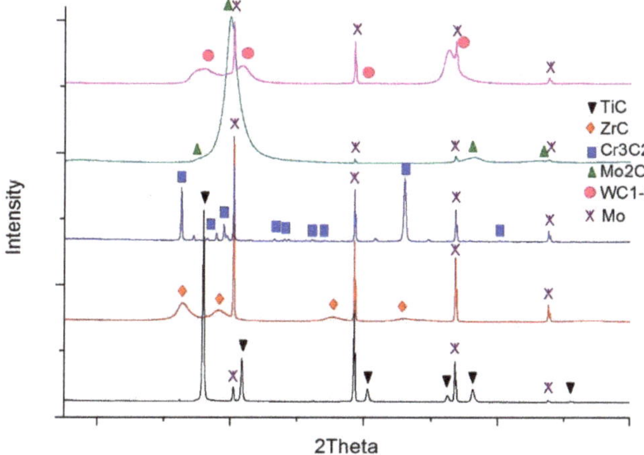

Figure 2. XRD patterns of the carbide layers over molybdenum surfaces.

2.1.3. Substrate Material for Plasma Surfacing

The base material applied was 1.0553 grade (S355J0) low-alloy steel in the form of 10 × 50 × 150 mm plates.

2.2. Methods

2.2.1. Wettability and Spreadability Test

The study was carried out by the sessile drop method in strictly controlled conditions. The substrate samples (18 × 18 × 1 mm Mo plates with deposited carbide layers) and NiBSi alloy (φ3 × 3 mm cylinders) were washed in an ultrasonic cleaner in acetone and were placed on a leveled measuring table in the chamber of the heating device. Due to the fact that plasma surfacing process was performed in an argon shield, the wetting and spreading tests were also carried out in an argon atmosphere with a high 99.999% purity. The working chamber of the device was pumped down to a pressure of 10 Pa, and then filled with argon. This treatment was repeated twice to thoroughly remove any residual air. Then, a constant argon flow of 0.5 L/min was established. The samples were heated from an ambient temperature to 900 °C at a rate of 100 °C/min, and then the rate was reduced to 20 °C/min. The measurement consisted of registering the shape of the drop contour from the time of melting every 10 °C as the temperature increased. The contact angles were measured on both sides of the droplets and the results were averaged. The processes were terminated after the temperature reached 1350 °C. The samples were cooled then and taken off the device. The areas of drop spread were measured. Two such tests were carried out for each NiBSi alloy–carbide substrate system.

2.2.2. Plasma Surfacing

Padding weld samples were made using the plasma transferring arc welding (PTAW) method. The NiBSi matrix powder was mixed with each carbide powder in a volume ratio of 60:40 prior to the process. Before starting, each of the substrate samples (1.0553 grade steel) was sandblasted and cleaned with acetone. The PTAW process was carried out at different current values of the main arc whereas all remaining parameters were constant (Table 3). There was no pre-heating. All of the overwelds had the same length of 60 mm.

Table 3. Parameters applied in PTAW surfacing.

Parameters		Value
Main arc current (welding current)		60, 70, 80, 90, 100, 110, and 120 A
Internal arc current		40 A
Plasma arc voltage		25 V
Powder output		6 g/min
Surfacing rate		50 mm/min
Gas flow (argon):	• Plasma generating (orifice) gas	1.5 L/min
	• Shielding gas	8 L/min
	• Powder transporting gas	5 L/min
Oscillation amplitude		8 mm
Oscillating speed		450 mm/min
Plasmatron-welded substrate distance		15 mm
Nozzle diameter		4 mm

The structural tests were carried out on specimens in the plane perpendicular to the weld axis. The samples were cut by means of an electrical discharge cutter at the same distance of 20 mm from the beginning of each weld.

3. Results and Discussion

3.1. Interaction in the Model System of Liquid NiBSi Alloy-Solid Carbide Surface

The results of the model tests are presented in Figures 3 and 4. The first shows the shapes of the liquid NiBSi droplets resting over the carbide substrates captured at 1290 °C, along with the plots of contact angles as a function of the temperature. The results indicate that all of the tested TMCs of the VIB group were more wettable than those of the IVB tranistion metals group. The wettability of the carbides increased (θ decreases) with the temperature increase, which is characteristic for the thermally activated processes [14]. The wettability of ZrC was slightly better than that of TiC. In the VIB group, at the initial 1290 °C temperature, the carbide wettability remained in the following sequence WC > Cr_3C_2 > Mo_2C, but all contact angles droped quickly to a value below 10° with the increase in temperature. However, it should be taken into account that the measurements of small contact angles were subjected to the relatively higher error.

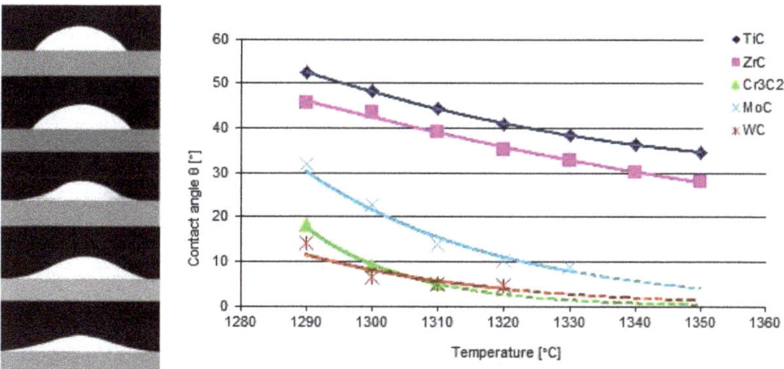

Figure 3. Wettability of carbides with a liquid NiBSi alloy: contours of droplets over flat TiC, ZrC, Cr$_3$C$_2$, Mo$_2$C, and WC substrates at an initial temperature of 1290 °C, and the temperature dependence of the contact angle (average of two trials). The broken line extrapolates the relationship after the completely spread of droplets ($\theta \approx 0°$).

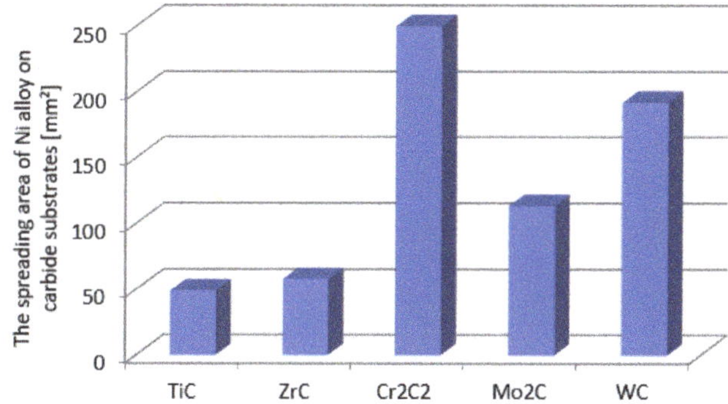

Figure 4. The surface areas of the NiBSi alloy droplets spread over the surface of particular carbides after the end of the entire wettability test (average of two trials).

Figure 4 presents the droplet spread surface areas after the whole test. The obtained results were consistent with those of the wettability, and indicated a more intense interaction of the liquid NiBSi alloy with the VIB group TMCs.

3.2. Interaction of Carbide Powder Particles with Liquid NiBSi during Surfacing

Studies on the production of SCLs were carried out in a wide range of parameters (Table 3) due to the differences in the physical properties of individual carbides (density, thermal capacity and conductivity, specific heat, etc.). However, only the layers selected for all of the comparative tests described in the article that were obtained at the welding current value shown in Table 4 were considered optimal for the given chemical composition of the padding weld. The criterion was the continuity of the layer along with the uniform distribution of the carbide without metallurgical discontinuities in the form of cracks and porosity in the bulk.

Table 4. Welding current values used to produce overwelds for comparative testing.

Transition Metals Group	Carbide	Welding Current [A]
IVB	TiC	90
	ZrC	90
VIB	Cr$_3$C$_2$	90
	Mo$_2$C	80
	WC	90

3.2.1. Overlays with Carbides of IVB TM Group

The instrumental analysis of the cross sections of the padding welds and their fractured surfaces revealed a composite structure with a specific morphology: diversified shapes and sizes of the reinforced particles were observed. Figure 5a shows large, irregular TiC grains surrounded with a fraction of much smaller particles on the matrix background. A fairly similar grain morphology was visible in the welds containing ZrC (Figure 5b). This was confirmed by the fractographic analysis of both composite layers. The trans-granular cleavage of the reinforcing particles and the relatively ductile inter-granular matrix fracture of the layers are visible in Figure 6.

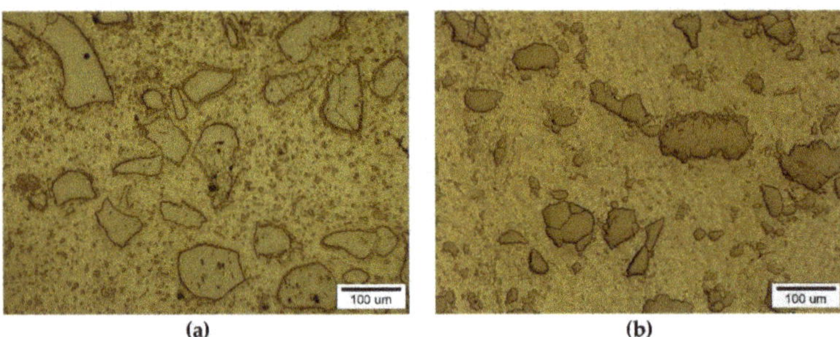

Figure 5. Optical microscope micrograph: Structure of PTAW obtained composite overlays on NiBSi alloy basis reinforced with TiC (**a**) and ZrC (**b**) particles.

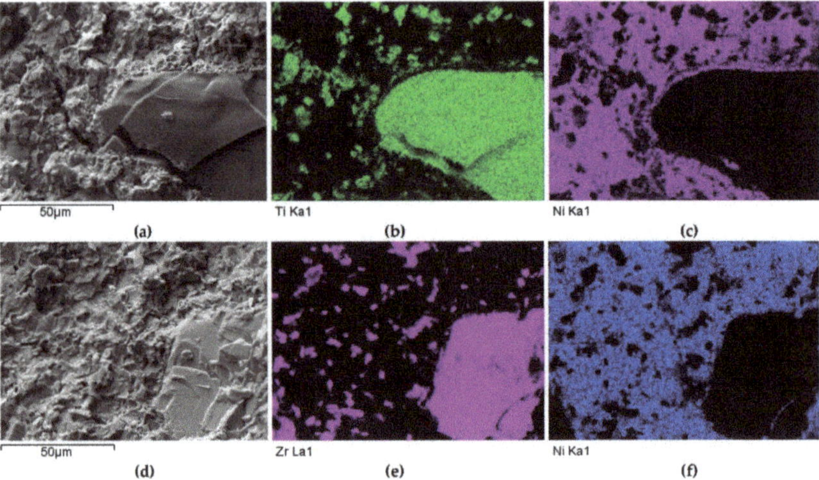

Figure 6. SEM/EDS micrographs of fractured surfaces of NiBSi-TiC (**a–c**) and NiBSi-ZrC (**d–f**) composite welds enabling phases identification.

The distributions of the main element concentrations comprising the tested system at the layers' cross-sections are presented in Figure 7 (for TiC) and Figure 8 (for ZrC). The maps for the dissemination of iron, which came from the partially melted steel substrates, are also included. The distribution maps are accompanied by a linear concentration of elements along the lines perpendicular to the interface between the carbide and matrix (Figures 8b and 9).

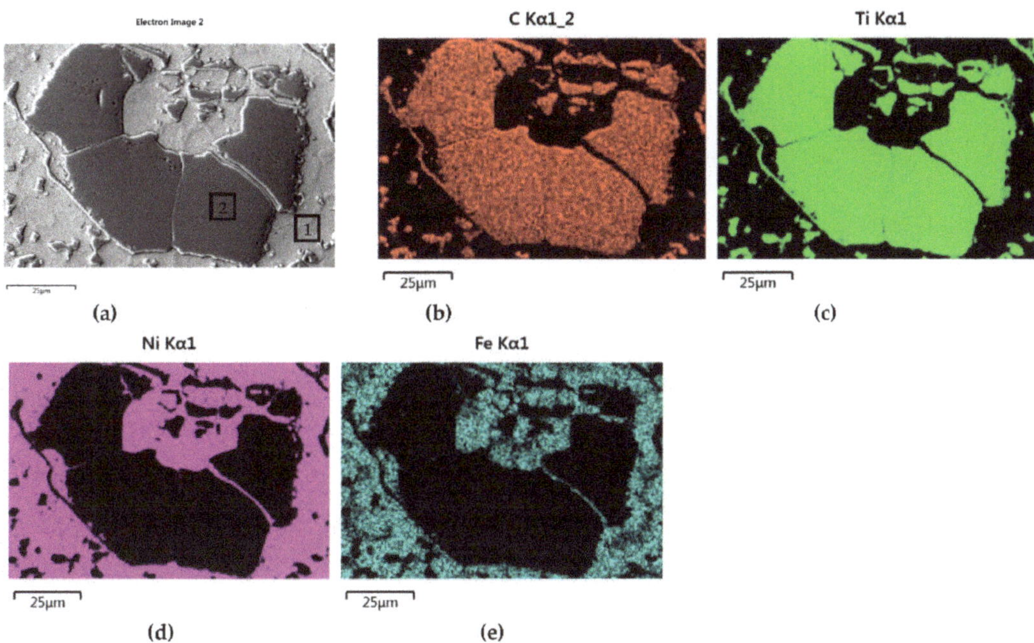

Figure 7. NiBSi-TiC layer: SEM micrograph with points of the quantitative analysis marked (**a**) and surface distributions of C (**b**), Ti (**c**), Ni (**d**), and Fe (**e**).

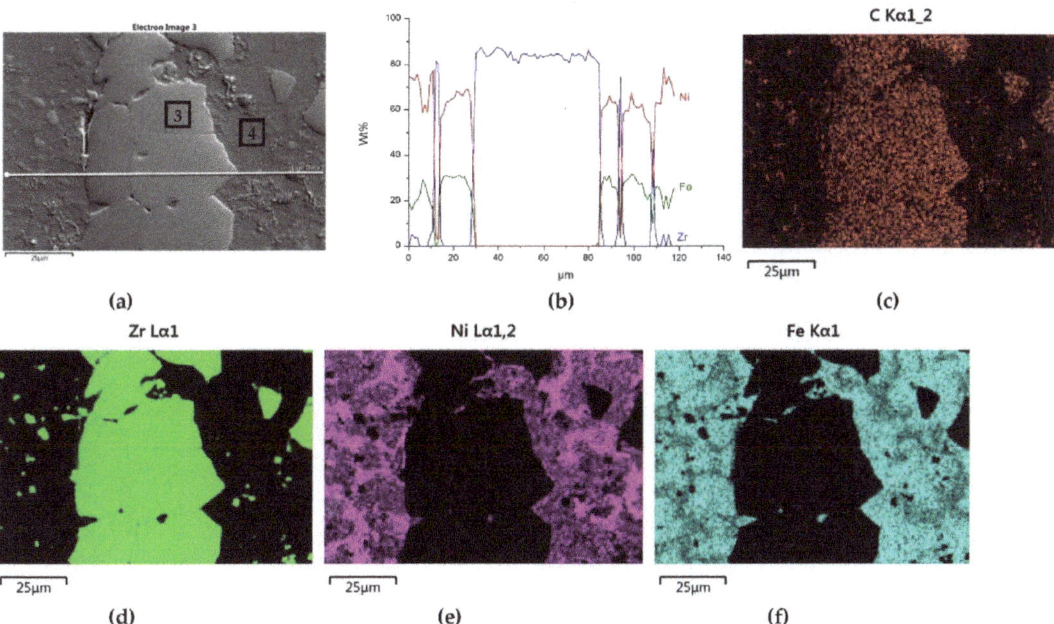

Figure 8. NiBSi-ZrC layer: SEM micrograph with points of the quantitative analysis marked (**a**), linear propagation of components along the marked line (**b**) and EDS maps of C (**c**), Zr (**d**), Ni (**e**), and Fe (**f**) distribution.

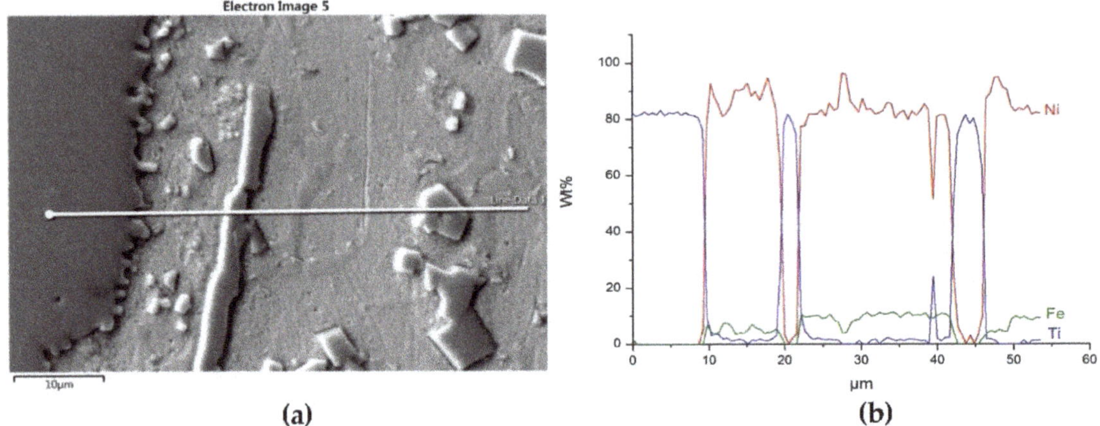

Figure 9. Structure close to the Ni alloy−TiC interfacial boundary (**a**) and concentrations of elements along the marked line (**b**).

Sound boundaries without any discontinuities between the matrix and strengthening phase were visible in both cases. No transition layer was visible at the interface of both carbides and the matrix (Figures 7–9). The linear distributions of the Ti, Zr, Ni, Fe, and C concentrations along the marked line (shown in Figures 8b and 9b) do not suggest this either, as evidenced by the steep transitions of the curves between the phases.

A disintegration of the large TMC (about 80 μm and above) agglomerates was observed, caused by the penetration of the liquid matrix alloy (i.e., Ni-B-Si-Fe) along the grain boundaries during the contact of the solid and liquid phases. This was clearly indicated by photos and the distribution of nickel in Figures 7 and 8. In addition to this breakdown of large TMCs agglomerates into several parts, a splitting of the outer layer of the strengthening phase grains into the much smaller particles is visible in Figure 10. As a result, some amounts of a fine micrometer-order fraction of titanium and zirconium carbides were formed. This surface disintegration was twofold: a narrow band of TiC and ZrC separated from the surface around the large carbide particles, which then broke down into the smaller parts. They dissipated over the entire volume of the liquid through its movement in the welding pool during the process.

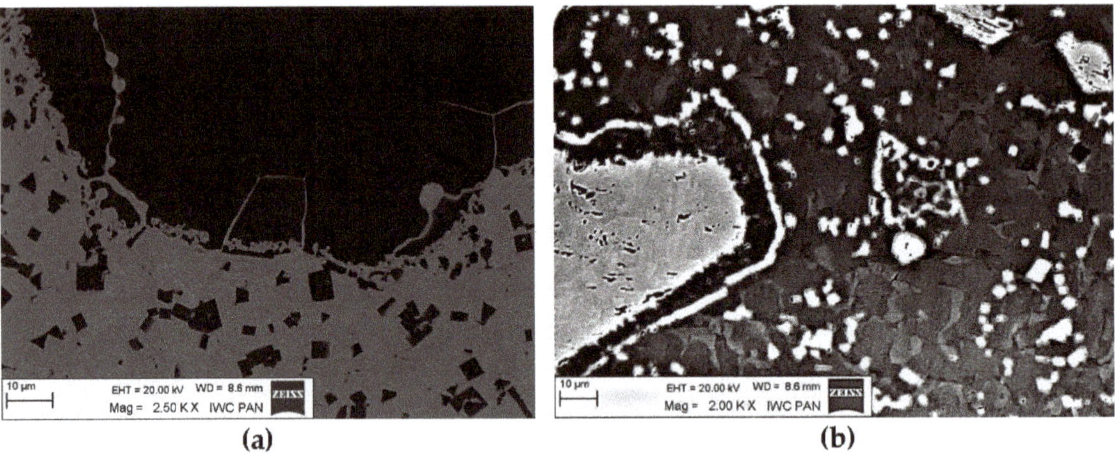

Figure 10. Disintegration of the carbide particle outer layer in the liquid matrix: (**a**) TiC and (**b**) ZrC.

The results of the point analysis of elements in the indicated areas (shown in Table 5) reveal that in both cases, except for nickel and silicon (present in the starting powder), a significant quantity of iron existed in the matrix of layers, coming from the partially molten steel substrate. Zr was not present in this area, contrary to a small amount of Ti noticed there. This proved a slight solubility of the last carbide.

Table 5. Point EDS analysis of nickel alloy–TiC and –ZrC regions indicated in Figures 7a and 8a.

Overlay	Region	Elements, Wt%					
		Ti	Zr	C	Ni	Fe	Si
NiBSi–TiC	1	80.8	-	19.2	-	-	-
	2	1.7	-	7.8	79.6	9.2	1.8
NiBSi–ZrC	3	-	82.6	17.4	-	-	-
	4	-	-	4.3	63.8	30.6	1.4

3.2.2. Overlays with Carbides of VIB TM Group

A different nature and intensity of interaction was observed in the SCLs reinforced with TMCs from the VIB group. Beside large irregular grains, much smaller elongated particles of the strengthening phase were visible in the padding welds with Cr_3C_2 (Figures 11a and 12a–c). These needle-like particles were presumably Cr_3C_2 carbides crystallized from a supersaturated solution. They formed characteristic star-shaped configurations. A little similar morphology was characterized by the molybdenum carbide (Figures 11b and 12d–f). However, the percentage of elongated particles in this case was lower, which may have indicated a reduced solubility of Mo_2C in the liquid weld pool. Large Mo_2C particles have clearly developed surfaces. Overlays with WC contain large, nodular carbide particles surrounded by a fine-grained fraction (Figures 11c and 12g–i).

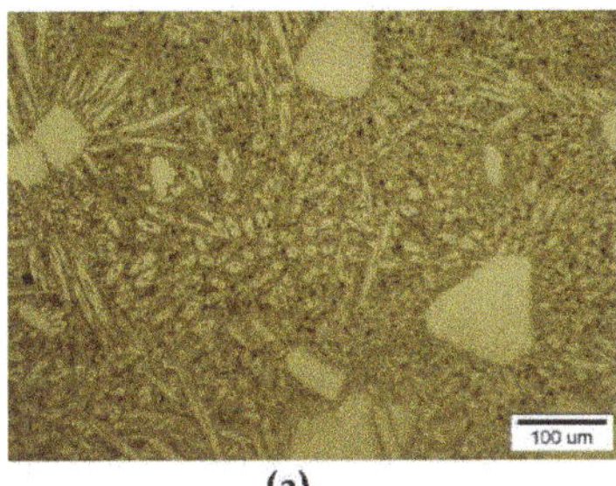 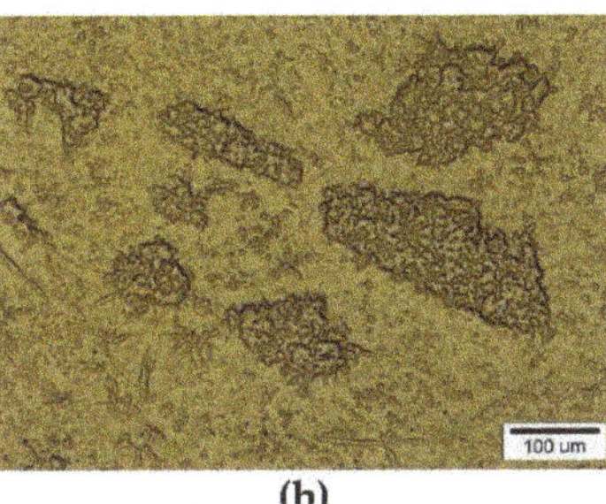 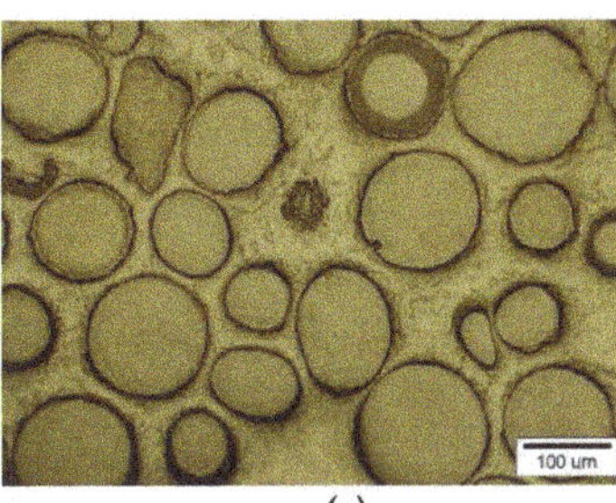

(a) (b) (c)

Figure 11. Structure of PTAW obtained composite overlays on NiBSi alloy basis reinforced with Cr_3C_2 (**a**), Mo_2C (**b**) and WC (**c**) particles.

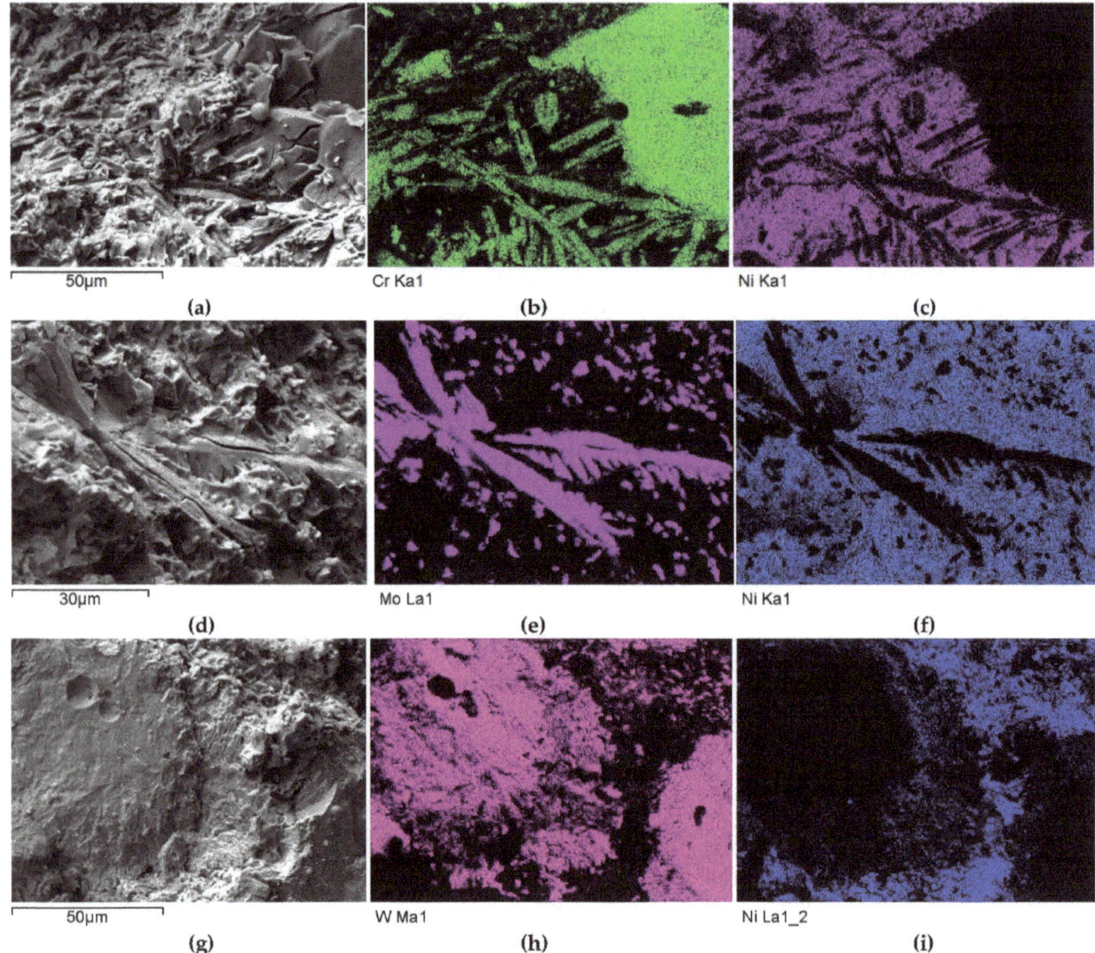

Figure 12. SEM/EDS micrographs of fractured surfaces of NiBSi–Cr$_3$C$_2$ (**a–c**), NiBSi–Mo$_2$C (**d–f**), and NiBSi–WC (**g–i**) composite welds enabling phases identification.

A set of distribution maps and linear concentrations of components, including the iron allocation, is presented in Figures 13–17, the same as for the SCLs with the IVB group TMCs. The results of the point quantitative analysis in selected areas marked in Figure 13a, Figure 14a, and Figure 16a are demonstrated in Table 6.

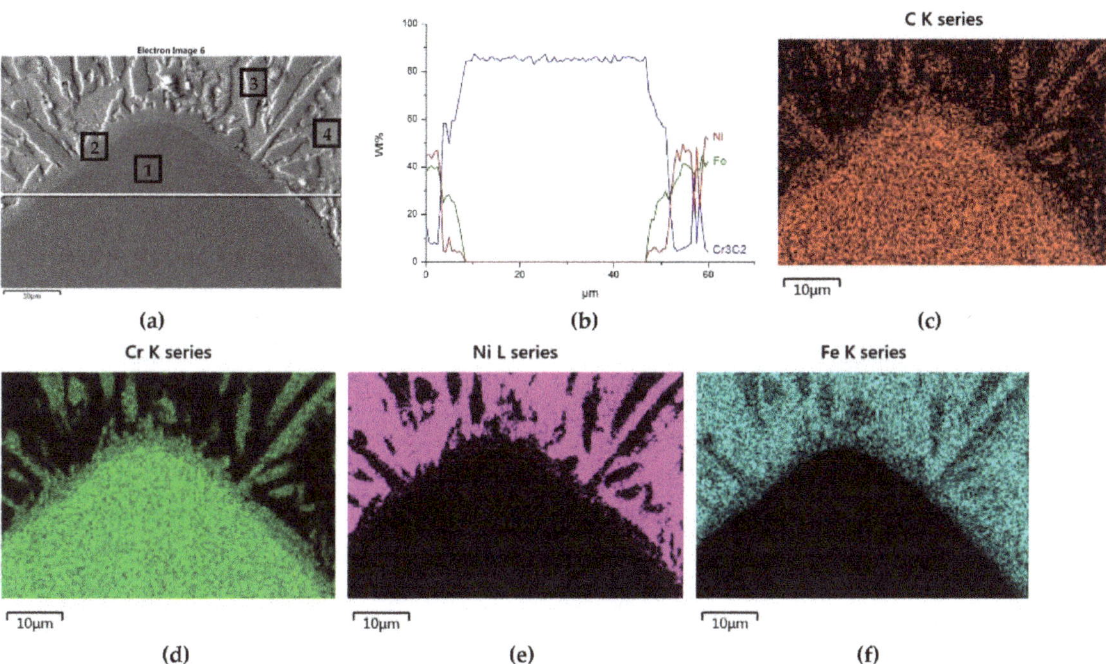

Figure 13. NiBSi-Cr_3C_2 layer: SEM micrograph with points of the quantitative analysis marked (**a**), the linear concentration of main components along the marked line (**b**) and the maps of C (**c**), Cr (**d**), Ni (**e**), and Fe (**f**) distribution.

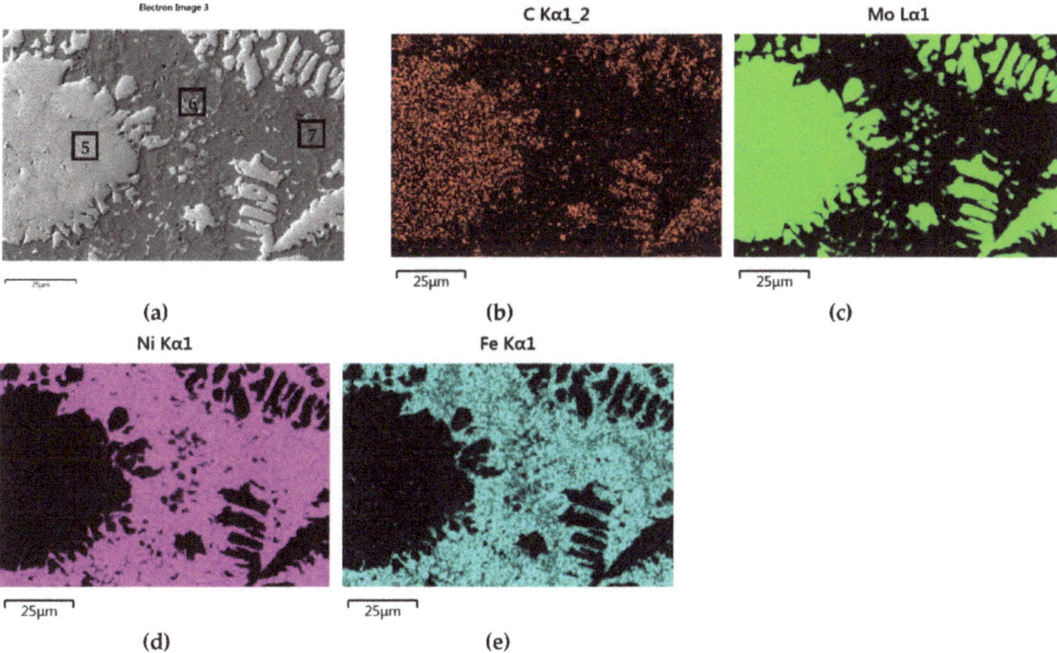

Figure 14. NiBSi–Mo_2C layer: SEM micrograph with points of the quantitative analysis marked (**a**), and surface distribution of: C (**b**), Mo (**c**), Ni (**d**), and Fe (**e**).

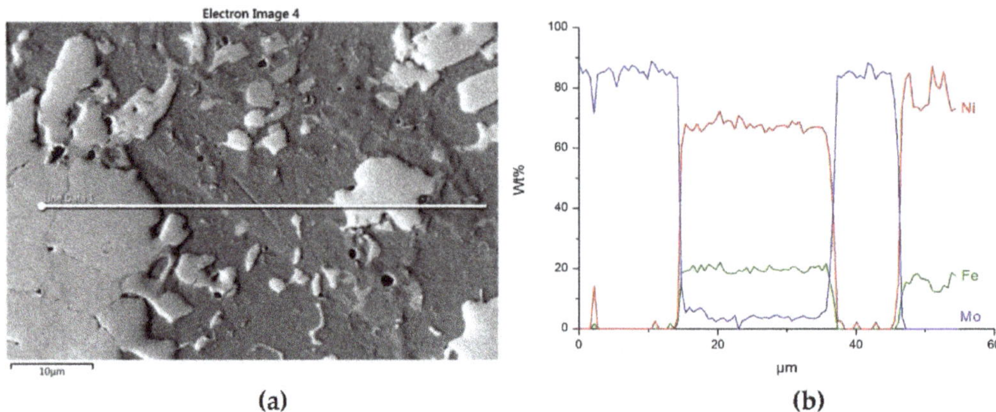

Figure 15. The structure close to the Ni alloy–Mo$_2$C interfacial boundary (**a**) and concentrations of elements along the marked line (**b**).

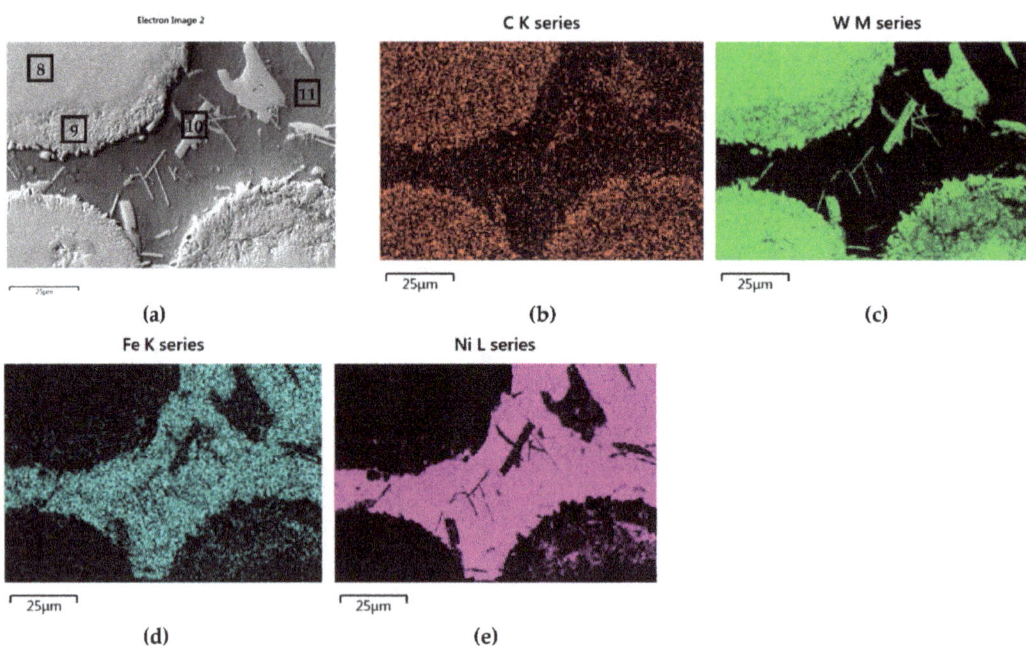

Figure 16. NiBSi–WC layer: SEM micrograph with points of the quantitative analysis marked (**a**), and surface distribution of C (**b**), W (**c**), Ni (**d**), and Fe (**e**).

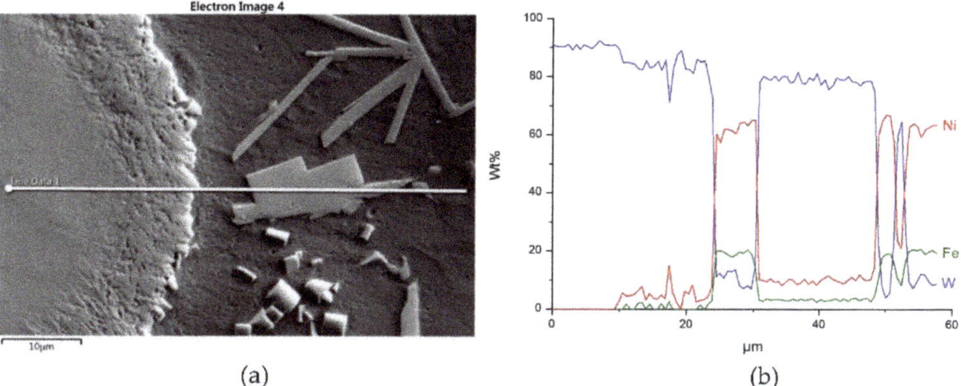

Figure 17. The structure close to the interfacial boundary (**a**) and concentration profiles of Ni, W, and Fe (**b**) in the NiBSi–WC overlay along the marked line.

Table 6. Results of the EDS quantitative analysis of SCLs with Cr_3C_2, Mo_2C, and WC in points marked in Figure 13a, Figure 14a, and Figure 16a.

Overlay	Region	Elements, Wt%						
		Cr	Mo	W	C	Ni	Fe	Si
NiBSi–Cr_3C_2	1	85.3	-	-	14.7	-	-	-
	2	60.2	-	-	10.6	4.3	24.9	
	3	33.1	-	-	9.2	22	35.5	0.3
	4	5	-	-	4.3	50	39.2	1.6
NiBSi–Mo_2C	5	-	84	-	16	-	-	-
	6	-	4.8	-	7.2	68.7	17.6	1.6
	7	-	1.2	-	7.5	80.7	10.5	
NiBSi–WC	8	-	-	90.6	9.4	-	-	-
	9	-	-	83	10.9	4.5	1.5	-
	10	-	-	79.7	8.2	8.9	3.2	-
	11	-	-	10.3	7.1	61.1	20.2	1.3

The mapping, linear elements concentration in the NiBSi–Cr_3C_2 coating in Figure 13 and the data from Table 6 exhibit the presence of nickel and iron in the transition zone at the matrix–carbide interface and in the bulk of needle-like precipitates. The matrix is rich in nickel and iron with a small amount of silicon and chromium. The gray shell around the Cr_3C_2 particles and precipitates in the matrix are composed of chromium, iron, and nickel with some carbon. The results prove the significant solubility of Cr_3C_2 in the liquid ground of the padding weld, from which the needles of the Cr–Fe carbides with the addition of Ni crystallize. The intensity of the phase interaction in the system is also indicated by the formation of wide intermediate zones. The concentration profiles of Ni and Fe suggest relatively long range of diffusion, with Fe being more active.

The structure of Mo_2C reinforced SCL is shown in Figures 14 and 15. The carbide–matrix interfaces are continuous. The disintegration of grains in the strengthening phase is visible. Precipitations of a new phase against the matrix are detectable. The distribution maps of the elements document the presence of molybdenum in the matrix, which proves the solubility of its carbide during the formation of the coating, confirmed also by the point quantitative analysis (Table 6). However, the Mo fraction in this solution is small.

Figures 16 and 17 show the structure of WC containing SCL and its component distribution. Large, spherical particles of the reinforced phase with continuous interfaces are visible on the matrix background. A wide transition zone is detected around all of them. The maps and linear distributions of the elements' concentration and the point analysis

(Table 6) demonstrate the existence of nickel and iron in both the transition zone and in the weld matrix. The presence of nickel and iron in the transition zone indicates their diffusion into WC. Tungsten is also dissolved in the matrix. Longitudinal secondary precipitations of tungsten carbides with Ni and Fe crystallized out of the matrix solution.

3.3. Strengthening Phase Fraction in Overlays

The steel substrate partially melts in the course of surfacing, and the resulting liquid stirs with the welding pool. The fractions (D) of the metal coming from the substrate into SCL were calculated according to Formula (1), where P and S are the cross-sectional areas of the molten substrate and the overlay, respectively. The results show that the fraction of the substrate material in the SCLs significantly increases with the increase of the welding current (Figure 18). The graphs are complex and no clear relationship is visible between D and the position of the given carbide-forming metal in the group of the periodic table of elements. The lowest values of composite matrix dilution with the substrate material are registered for the layers containing titanium and molybdenum carbides.

$$D = \frac{P}{P + S} \, 100\% \tag{1}$$

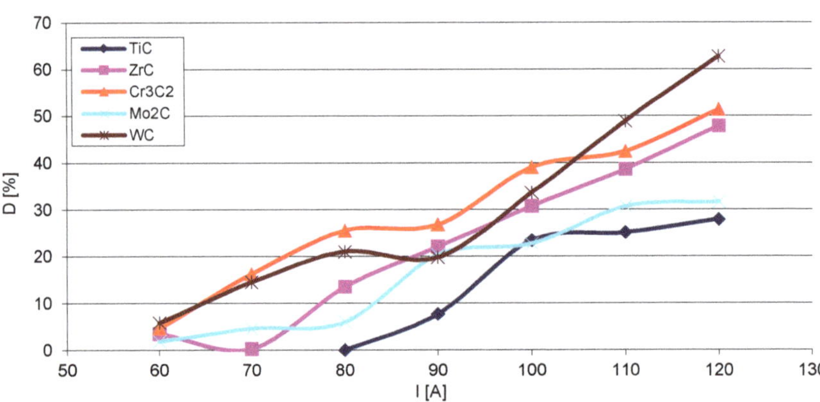

Figure 18. Volume fraction (D) of the substrate material in overlays with particular carbides as a function of the welding current I.

The volume fraction of the strengthening phase in SCLs as a function of the welding current was also analyzed. The measurements were carried out at $100\times$ magnification applying the Cavalieri–Hacquert principle, according to which the ratio of the given phase volumes in the composite is equal to the ratio of the surface areas of these phases at the cross-sections. The obtained results are presented graphically in Figure 19a,b. In the coatings reinforced with TiC and ZrC (group IVB), a systematic increase of fraction with the increasing current intensity was observed, despite the increasing volume of the padding weld due to the increasing fusion of the substrate. The opposite tendency was noted for surfacing welds with carbides of VIB group metals (Cr_3C_2, Mo_2C, and WC). This can be explained by the inferior wettability (less intense interaction with the matrix) group IVB TMCs, which impact their introduction into the liquid. The increase of welding current promotes the thermal activation of the wetting process to maintain larger amounts of TiC and ZrC particles in the liquid. In SCLs reinforced by TMCs of the VIB group, the increase in the linear energy of the welding intensifies the dissolution of carbides, so their fraction decreases with its increase.

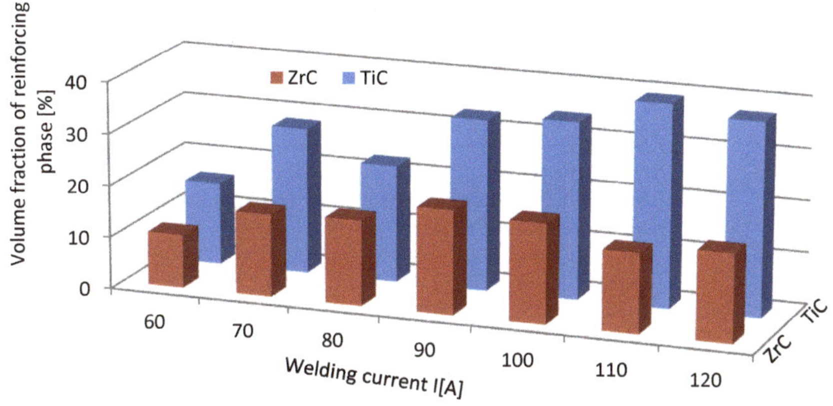

(**a**)

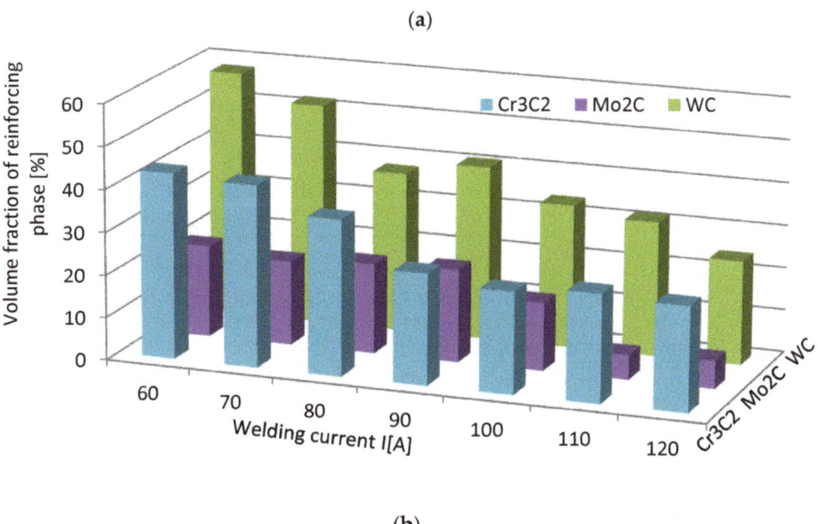

(**b**)

Figure 19. Impact of welding current on the volume fraction of the strengthening phase in overlays with TMs carbides of IVB (**a**) and VIB groups (**b**).

3.4. Metallurgical Considerations

The results of the experiments show the intense interaction of all TMCs with liquid nickel modified with B and Si additives, both under the stable conditions of the contacting phases and during the dynamic technological process of PTAW, including the melting of powders with a plasma beam, the formation of a liquid solution with the partially molten substrate material, and the crystallization of SCL. As a measure of such an activity, the equilibrium contact angle along with the spread area on a flat surface for the first group of tests was taken, while the formation of SCLs and their structure, with particular emphasis on the vicinity of the interfacial boundary, was taken for the second. The wettability of all of the tested TMCs with the NiBSi alloy was generally good from the initial temperature of 1290 °C, and improved with its increase, as expected. The wettability of group VIB TMCs was better than that of IVB: the contact angles in both groups were close to 0° and in the range 30–40°, respectively (Figure 3). This was sufficient for the technological process. The trend of improved wettability from top to bottom in both TMC groups was observed. The wettability of TMCs had been tested previously, but only for pure Ni [30,31], necessarily at much higher temperatures (T_{fNi} = 1455 °C). The spreadability of the liquid NiBSi alloy over the surface of particular TMCs was consistent with the results of the wettability tests (Figure 4). The results obtained in the study indicate the possibility of producing Ni matrix composites with B and Ni temperature depressants in a much lower temperature range, which is important in massive scale manufacturing.

The structures of the PTAW obtained SCLs were complex and varied for both of TMCs groups studied. The research confirms the different nature of the interaction of the liquid Ni alloy matrix with each TMC. Penetration of the liquid matrix along the grain boundaries of TiC and ZrC took place, resulting in a disintegration of large carbide agglomerates into smaller particles. This is a known mechanism that utilizes the energy surplus of grain boundaries in relation to the interior of the solid phase [14]. If the condition

$$\sigma_{GB} > \sigma_{SL} \cos (\psi/2) \tag{2}$$

is met (where σ_{GB}, σ_{SL}, and ψ are the solid–solid grain boundary energy, solid–liquid interfacial energy, and dihedral contact angle, respectively), the equilibrium in the system was not reached and the liquid flows into the solid along grain boundaries.

The interfacial boundaries in the SCLs of both groups are continuous, with no discrepancies. This means there was compliance with the results of the phase interaction tests in the model studies for small objects with curved surfaces. This also confirms the correctness of the PTAW process carried out, as the contact time of the phases in the liquid pool was sufficient to wet all the surfaces of the solid particles. In the SCLs with the IVB group TMCs there were no transition zones at the boundaries between the carbides and the matrix. TiC is very slightly soluble in liquid, while in the SCLs reinforced with ZrC the presence of Zr in the matrix was not revealed, which proves the complete insolubility of this carbide under the process conditions. In SCLs fortified with VIB TMCs, the Cr_3C_2, Mo_2C, and WC particles partially dissolved and crystallized from the supersaturated solution. In addition, transition zones were formed in these SCLs at the carbide–matrix boundaries, especially in the Cr_3C_2- and WC-reinforced coatings. The iron from the molten substrate played an active role in the phases interaction: its presence was noted both in the matrix, including the part separating carbide agglomerates (group IVB), and in transition layers at the interphase boundaries (group VIB). Therefore, during PTAW we are dealing with the active role of three elements: boron, silicon, and iron. A similar mechanism was observed in earlier works [32,33]. The analysis of the double phase equilibrium systems for the combinations of tested TMs with B, Si, and Fe revealed liquid solutions in the entire concentration range and the presence of numerous intermetallic phases for all cases [34]. This shows a tendency towards their mutual interaction.

When comparing the behavior of TMCs during the production of SCLs, one should take into account their electronic structure, physicochemical properties, and the position of the parent TM in the periodic table of elements. Table 7 presents the data relevant from this point of view from many sources. These data are referred to the TMCs of IVB and VIB groups, and also to the not experimentally tested VB group.

Table 7. TMCs of the IVB–VIB groups: melting point T_f [35], electrical resistivity ρ [35], standard free enthalpy of formation ΔG_F [36], formation energy E_{TMC} [37], bond dissociation energy E_{BD} [38], and enthalpy of parent metal solubility in Ni $\Delta H_{Ni\ mix}$ [39].

TM Group	Carbide	T_f [°C]	ρ [μΩcm]	ΔG_F [kJ/mole C]	E_{TMC} [eV]	E_{BD} [eV]	$\Delta H_{Ni\ mix}$ [kJ/mole]
IVB	TiC	3067 [40]	100	−164.73	−1.62	3.857	−154
	ZrC	3572 [40]	75	−186.11	−1.63	4.892	−236
	HfC	3982 [40]	67	−214.16	−1.88	4.426	−250 [41]
VB	$VC_{0.88}$	2650	69	−91.40	−0.83	4.109 [42]	−75
	NbC	3610	20	−133.27	−1.06	5.620	n.a.
	TaC	3985	15	−140.70	−1.17	4.975	−133
VIB	Cr_3C_2	1810	75	−114.32	n.a.	n.a.	−27
	Mo_2C	2520	57	−62.13	−0.18 *	n.a.	−32
	WC	2776	17	−34.01	−0.24	4.289 [43]	−14

* value for MoC.

TMCs have generally low electrical resistance, decreasing with the group number and from top to bottom inside. This proves the significant participation of TM–TM metallic bonds, in addition to the existing TM–C bonds in these compounds. The structure of TMCs is a metallic lattice structure (fcc for monocarbides, orthorhombic for Cr and Mo, and hexagonal for WC) with the C atoms in the interstitial positions [41]. Therefore, the energy necessary to create the network is reduced due to the inclusion of carbon atoms [44]. TM–C bonds are weakened with the increasing filling of the d-band. This applies to all parent TMs: 3d, 4d, and 5d [37,45]. Thus, the tendency of the metallic character of TMCs increases from left to right and from top to bottom in groups IVB–VIB of the periodic table.

The thermodynamic stability of the TMCs of groups IVB and VB is high, decreasing with the group number from very high for IVB, high for VB, and somewhat softer for the TMCs of group VIB under consideration. The trend of the formation of parent TMs solutions with liquid nickel is similar. This is evidenced by the ΔG_F values of the carbide formation and the bond dissociation energies E_{BD}, as well as the heat of formation of $\Delta H_{Ni\ mix}$ solutions quoted in Table 7.

In summarizing, it can be stated that the interaction of TMCs with the NiBSi matrix during the production of SCLs by the PTAW method tends to increase with the group number of the periodic table, and from the top to bottom inside individual groups. It is presented schematically in Figure 20. The statement is based on the correlations obtained for the two border groups, assuming a specific "interpolation", and would require verification for the "missing" VB TMCs, which will be the subject of further research.

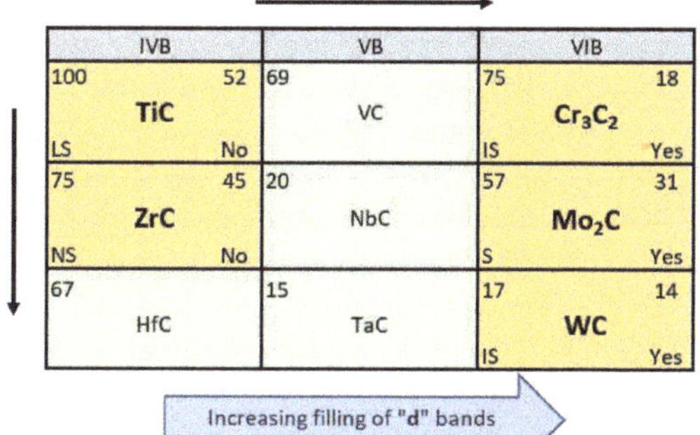

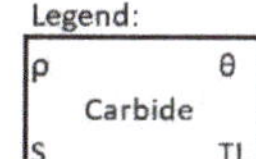

ρ – electrical resistivity [μΩcm],
θ – wetting angle at 1290°C [°],
S – solubility of carbides: LS – low solubility,
NS – no solubility; IS – intense solubility,
TL – transition layer between matrix and carbide: No/Yes.

Figure 20. Intensity trends of TMC interactions with the NiBSi matrix during the PTAW formation of SCLs. Each carbide position corresponds to its parent metal location on the periodic table. Carbides subjected to the research are marked.

4. Conclusions

Based on the experimental research carried out in the model system and in the real technological process of PTAW, and in the discussion on results, the following can be stated:

1. Carbides of the IVB and VIB TM groups of the periodic table interact with NiBSi alloy, with the increasing intensity related to the group number and the TM location in it. Its measure is the wettability and spreadability, as well as selected aspects of the formation of SCLs: the mechanisms of the disintegration of the strengthening phase in the liquid pool, the solubility of TMCs in the matrix and secondary crystallization, and the formation of transition zones at the interfacial boundaries.

    (a) TMCs of the IVB group are good and these of VIB are perfectly wettable with liquid Ni alloy.

    (b) The fraction of the strengthening phase particles in SCLs and their distribution are related to the interaction of TMCs with the matrix, which is more intense for the VIB group TMCs.

    (c) Particles of TiC and ZrC are decomposed by the penetration of the liquid phase along the grain boundaries. As a result, decomposition of agglomerates on smaller parts occurs and a significant amount of a fine fraction is formed. In contrast with this, Cr_3C_2, Mo_2C, and WC particles are dissolved partially or completely, enriching the matrix in Cr, Mo, and W, respectively. New phases are crystallized from the supersaturated solution when cooling.

    (d) No transition zones are formed at the interfacial boundaries of the IVB group TMCs, unlike in the case of the VIB group TMCs.

    (e) There is a tendency for both of the studied groups of TMCs to intensify the interaction with the Ni alloy matrix, with an increase in the atomic number of the parent metal forming the given carbide.

2. Fe coming from the partially PTAW melted substrate plays an active role in the system, along with temperature depressants B and Si present in the matrix of the composite.

3. The obtained experimental results can be successfully interpreted in light of the electronic structure of TMCs and their physicochemical properties.

Author Contributions: Conceptualization, M.B. and J.S.; methodology, J.S. and M.B.; validation, M.B., J.S. and H.L.; formal analysis, M.B., J.S. and H.L.; investigation, M.B.; writing—original draft preparation, M.B. and J.S.; writing—review and editing, J.S., M.B. and H.L.; visualization, M.B.; supervision, M.B. and J.S. All authors have read and agreed to the published version of the manuscript.

Funding: This research received no external funding.

Institutional Review Board Statement: Not applicable.

Informed Consent Statement: Not applicable.

Data Availability Statement: Experimental methods and results are available from the authors.

Conflicts of Interest: The authors declare no conflict of interest.

References

1. Berger, L.M. Application of Hardmetals as Thermal Spray Coatings. *Int. J. Refract. Met. Hard Mater.* **2015**, *49*, 350–364. [CrossRef]
2. Pawlowski, L. Thick Laser Coatings: A Review. *J. Therm. Spray Technol.* **1999**, *8*, 279–295. [CrossRef]
3. Zhong, M.; Liu, W. Laser Surface Cladding: The State of the Art and Challenges. *Proc. Inst. Mech. Eng. Part C J. Mech. Eng. Sci.* **2010**, *224*, 1041–1060. [CrossRef]
4. Praedeep, G.R.C.; Ramesh, A.; Prasad, B.D. A Review Paper on Hardfacing Processes and Materials. *Int. J. Eng. Sci. Technol.* **2010**, *11*, 6507–6510.
5. Czupryński, A. Microstructure and Abrasive Wear Resistance of Metal Matrix Composite Coatings Deposited on Steel Grade AISI 4715 by Powder Plasma Transferred Arc Welding Part 2. Mechanical and Structural Properties of a Nickel-Based Alloy Surface Layer Reinforced with Particles of Tungsten Carbide and Synthetic Metal–Diamond Composite. *Materials* **2021**, *14*, 2805. [CrossRef]

6. Meng, Q.W.; Geng, L.; Zhang, B.Y. Laser cladding on Ni-base Composite Coatings onto Ti6Al4V Substrates with Pre-Placed B4C+NiCrBSi Powders. *Surf. Coat. Technol.* **2006**, *200*, 4923–4928. [CrossRef]

7. Li, C.-W.; Chang, K.-C.; Yeh, A.-C.; Yeh, J.-W.; Lin, S.-J. Microstructure Characterization of Cemented Carbide Fabricated by Selective Laser Melting Process. *Int. J. Refract. Met. Hard Mater.* **2018**, *75*, 225–233. [CrossRef]

8. Chao, M.-J.; Niu, X.; Yuan, B.; Liang, E.-J.; Wang, D.-S. Preparation and Characterization of in Situ Synthesized B4C Particulate Reinforced Nickel Composite Coatings by Laser Cladding. *Surf. Coatings Technol.* **2006**, *201*, 1102–1108. [CrossRef]

9. Liu, Y.; Wang, H. A Novel Load-Insensitive Co3Mo2Si Reinforced In-Situ Metal-Matrix Composite Coating for Wear Resistance Application. *Mater. Lett.* **2010**, *64*, 2494–2497. [CrossRef]

10. Betts, J.C. The Direct Laser Deposition of AISI316 Stainless Steel and Cr3C2 Powder. *J. Mater. Process. Technol.* **2009**, *209*, 5229–5238. [CrossRef]

11. Asthana, R.; Sobczak, N. Wettability, Spreading, and Interfacial Phenomena in High-Temperature Coatings. *JOM-E* **2000**, *52*, 1–19.

12. Oukach, S.; Pateyron, B.; Pawłowski, L. Physical and chemical Phenomena Occurring Between Solid Ceramics and Liquid Metals and Alloys at Laser and Plasma Composite Coatings Formation: A review. *Surf. Sci. Rep.* **2019**, *74*, 213–241. [CrossRef]

13. Huebner, J.; Kata, D.; Rutkowski, P.; Petrzak, P.; Kusiński, J. Grain-Boundary Interaction between Inconel 625 and WC during Laser Metal Deposition. *Materials* **2018**, *11*, 1797. [CrossRef] [PubMed]

14. Eustathopoulos, N.; Nicholas, M.G.; Drevet, B. *Wettability at High Temperatures*, 1st ed.; Pergamon Press: Amsterdam, The Netherlands; Lausanne, NY, USA; Oxford, UK, 1999; Volume 3.

15. Bourell, D.; Kruth, J.P.; Leu, M.; Levy, G.; Rosen, D.; Beese, A.; Clare, A. Materials for Additive Manufacturing. *CIRP Ann.* **2017**, *66*, 659–681. [CrossRef]

16. Rojasa, J.G.M.; Wolfeb, T.; Flecka, B.A.; Qureshi, A.J. Plasma Transferred Arc Additive Manufacturing of Nickel Metal Matrix Composites. *Manuf. Lett.* **2018**, *18*, 31–34. [CrossRef]

17. Zhong, Y.; Xia, X.; Shi, F.; Zhan, J.; Tu, J.; Fan, H.J. Transition Metal Carbides and Nitrides in Energy Storage and Conversion. *Adv. Sci.* **2016**, *3*, 1500286. [CrossRef] [PubMed]

18. Dinh, K.N.; Liang, Q.; Du, C.-F.; Zhao, J.; Tok, A.; Mao, H.; Yan, Q. Nanostructured Metallic Transition Metal Carbides, Nitrides, Phosphides, and Borides for Energy Storage and Conversion. *Nano Today* **2019**, *25*, 99–121. [CrossRef]

19. Xiao, P.; Ge, X.; Wang, H.; Liu, Z.; Fisher, A.C.; Wang, X. Novel Molybdenum Carbide-Tungsten Carbide Composite Nanowires and Their Electrochemical Activation for Efficient and Stable Hydrogen Evolution. *Adv. Funct. Mater.* **2015**, *25*, 1520–1526. [CrossRef]

20. Sebakhy, K.O.; Vitale, G.; Hassan, A.; Pereira-Almao, P. New Insights into the Kinetics of Structural Transformation and Hydrogenation Activity of Nano-crystalline Molybdenum Carbide. *Catal. Lett.* **2018**, *148*, 904–923. [CrossRef]

21. Li, T.; Liu, B.; Liu, Y.; Guo, W.; Fu, A.; Li, L.; Yan, N.; Fang, Q. Microstructure and Mechanical Properties of Particulate Reinforced NbMoCrTiAl High Entropy Based Composite. *Entropy* **2018**, *20*, 517. [CrossRef]

22. Zhu, T.; Wu, H.; Zhou, R.; Zhang, N.; Yin, Y.; Liang, L.; Liu, Y.; Li, J.; Shan, Q.; Li, Q.; et al. Microstructures and Tribological Properties of TiC Reinforced FeCoNiCuAl High-Entropy Alloy at Normal and Elevated Temperature. *Metals* **2020**, *10*, 387. [CrossRef]

23. Zhang, H.; Akhtar, F. Processing and Characterization of Refractory Quaternary and Quinary High-Entropy Carbide Composite. *Entropy* **2019**, *21*, 474. [CrossRef]

24. Zhang, Y.; Li, R. New Advances in High-Entropy Alloys. *Entropy* **2020**, *22*, 1158. [CrossRef]

25. Wang, Z.H.; Wang, H.; He, D.Y.; Cui, L.; Jiang, J.M.; Zhou, Z.; Zhao, Q.Y. Microstructure Characterization of In Situ NbC/high Entropy Alloys by Plasma Cladding. *Rare Met. Mater. Eng.* **2015**, *44*, 3156–3160.

26. Hardfacing Alloys. Available online: www.deloro.com (accessed on 8 September 2021).

27. Tokunaga, T.; Nishio, K.; Ohtani, H.; Hasebe, M. Phase Equilibria in the Ni-Si-B System. *Mater. Trans.* **2003**, *44*, 1651–1654. [CrossRef]

28. Rumble, J.R. *CRC Handbook of Chemistry and Physics*, 99th ed.; Taylor&Francis: Boca Raton, FL, USA; London, UK, 2018.

29. Bober, M.; Senkara, J.; Wendler, B. Persistence of the Thin Layers of Transition Metal Carbides in Contact with Liquid NiBSi alloy. *Weld. Technol. Rev.* **2021**, *93*, 5–12. [CrossRef]

30. Ramqvist, L. Wetting of Metallic Carbides by Liquid Copper, Nickel, Cobalt and Iron. *Int. J. Powder Met.* **1965**, *1*, 2–21.

31. Naidich, Y.V. The Wettability of Solids by Liquid Metals. *Progress in Surface and Membrane Sciences* **1981**, *14*, 353–484.

32. Bober, M.; Senkara, J. Comparative Tests of Plasma-Surfaced Nickel Layers with Chromium and Titanium Carbides. *Weld. Int.* **2016**, *30*, 107–111. [CrossRef]

33. Bober, M. Formation and Structure of the Composite Coatings Ni-NbC deposited by plasma transferred arc. *Weld. Technol. Rev.* **2018**, *90*, 65–69.

34. Okamoto, H.; Massalski, T.B. Binary Alloy Phase Diagrams Requiring Further Studies. *J. Phase Equilibria Diffus.* **1994**, *15*, 500–521. [CrossRef]

35. Lengauer, W. Transition Metal Carbides, Nitrides and Carbonitrides. In *Handbook of Ceramic Hard Materials*; Riedel, R., Ed.; Wiley-VCH Verlag GmbH&Co: Weinheim, Germany, 2000.

36. Pankratz, L.B. *Thermodynamic Properties of Carbides, Nitrides, and Other Selected Substances*; United States Bureau of Mines Reports: Washington, DC, USA, 1994; Available online: https://digital.library.unt.edu/ark:/67531/metadc12836/ (accessed on 31 August 2021).

37. Wang, Q.; German, K.E.; Oganov, A.R.; Dong, H.; Feya, O.D.; Zubavichus, Y.V.; Murzin, V.Y. Explaining Stability of Transition Metal Carbides—and Why TcC Does Not Exist. *RSC Adv.* **2016**, *6*, 16197–16202. [CrossRef]

38. Sevy, A.; Matthew, D.J.; Morse, M.D. Bond Dissociation Energies of TiC, ZrC, HfC, ThC, NbC, and TaC. *J. Chem. Phys.* **2018**, *149*, 044306. [CrossRef]

39. Niessen, A.; de Boer, F.; Boom, R.; de Châtel, P.; Mattens, W.; Miedema, A. Model Predictions for the Enthalpy of Formation of Transition Metal Alloys II. *Calphad* **1983**, *7*, 51–70. [CrossRef]

40. Ushakov, S.V.; Navrotsky, A.; Hong, Q.-J.; Van De Walle, A. Carbides and Nitrides of Zirconium and Hafnium. *Materials* **2019**, *12*, 2728. [CrossRef]

41. Gubanov, V.A.; Ivanovsky, A.L.; Zhukov, V.P. *Electronic Structure of Refractory Carbides and Nitrides*; Cambridge University Press: Cambridge, UK, 1994.

42. Johnson, E.L.; Davis, Q.C.; Morse, M.D. Predissociation Measurements of Bond Dissociation Energies: VC, VN, and VS. *J. Chem. Phys.* **2016**, *144*, 234306. [CrossRef] [PubMed]

43. Sevy, A.; Huffaker, R.F.; Morse, M.D. Bond Dissociation Energies of Tungsten Molecules: WC, WSi, WS, WSe, and WCl. *J. Phys. Chem. A* **2017**, *121*, 9446–9457. [CrossRef]

44. Häglund, J.; Guillermet, A.F.; Grimvall, G.; Körling, M. Theory of Bonding in Transition-Metal Carbides and Nitrides. *Phys. Rev. B* **1993**, *48*, 11685–11691. [CrossRef] [PubMed]

45. Greczynski, G.; Primetzhofer, D.; Hultman, L. Reference Binding Energies of Transition Metal Carbides by Core-Level X-Ray Photoelectron Spectroscopy Free from Ar+ Etching Artefacts. *Appl. Surf. Sci.* **2018**, *436*, 102–110. [CrossRef]

Article

The Phenomena and Criteria Determining the Cracking Susceptibility of Repair Padding Welds of the Inconel 713C Nickel Alloy

Katarzyna Łyczkowska * and Janusz Adamiec

Faculty of Materials Engineering, Silesian University of Technology, ul. Krasińskiego 8, 40-019 Katowice, Poland; janusz.adamiec@polsl.pl
* Correspondence: katarzyna.lyczkowska@polsl.pl

Abstract: The creep-resistant casting nickel alloys (e.g., Inconel 713C) belong to the group of difficult-to-weld materials that are using for precise element production; e.g., aircraft engines. In precision castings composed of these alloys, some surface defects can be observed, especially in the form of surface discontinuities. These defects disqualify the castings for use. In this paper, the results of technological tests of remelting and surfacing by the Tungsten Inert Gas method (TIG) in an argon shield and TecLine 8910 gas mixture are presented for stationary parts of aircraft engines cast from Inconel 713C alloy. Based on the results of metallographic studies, it was found that the main problem during remelting and pad welding of Inconel 713C castings was the appearance of hot microcracks. This type of defect was initiated in the partial melting zone, and propagated to the heat affected zone (HAZ) subsequently. The transvarestraint test was performed to determine the hot-cracking criteria. The results of these tests indicated that under the conditions of variable deformation during the remelting and pad welding process, the high-temperature brittleness range (HTBR) was equal 246 °C, and it was between 1053 °C and 1299 °C. In this range, the Inconel 713C was prone to hot cracking. The maximum deformation for which the material was resistant to hot cracking was equal to 0.3%. The critical strain speed (CSS) of 1.71 1/s, and the critical strain rate for temperature drop (CST), which in this case was 0.0055 1/°C, should be used as a criteria for assessing the tendency for hot cracking of the Inconel 713C alloy in the HTBR. The developed technological guidelines and hot-cracking criteria can be used to repair Inconel 713C precision castings or modify their surfaces using welding processes.

Keywords: high-temperature brittleness range; hot cracking; TIG welding; transvarestraint test; Inconel 713C; nickel alloy

Citation: Łyczkowska, K.; Adamiec, J. The Phenomena and Criteria Determining the Cracking Susceptibility of Repair Padding Welds of the Inconel 713C Nickel Alloy. *Materials* **2022**, *15*, 634. https://doi.org/10.3390/ma15020634

Academic Editor: Jordi Sort

Received: 21 December 2021
Accepted: 10 January 2022
Published: 14 January 2022

Publisher's Note: MDPI stays neutral with regard to jurisdictional claims in published maps and institutional affiliations.

1. Introduction

Nickel-based casting alloys are widely used; e.g., in the aviation industry as materials for engine elements such as high- and low-pressure turbine blades, control segments, etc. [1–3]. Such components are manufactured by precision casting, which enables castings of a high dimensional accuracy and with the correct shape to be obtained without the need for further mechanical treatment. Analysis of the literature data indicated a considerable proportion of castings are disqualified for use due to identified casting defects in the form of pores, blowholes, shrinkage porosities, or cracks [4,5]. In the industry, these types of defects are commonly repaired by welding techniques.

Analysis of the present knowledge on weldability of nickel-based casting alloys indicated that the main limitation of the repair and remanufacturing of such precision castings is the hot-cracking effect. Hot cracks most often run along the weld/pad weld axis, or as intercrystalline cracks [6].

The authors of [7–11] pointed out that the most common cause of low resistance to hot cracking was plastic deformation in the material during weld crystallisation, leading to the rupture of the liquid film along dendrite boundaries, as well as the deformation growth rate and the temperature brittleness range. Cracks the form during welding (crystallisation and liquation cracks) initiate within the high-temperature brittleness range (HTBR), whereas cracks occurring below the solidus temperature—ductility-dip cracking (DDC) cracks (Figure 1)—are related to the ductility-dip temperature range (DTR) [6,12–15]. The HTBR is defined as the range between the nil strength temperature (NST) upon heating and the ductility recovery temperature (DRT) upon cooling [16,17]. The types of hot cracks that form in the HTBR or the DTR depending on the welding temperature are shown in Figure 2.

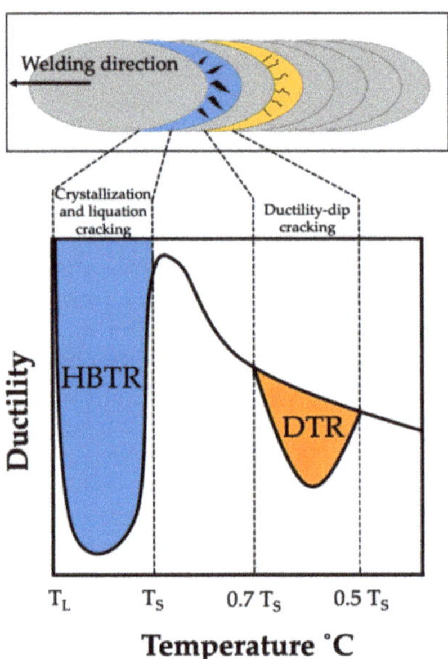

Figure 1. Areas of hot-crack initiation in the weld and the heat affected zone.

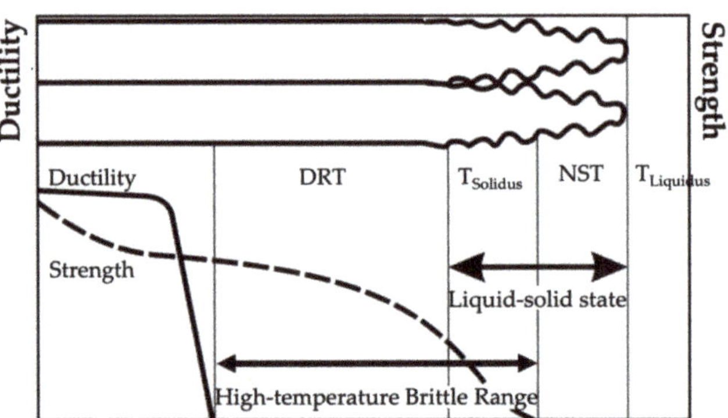

Figure 2. The high-temperature brittleness range (HTBR) determining crystallisation cracking in welded joints and padding welds. DRT—ductility recovery temperature; NST—nil strength temperature [16].

The most frequently occurring type of hot crack is the crystallisation crack. During the final phase of crystallisation, nickel-based alloys display a tendency towards the segregation of alloying elements along the solidification grain boundary, which leads to the formation of a liquid film. The liquid film is characterised by poor mechanical properties and ruptures as a result of local tensile stresses related to weld shrinkage, which in turn leads to the initiation of a crack [18].

The number of crystallisation cracks depends i.a. on the number and nature of intermetallic phases formed during solidification, the surface tension of the liquid metal, the distribution of the liquid at the final phase of crystallisation, the solidification temperature range, the weld's tendency towards shrinkage, etc. The process of crystallisation cracking is presented in Figure 3.

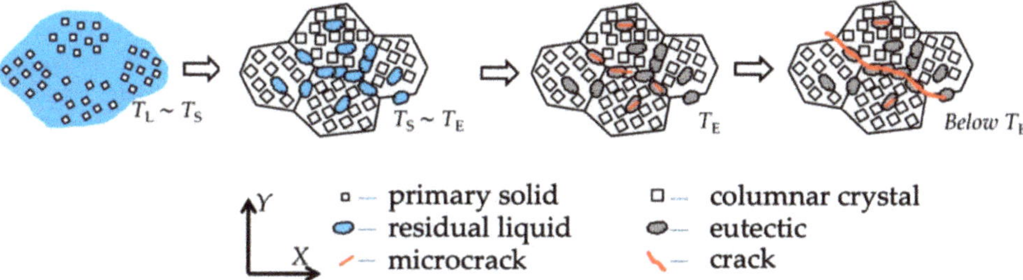

Figure 3. Schematic diagram of the crystallisation cracking mechanism.

According to the theory described by J. F. Lancaster in [19], materials having a wide HTBR are characterised by a low strength/temperature gradient, and thus are susceptible to crystallisation cracking, whereas materials with a narrow HTBR are resistant to crystallisation cracking. The work also demonstrated that a major factor affecting the crystallisation cracking susceptibility of a material is its ductility. The higher the ductility, the better the cracking resistance [19].

However, the basic theory describing crystallisation cracking within the high-temperature brittleness range was presented by N.N. Prokhorov [20]. In his work, he assumed that there was a certain reserve of material plasticity (A = CST·HTBR (%), being the product of the HTBR width (°C) and a parameter referred to as the critical strain rate for temperature drop (CST) (%/°C). He claimed that the main measure of crystallisation-cracking susceptibility was the material's plasticity within the HTBR (Figure 4).

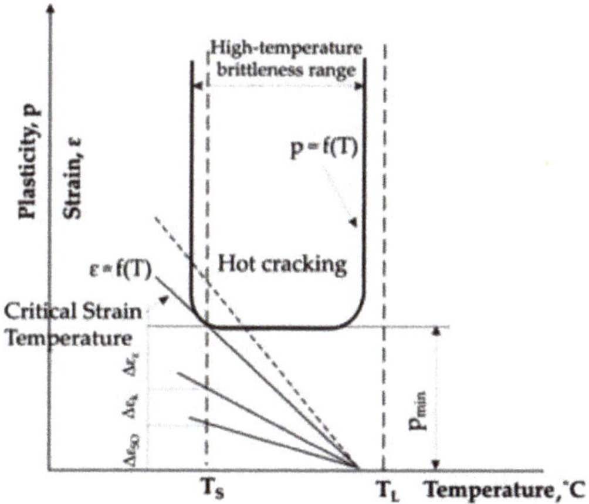

Figure 4. Dependence of alloy ductility within the HTBR and the strain rate [20].

During weld crystallisation, weld ductility drops to a value referred to as p_{min}. Cracking occurs if the built-up strain during weld crystallisation exceeds the HTBR; accordingly, if the accumulated strain related to free shrinkage and the change in the weld shape is lower than p_{min} and falls within the reserve of plasticity, no cracking will occur in the welded joint.

The research results published in [21–25], concerning crystallisation cracking in nickel-based casting alloys, also confirmed that such cracking was caused by the contamination of the material with low-melting phases. During weld crystallisation, they segregated towards grain boundaries, and thus reduced the material's ductility within the HTBR. It was found that the materials described had a wide HTBR, which resulted in crystallisation cracking.

A second type of hot crack is the liquation crack, which forms most frequently in nickel-based alloys with a high Al + Ti content. The literature points to the presence of the γ' phase—Ni_3(Al, Ti)—as their main cause [6,26]. They form due to the recrystallisation of low-melting eutectic mixtures based on partially melted γ' phase, which leads to the formation of a thin liquid film along dendrite boundaries. Such cracks are usually identified along grain boundaries within the partially melted zone [14]. Elements such as B, S, and P, which segregate towards grain boundaries, also contribute to higher susceptibility to such cracks [11]. The mechanism of liquation crack formation is shown in Figure 5.

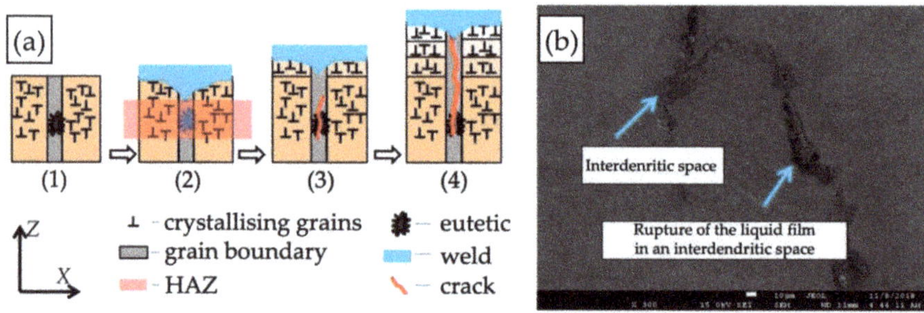

Figure 5. Liquation cracking mechanism: (**a**) diagram; (**b**) liquation cracks between dendrites [6].

Cracks of this type have been described in the literature mainly with regard to austenitic steels and a number of nickel-based alloys; however, there are no precise and exhaustive descriptions of the liquation-cracking mechanism in welded joints and padding welds of nickel-based casting alloys, including Inconel 713C.

A third type of hot crack is the DDC crack. Such cracks occur within $0.5 \div 0.7$ of the solidus temperature; i.e., within the ductility-dip temperature range (DTR) in the solid state. It is deemed that the main cause of this type of cracking is the formation of microvoids along the boundaries of crystallising grains (Figure 6, Type 1) or the partial melting of carbides (Figure 6, Type 2), as well as thermal stresses during crystallisation and low metal ductility within the DTR. This leads to plastic deformations in the material, which depend i.a. on the material's thermal conductivity, the crystallisation rate, the presence of impurities in the welded joint, and interdendritic microporosity. If the strain exceeds the limit values, cracks will initiate in the material [6].

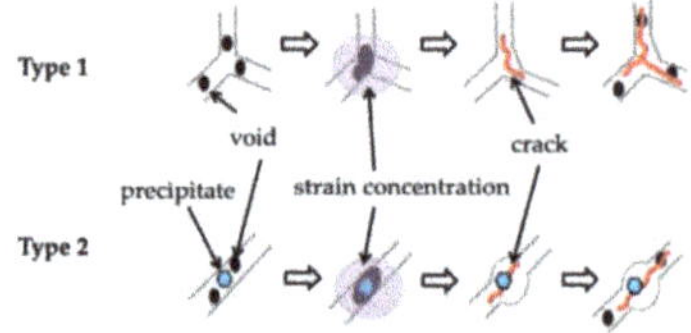

Figure 6. Schematic diagram of the ductility dip cracking mechanism.

Published articles have described the DDC phenomenon mainly for wrought nickel-based alloys; e.g., Alloy 690 [27], Inconel 625, and Inconel 600 [28]. Some works on casting alloys such as Inconel 738 [29] have also been published. The research indicated that the main cause of such cracking was the partial melting of carbides—especially NbC.

The main problems identified in connection with the joining of nickel-based casting alloys, according to the strengthening type, are presented in Table 1 [17,30].

Table 1. The parameters of main weldability concerns for nickel-based alloys [17].

Material Type	Strengthening Type	Main Components	Examples of Alloys	Main Weldability Concerns
Heatproof	Solution Strengthened	Ni–Cu	Monel 400, Monel K-500 (New York, NY, USA)	Weld porosity, crystallisation cracking
		Ni–Mo	Hastelloy B-2 (Kokomo, IN, USA)	Weld and HAZ corrosion
		Ni–Cr–Mo	Hastelloy G-35 (Kokomo, IN, USA)	Weld and HAZ corrosion
		Ni–Cr–Mo–W	Hastelloy C-22 (Kokomo, IN, USA), Inconel 686 (New York, NY, USA)	Weld and HAZ corrosion
		Ni–Cr–Mo–Cu	Hastelloy C-2000 (Kokomo, IN, USA)	Weld and HAZ corrosion
Creep-resistant	Solution Strengthened	Ni–Fe–Cr	Incoloy 800H (New York, NY, USA), RA330 (Temperance, MI, USA)	Liquation cracking
		Ni–Cr–Fe	Inconel 600, Inconel 690 (New York, NY, USA)	DDC
		Ni–Cr–Fe–Mo	Hastelloy X (Kokomo, IN, USA)	Liquation cracking
		Ni–Cr–Mo–Nb	Inconel 625 (New York, NY, USA), Haynes 625SQ (Kokomo, IN, USA)	Crystallisation cracking
		Ni–Cr–Co–Mo	Inconel 617 (New York, NY, USA)	Liquation cracking
		Ni–Cr–W–Mo	Haynes 230 (Kokomo, IN, USA)	Crystallisation and liquation cracking
		Ni–Co–Cr–Si	Haynes R-160 (Kokomo, IN, USA)	Crystallisation cracking
	Precipitation-Strengthened	γ' phase	Rene 41 (Boston, MA, USA), Waspaloy (Hartford, CT, USA), Inconel 713C (New York, NY, USA)	Annealing, crystallisation, and liquation cracking
		γ'' phase	Allvac 718Plus (Pittsburgh, PN, USA)	Crystallisation and liquation cracking
		Ni_3Al	IC-218, IC-25 (Ohio, OH, USA)	Crystallisation and liquation cracking
	Dispersion Strengthened	Y_2O_3	Inconel MA754, Inconel MA6000 (New York, NY, USA)	Metal oxidation

Despite numerous attempts to determine the HTBR and identify the hot-cracking criteria, mainly for wrought nickel-based alloys, there were no unambiguous research results that described cracking mechanisms in precipitation-strengthened nickel-based casting alloys and that evaluated and critiqued their weldability.

The information available in the literature indicated that most nickel-based casting alloys, especially plastically deformed ones, belong to the weldable materials. However, nickel-based casting alloys, in particular those containing aluminum and titanium, are hard to weld, or even unweldable.

An example of a precipitation-strengthened nickel-based casting alloy is Inconel 713C, which is used for vital components of aircraft engines designed to operate at above 700 °C, such as turbine blades or vane clusters.

The attempts at joining and repairing IN713C cast alloys by welding methods described in the literature to date have mainly concerned model components, whereas the translation of the technologies described into actual castings of complex shapes and various wall thicknesses has ended in failure, mainly due to hot cracking on the surface or inside the casting.

Analysis of the literature data showed that due to its content of aluminum (approx. 6%) and titanium (up to 1%), Inconel 713C is classified as hard to weld or unweldable [31]. Thus, it is necessary to explore the mechanisms determining its hot-cracking susceptibility and to analyse the structural phenomena occurring during the crystallisation of remelted areas and padding welds in the casting repair process.

The main purpose of conducted technological tests of remelting and pad welding for Inconel 713C precision castings and the performed remelting test under variable deformation conditions (transvarestraint test) was to assess the possibility of repairing or modifying the casting surface, and determine the criteria for hot remelting cracking. The determined range of technological parameters and hot-cracking criteria are the basis for the development of repair technology or even for the regeneration and modification of surface of Inconel 713C precision cast alloy. The performed structural tests presented an opportunity to describe HTBR and the mechanism of hot melt cracking for a remelted and pad-welded surface using the TIG method.

2. Materials and Methods

The material used in the tests was the nickel-based casting alloy Inconel 713C (New York, NY, United States), which is a polycrystalline, precipitation-strengthened material. The test material was delivered in the form of 5 mm thick plates and castings having a rectangular cross-section measuring $100 \times 80 \times 8$ mm^3. The test castings were made by precision casting. The vacuum induction melting (VIM) method was used to melt the charge material.

The metallographic examinations were conducted using an Olympus GX71 (Warsaw, Poland) light microscope (LM) at magnifications of up to $500\times$. The surface structure after the tests was examined under scanning electron microscopes (SEM): a ZEISS Merlin Gemini II (Oberkochen, Germany) and a JEOL JCM-6000 Neoscope II (Tokyo, Japan). Images were recorded in the secondary electron mode at a magnification of $80,000\times$ and at a voltage accelerating the electron beam to 15 keV.

The structural examinations of the Inconel 713C precision castings revealed that they had a dendritic structure (Figure 7a) with primary MC carbide precipitates (the main precipitate product of carbon) and eutectic mixture areas in interdendritic spaces (Figure 7b). The dendrites were built of the γ phase, being the matrix for γ' phase precipitates. This is a typical structural arrangement for precision castings of IN713C, which was also confirmed by an analysis of the literature data [1,4,32]. The carbides observed were most frequently arranged in the "Chinese script" morphology.

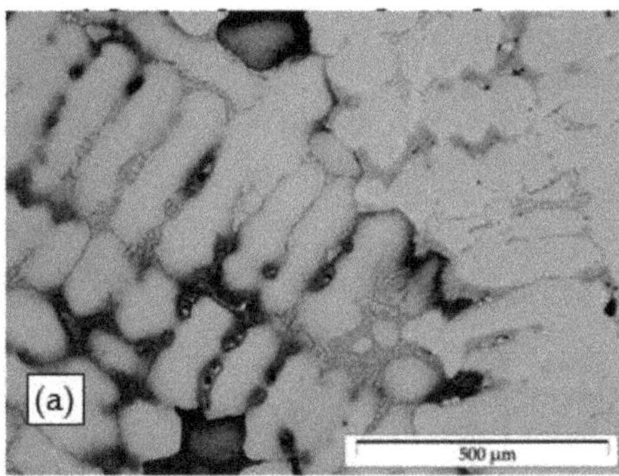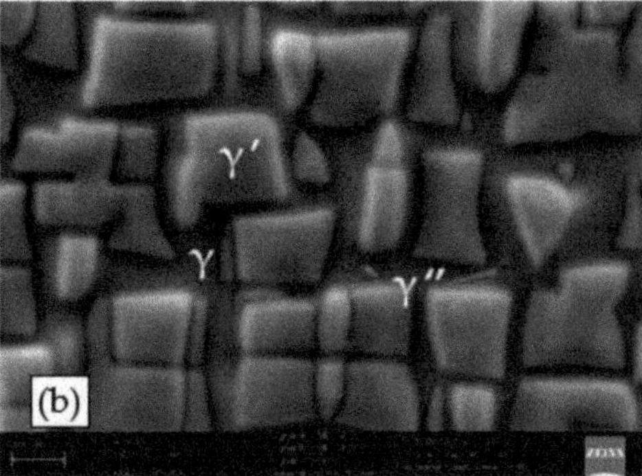

Figure 7. Structure of the Inconel 713C castings: (**a**) dendritic structure with visible eutectic mixtures and carbides (LM); (**b**) γ' phase in the γ phase matrix (SEM).

Based on the literature data and a microanalysis of their chemical composition, it was confirmed that they were complex carbides containing Nb and Mo. Some fine-sized precipitates, which could be identified as the γ'' phase, were also observed in the matrix.

Subsequently, technological trials to repair simulated defects on the side surface of Inconel 713C precision castings were conducted using the TIG welding process. The TIG remelting and pad-welding tests were aimed at developing a technology for repairing surface defects in castings. The tests were performed using two gas shield variants: technically pure argon and a special gas mixture.

The TIG remelting and pad welding in a pure argon atmosphere (99.995) by Messer (Bad Soden, Germany) was conducted using an Esab Aristotig 200 DC power supply (Gothenburg, Sweden), and a WT20 tungsten electrode by ESAB (Gothenburg, Sweden)with a diameter of 2.4 mm according to PN EN ISO 6848. The technological parameters of the processes are shown in Table 2. Thermanit 625 welding wire (EN ISO 18274–S Ni6625 (NiCr22Mo9Nb, AWS A5.14:ERniCrMo-3)) by Böhler Schweisstechnik GmbH (Linz, Austria), 1.0 mm in diameter, was used as filler material in the pad-welding tests.

Table 2. Parameters of the Tungsten Inert Gas remelting and pad-welding processes in an argon atmosphere.

	Specimen Designation	Current (A)	Arc Voltage (V)	Remelting /Pad-Welding Rate (mm/s)	Arc Linear Energy (kJ/mm)	Gas Flow Rate (l/min)	Visual Assessment of the Weld Face According to EN ISO 5817
Remelting	15	25	12	1.20	0.15	12	C
	16	30	12	1.20	0.18	12	C
	17	35	12	1.20	0.21	12	C
	18	40	15	1.20	0.30	12	C
	19	45	15	1.20	0.34	12	C
	20	50	15	1.20	0.38	12	B
Pad Welding	625.1	30	15	1.03	0.26	7	B
	625.2	35	15	1.03	0.31	7	C
	625.3	40	15	1.03	0.35	12	B

In the other test variant, the TIG remelting and pad-welding processes were conducted in the TecLine 8910 gas mixture by Messer (Bad Soden, Germany) (15% He, 2% H_2, 0.015% N_2, Ar-balance). A Lincoln Electric Bester Invertec V405-T Pulse power supply (Cleveland,

OH, USA) and a tungsten electrode by ESAB (Gothenburg, Sweden) (WT20 according to the AWS classification), 2.4 mm in diameter, were used. The pad welding was performed using the same welding wire The pad welding was performed using the same welding wire by Böhler Schweisstechnik GmbH (Linz, Austria) (Thermanit 625, Ø1.0 mm) as in the case of the pad welding in an argon atmosphere. The parameters of the TIG remelting and pad-welding processes are set out in Table 3.

Table 3. Parameters of the TIG remelting and pad-welding processes in a TecLine 8910 gas mixture atmosphere.

	Specimen Designation	Current (A)	Arc Voltage (V)	Remelting/Pad-Welding Rate (mm/s)	Arc Linear Energy (kJ/mm)	Gas Flow Rate (l/min)	Visual Assessment of the Weld Face According to EN ISO 5817
Remelting	1	25	12	1.30	0.15	12	B
	2	30	12	1.30	0.17	12	B
	3	35	12	1.30	0.19	12	B
	4	40	15	1.30	0.28	12	B
	5	45	15	1.30	0.31	12	B
	6	50	15	1.30	0.35	12	B
Pad Welding	7	25	10	1.15	0.13	7	B
	8	30	12	1.15	0.17	7	B
	9	35	12	1.15	0.22	12	B
	10	40	15	1.15	0.31	12	B
	11	45	15	1.15	0.35	12	B
	12	50	15	1.15	0.39	12	B

The influence of factors determining the viability of remelting of Inconel 713C was evaluated based on the results of the assessment of the HTBR under forced deformation conditions (transvarestraint test). The transvarestraint test consisted of fast bending of flat samples on a cylindrical die block, perpendicular to the direction of remelting [33]. The strain inflicted was related to the radius of the die block, and depended on the thickness of the bent specimen.

Cast plates of Inconel 713C measuring $100 \times 80 \times 5$ mm^3 were prepared for the tests. The remelting was performed with a direct current of 40 A, at a rate of approx. 1 mm/s. The remelting parameters were selected based on technological tests, so as to obtain full penetration. The strain inflicted in particular tests was calculated using the following Equation (1):

$$\varepsilon = \frac{g}{2R} \cdot 100\% \tag{1}$$

where: ε—strain (%), g—specimen thickness (mm), and R—radius of die block curvature (mm) [34,35].

Following the remelting tests, the length of the longest crack in the remelted area axis (L_{max}) and the total length of all cracks classified as hot cracks were determined. With the individual strain value during remelting (Equation (1)) and the welding rate (v_s) being known, the crack growth time (t_{max}) was calculated based on the following Equation (2):

$$t_{max} = \frac{L_{max}}{v_s} \tag{2}$$

where: t_{max}—crack growth time (s), L_{max}—longest crack (mm), and v_s—welding rate (mm/s) [34].

With the welding heat cycle and the crack growth time during remelting being known, the temperature at the end of the longest crack was determined, which enabled identification of the HTBR for the Inconel 713C precision castings under variable strain conditions; i.e., under crystallisation conditions typical of welding processes. The schematic methodology is shown in Figure 8.

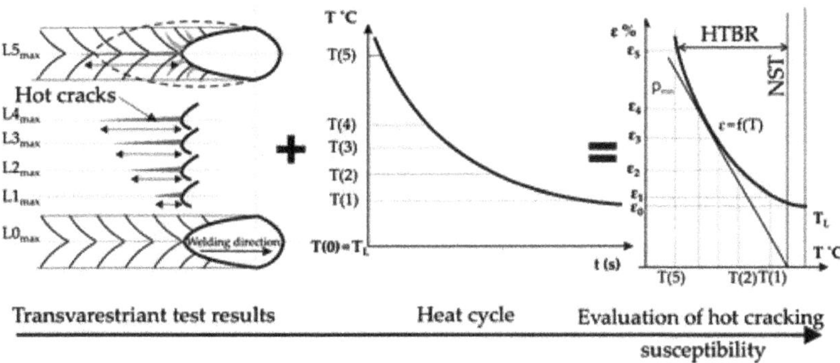

Figure 8. Methodology for determining the HTBR based on the results of the transvarestraint test.

3. Results

Visual examinations of the remelted area surfaces on the Inconel 713C precision castings obtained by TIG in an argon atmosphere revealed no cracks (Figure 9a,c,e). The surfaces obtained at an arc linear energy below 0.3 kJ/mm were uneven, with visible ripples (Figure 9a). Remelting at a higher linear energy (more than 0.3 kJ/mm) yielded an even and smooth surface (Figure 9c,e). Based on the visual examinations of the remelted area surfaces, they were classified as quality level C according to EN ISO 5817 (Table 2).

Visual examinations of the microstructure of the padding weld shown in Figure 9d revealed that the area of the padding weld material was built of narrow columnar dendrites that grew perpendicularly to the heat-dissipation direction. Partially melted dendrites of the base material were observed in the partially melted zone (Figure 9a,d,e).

Examinations of the macrostructure of the remelted areas revealed that their width and depth increased with increasing arc energy (Figure 9b,d). Remelting at a linear energy of more than 0.3 kJ/mm resulted in the entire casting edge being remelted, which is important in the case of through-casting defects (Figure 9d).

As for the pad welding performed with the use of Inconel 625 wire as the filler material, the padding-weld faces were correct (Figure 9e). They had a regular shape with no visible ripples on the surface. Pad welding with a linear energy of 0.3 kJ/mm resulted in the entire casting edge being remelted, and defects could be filled by filler material, depending on their size. Examinations showed that the padding welds had correct macrostructures. No cracks or other welding defects were identified in the padding welds or the HAZ. On this basis, the padding welds made at a linear energy of 0.35 kJ/mm could be classified as quality level B according to EN ISO 5817.

Examinations of the microstructure of the remelted areas obtained at a low linear energy (below 0.15 kJ/mm) confirmed that their surfaces were flat, with the weld lines being distinctly visible (Figure 9b). A broad partially melted zone was revealed (approx. 300 μm), in which the interdendritic zone was partially melted (Figure 9b).

In the case of the remelting process conducted at a linear energy of more than 0.21 kJ/mm, interdendritic cracks (Figure 10a) that disqualified the remelted areas for use were identified in the HAZ and the partially melted zone. Such cracks initiated along MC carbide boundaries in the partially melted zone. They formed as a consequence of the partial melting of dendrite branches and the loss of cohesion by the interdendritic liquid, which resulted in decreased adhesion to the base material.

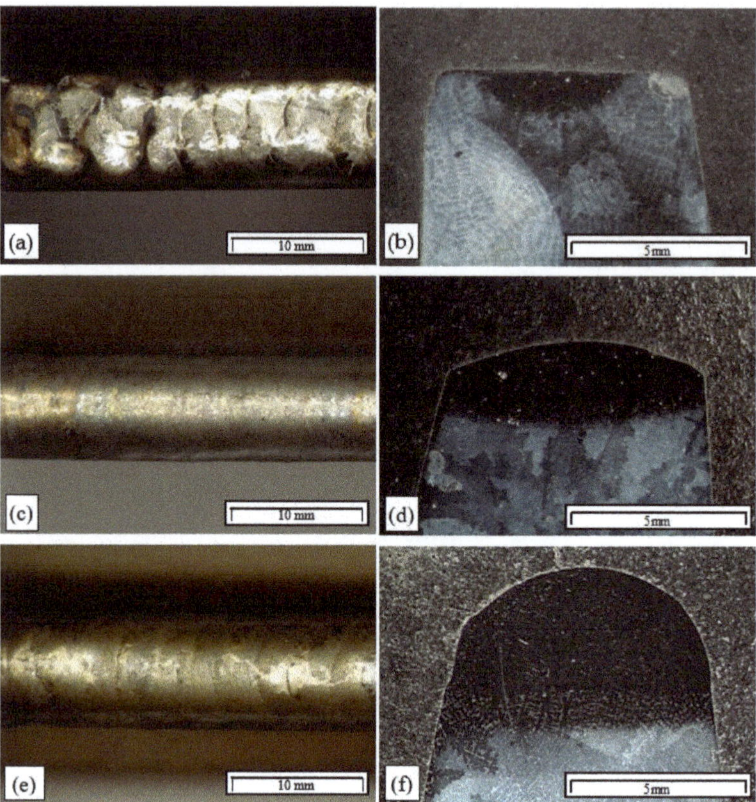

Figure 9. Surfaces and macrostructures of a remelted area and two padding welds made by TIG: (**a**,**b**) remelting of the base material with no filler, arc linear energy: 0.18 kJ/mm; (**c**,**d**) padding weld, arc linear energy: 0.34 kJ/mm; (**e**,**f**) padding weld made with Thermanit 625 wire, arc linear energy: 0.35 kJ/mm.

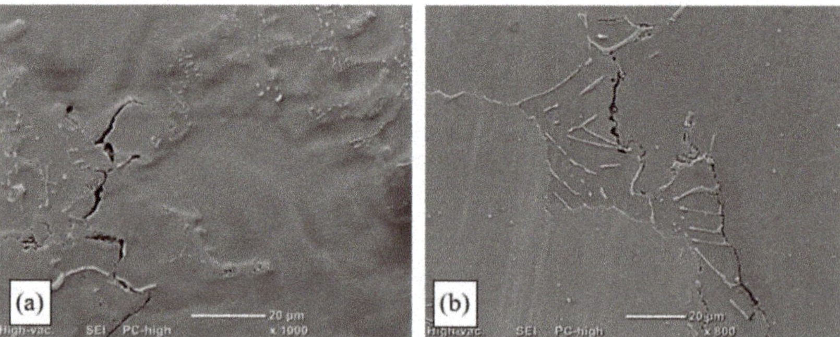

Figure 10. Structure of a remelted area on an Inconel 713C precision casting, obtained by TIG welding with no filler material (E_l = 0.38 kJ/mm): (**a**) crack in the partially melted zone; (**b**) cracks along dendrite boundaries in the area of "Chinese script" carbides.

Dendrites were observed that had been separated from the base material and had not melted in the welding pool. This confirmed that deep penetration by liquid metal occurred in interdendritic spaces in the partially melted zone. The fragmentation of primary carbides was observed in those spaces, which was related to their partial melting and coagulation (Figure 11a). Numerous microcracks were also identified that ran along primary carbide precipitates, along dendrite boundaries (Figure 11b). Analysis of the crack trajectory confirmed that depending on the heat cycle of the pad-welding process, the cracks were related to the partial melting of dendrite edges (Figure 10a), eutectic mixture areas, and carbides (Figure 11b). Cracks initiated in the partially melted zone due to the rupture of the liquid film, which was stretched during padding-weld crystallisation [36,37].

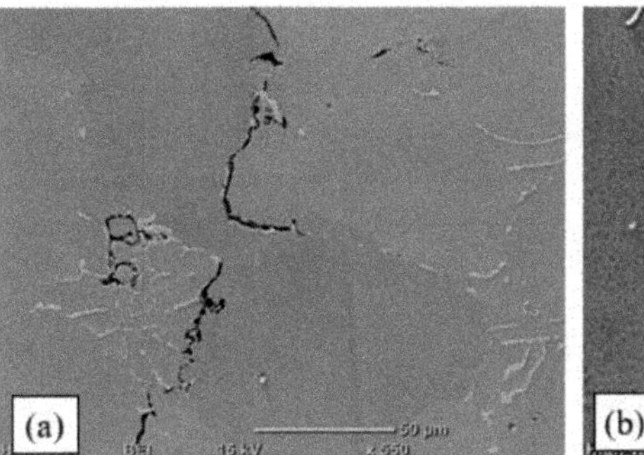
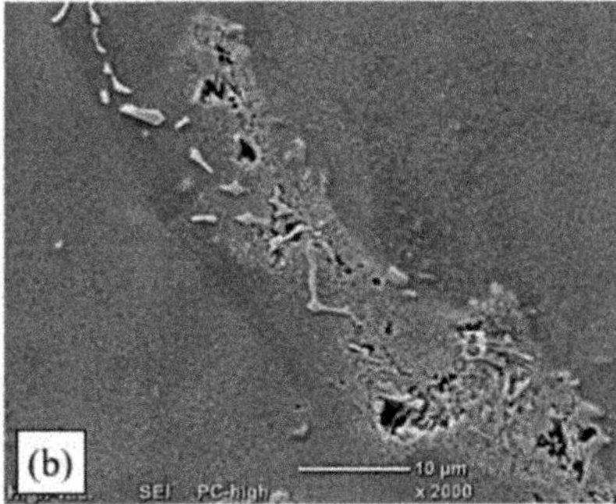

Figure 11. Structure of an Inconel 713C padding weld obtained by TIG welding with Inconel 625 as the filler material (E_l = 0.35 kJ/mm): (**a**) crack in the area of Chinese script carbide precipitates, SEM; (**b**) material discontinuities in the HAZ, in the area of the γ-γ' eutectic mixture and carbides.

The use of the TecLine 8910 mixture increased the welding rate and improved the stability of electric arc discharges. An important technological measure affecting the remelting process was to increase molten metal liquidity by lowering the surface tension. This enabled filling developing cracks with liquid metal [16]. The process parameters are presented in Table 3, and examples of padding-weld faces and macrostructures are shown in Figure 12. Photographs of the microstructures of the remelted areas and padding welds obtained are shown in Figures 13 and 14.

Visual examinations of the remelted area surfaces obtained by TIG remelting in a TecLine 8910 atmosphere revealed that in all cases, the surface was even and smooth, and free of welding defects (Figure 12a,c). Remelting with a linear energy of less than 0.17 kJ/mm led to the formation of ripples, caused by the gradual crystallisation of the molten pool (Figure 12a). Increasing the linear energy to more than 0.2 kJ/mm resulted in a smooth surface without visible ripples (Figure 12c).

Examinations of the macrostructure revealed a correct remelted area geometry with distinctly marked zones; i.e., the melted metal, with visible dendrites growing in the heat-dissipation direction, a wide partially melted zone, and the HAZ. The remelting parameters applied enabled the melting of the entire casting edge (Figure 12b,d). Based on the visual examinations of the surfaces of the remelted areas and the assessment of their macrostructures, it was determined that the remelted areas met the requirements of quality level B according to EN ISO 5817.

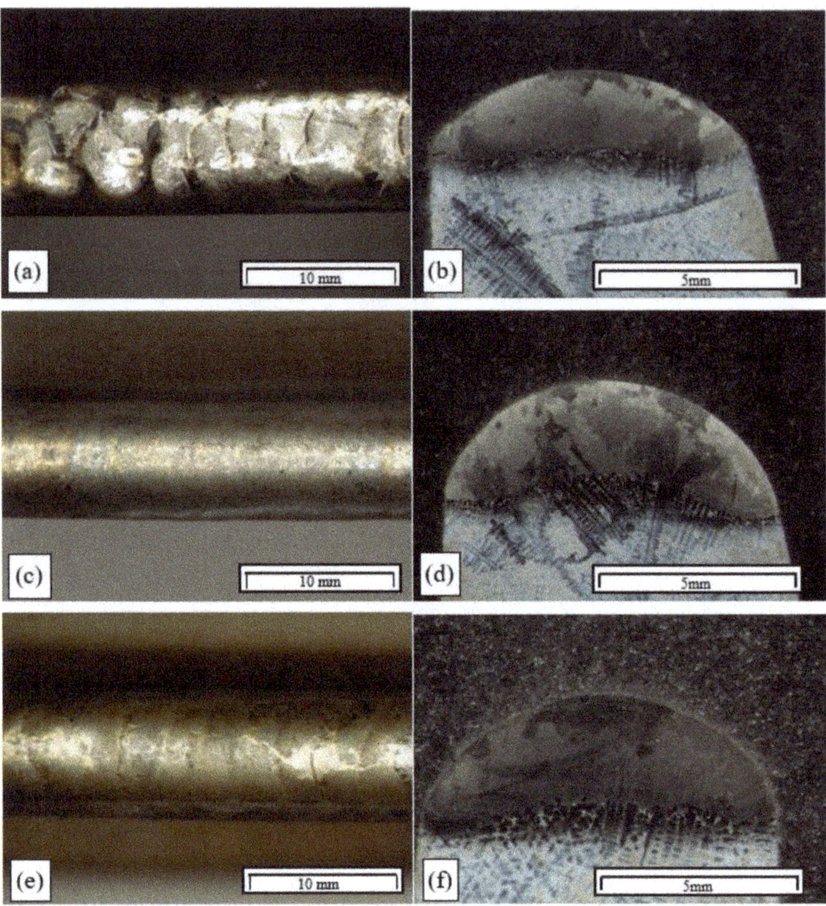

Figure 12. Surfaces and macrostructures of a remelted area and two padding welds made by TIG in a TecLine 8910 atmosphere: (**a**,**b**) remelting of the base material with no filler, arc linear energy: 0.15 kJ/mm; (**c**,**d**) padding weld, arc linear energy: 0.31 kJ/mm; (**e**,**f**) padding weld made with Thermanit 625 wire, arc linear energy: 0.35 kJ/mm.

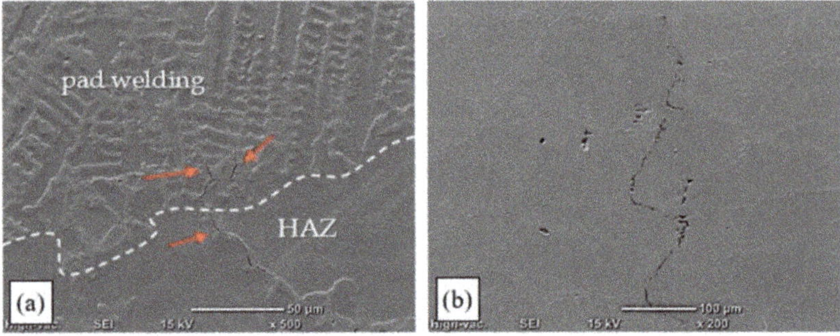

Figure 13. Structure of a remelted area in TIG remelting of Inconel 713C in a TecLine 8910 atmosphere (E_l = 0.35 kJ/mm): (**a**) crack along crystal boundaries in the HAZ; (**b**) crack along dendrite boundaries in the carbide area (SEM).

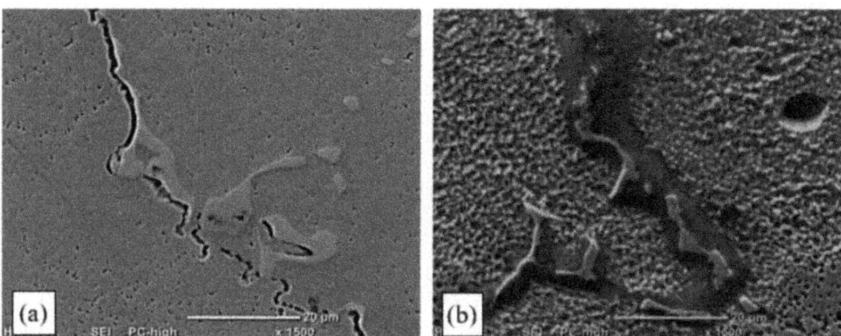

Figure 14. Padding-weld structure in TIG pad welding of Inconel 713C in a TecLine 8910 atmosphere (E_l = 0.35 kJ/mm) with filler material: (**a**) crack at a dendrite/carbide interface; (**b**) microcrack along dendrite boundaries with a visible privileged trajectory determined by carbides.

Visual assessment of the padding-weld faces made by TIG in a TecLine 8910 atmosphere with the addition of Inconel 625 wire revealed that pad welding with a linear energy of up to 0.15 kJ/mm led to the formation of ripples on the surface. This was related to the feeding of filler material into the molten pool and the process of padding-weld crystallisation. Increasing the linear energy to more than 0.15 kJ/mm resulted in a smooth and even weld face (Figure 12e). Examinations of the padding-weld macrostructures revealed no welding defects. The shapes of the padding welds were found to be correct, with a clearly outlined fusion zone and an approx. 1 mm wide HAZ (Figure 12f). Examinations conducted in accordance with EN ISO 17637 enabled qualifying the padding welds as quality level B according to EN ISO 5817 (Figure 12e,f, Table 3).

Analysis of the microstructure of the remelted areas obtained with a linear energy of less than 0.17 kJ/mm in a TecLine 8910 atmosphere revealed no cracks or other welding defects. A small number of hot cracks were only present in the HAZ of the remelted areas obtained with a linear energy of more than 0.17 kJ/mm. The cracks were found along dendrite boundaries, and their trajectories were determined by MC carbides (Figure 13).

The structure of the melted metal area was made up of fine columnar crystals, between which fine carbides, probably of the MC type, were revealed. In the fusion zone, the partial melting of dendrite boundaries was observed in the base material, as well as the partial melting of primary carbides, which had undergone fragmentation. On this basis, it can be stated that due to the identification in the interdendritic spaces of microcracks that were impossible to detect by nondestructive tests, this technology may be deemed acceptable, but is recommended only if the remelting is conducted with a linear energy of less than 0.17 kJ/mm.

The padding welds had a complex dendritic structure with carbides located in interdendritic spaces. This arrangement is typical of padding welds made on nickel-based casting alloys. The partial melting of carbides, leading to their coagulation and fragmentation, was also observed in the partially melted zone. The use of a gas mixture containing hydrogen and helium, increasing the arc linear energy and molten metal liquidity, resulted in a wider partially melted zone (approx. 300 μm), and thus enhanced the penetration of molten metal into interdendritic spaces.

In addition, in the case of the TIG pad welding in a TecLine 8910 atmosphere, microcracks were identified in the HAZ that had formed during pad welding at less than 0.17 kJ/mm. The cracks identified initiated at the weld line, where dendrites were partially melted. They grew as interdendritic cracks in the areas where MC primary carbides were present (Figure 14a).

The partial melting of carbides and dendrites was also observed in interdendritic spaces, which led—due to the ongoing crystallisation process—to the development of a network of fine material discontinuities that constituted DDC initiation spots (Figure 14b).

During the pad welding, similar to in the case of the remelting process, liquation cracks were identified in the HAZ. Although they were partially filled with metal, TIG pad welding should be deemed an acceptable technology only if the linear energy applied is below 0.17 kJ/mm, and if special production supervision and control conditions are satisfied.

The measurements and calculations presented in Table 4 enabled the determination of the high-temperature brittleness threshold; i.e., the strain value at which no cracking occurred. The high-temperature brittleness threshold (ε_p) adapted for the castings tested was 0.3%. This parameter can be adopted as a criterion for assessing the hot-cracking susceptibility of Inconel 713C.

Table 4. Results of the measurements and calculations of the indicators used to assess the high-temperature brittleness range of the Inconel 713C precision castings.

No.	Strain ε (%)	Longest Crack Length L_{max} (mm)	Crack Growth Time t_{max} (s)	Critical Strain Speed (1/s)	Critical Strain Temperature (1/°C)	ΔHTBR ** (°C)	HTBR ** (°C)
1	0.56	4	0.8				
2	0.77	6	1.2				
3	1.12	8	1.6	1.71	0.0055	1053–1299	246
4	1.67	12	2.4				
5	2.50	12.5	2.5				
6	5.00	18	3.6				

** The results are presented in [38].

With the welding heat cycle and the crack growth time during remelting (Figure 8) being known, the temperature at the end of the longest crack was determined, which enabled the identification of the HTBR for the Inconel 713C precision castings under variable strain conditions; i.e., under crystallisation conditions typical of welding processes.

Determination of the relation of $t_{max} = f(\varepsilon)$ also enabled the determination of the value of the critical strain speed (CSS) parameter, understood as the tangent of the inclination angle between the tangent to the crack growth curve and the deformation axis (Figure 15).

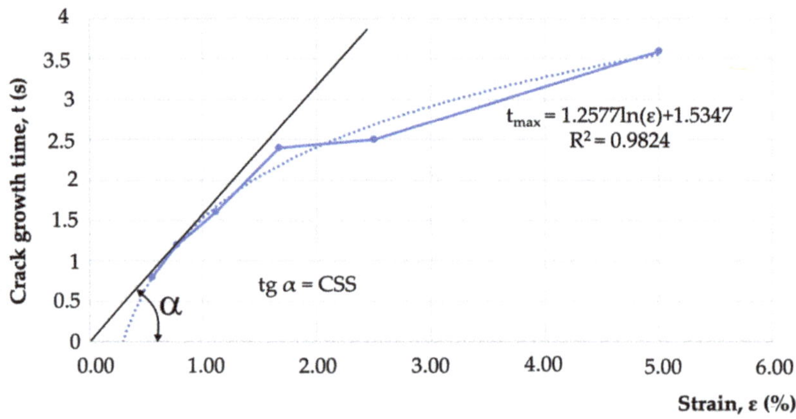

Figure 15. Hot-crack growth time as a function of specimen deformation in the transvarestraint test. The CSS value for the case in question was 1.71 1/s, which indicated that the alloy was highly susceptible to hot cracking during remelting. The results obtained enabled the determination of exponential ductility curves using $\varepsilon = f(T)$ (Figure 16).

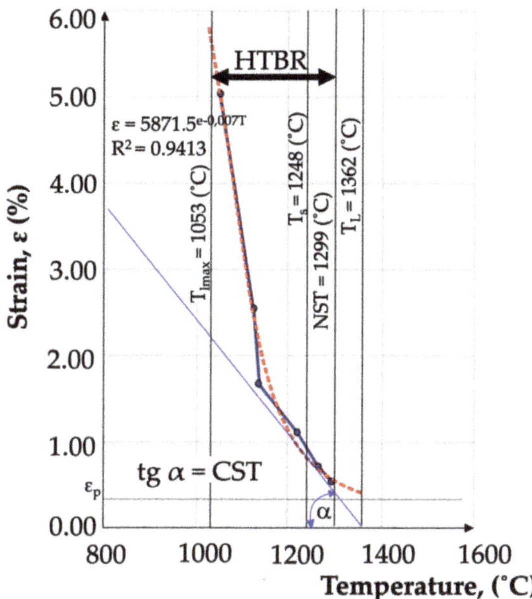

Figure 16. Strain as a function of temperature for Inconel 713C precision castings, determined based on transvarestraint tests.

The results of the tests enabled the determination of the maximum crack length in the padding-weld axis (L_{max}), the total crack length ($L1_{max}$), the cracking threshold (ε_p), the HTBR during welding, the critical strain rate for temperature drop (CST), and the critical strain speed (CSS). The results obtained made it possible to describe the phenomena occurring during padding-weld crystallisation and the factors affecting hot-cracking susceptibility within the HTBR, and thus to assess the weldability of Inconel 713C and the possibility of repairing defects in Inconel 713C castings.

Based on the regression and correlation analysis of a single variable function (nonlinear), it was found that the relationship determined was valid. The relation described enabled the determination of the HTBR under remelting conditions. The HTBR is defined as the difference between the NST and the temperature at the end of the longest crack. The relation also enabled the determination of certain hot-cracking criteria, including the critical strain rate for temperature drop (CST), which is the tangent of the angle between the tangent to the ductility curve $\varepsilon = f(T)$ and the temperature axis (Figure 16). The value of this parameter was 0.0055 1/°C.

Figure 17a shows the weld face on a specimen that was subjected to maximum deformation during the transvarestraint test ($\varepsilon = 5\%$). It was found that the hot crack caused by specimen deformation ran along the axis of the padding weld and across its entire melted part, which indicated its brittleness. Fractographic examinations confirmed that fine columnar dendrites grew perpendicularly to the remelted area surface, in the heat-dissipation direction.

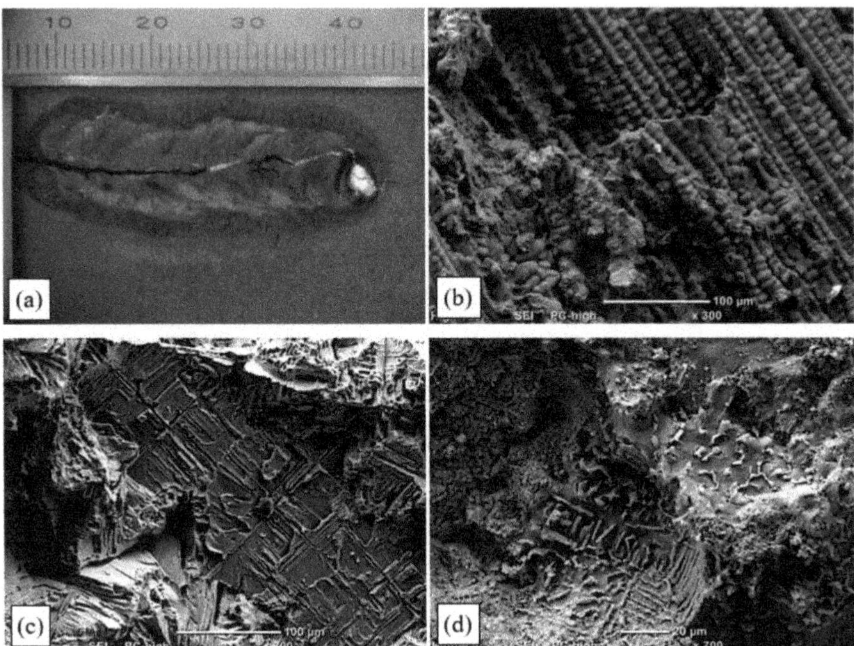

Figure 17. Results of fractographic examinations of the surface of a hot crack that developed during a transvarestraint test in a specimen subjected to 5% strain during remelting: (**a**) general view; (**b**) ruptured dendrites and interdendritic bridges in the remelted area; (**c**) crack surface with visible brittle transcrystalline fracture and areas of liquid film rupture; (**d**) partially melted carbides in the partially melted zone.

The crack initiation site was the molten pool, where the interdendritic liquid film lost cohesion at the NST due to tensile stresses involved in the crystallisation process. The rupture of "bridges" that formed the rigid structure of the liquid–solid state was also observed there (Figure 17b). The number of bridges was relatively small, and the dominant crack-initiation mechanism was the loss of continuity by the liquid film covering the crystallising dendrites. As the temperature dropped, the solid body lattice expanded, and thus the number of ruptured bridges between dendrite branches increased (Figure 17b). Near the solidus temperature, the inflow of liquid metal into the crystallising area of the padding weld stopped, leading to the formation of local voids, which—with the material's ductility dropping in the HTBR—reinforced the tendency for cracks to propagate (Figure 17c). Partially melted interdendritic spaces with distinctly visible carbides were observed in the partially melted zone (Figure 17d).

Brittle transcrystalline fracture surfaces were also observed (Figure 17d). They were ruptured base material dendrites that were partially melted. The fractographic examinations of hot-crack surfaces confirmed the same hot-cracking mechanism for all cases (irrespective of the strain degree).

4. Discussion

The analysis of the results of the technological TIG remelting and pad-welding tests in an argon atmosphere showed that the process could not be used for repairing precision castings. Despite correct surfaces having been obtained (particularly in the pad-welding process) (Figure 9), the examinations of the microstructure revealed numerous cracks in the heat-affected zone and the partially melted zone (Figures 10 and 11). The areas that were the most susceptible to hot cracking were the interdendritic spaces of the base material that underwent partial melting. As a result of the plastic strains at work, the liquid metal lost cohesion. It was found that the areas privileged for the appearance of cracks were sites with carbides in the Chinese script morphology (Figures 10b and 11a).

In order to enhance electric arc stability and increase metal liquidity during the TIG welding tests, some of the tests were conducted with a new gas mixture—TecLine 8910, containing approx. 15% He and 2% H_2. The gas mixture considerably improved the quality of the surfaces obtained (Figure 12); however, correct results were only obtained for remelted areas and padding welds made with a linear energy of less than 0.17 kJ/cm. Remelting and pad welding at a higher energy resulted in the formation of interdendritic cracks, which was related to strain occurring in the HAZ and resulting from the welding heat cycle (Figure 13).

The hot cracks revealed on metallographic specimens were most often located under the remelted and pad-welded surface, which made it impossible to identify defects by nondestructive tests. However, due to the need to ensure the safe use of the repaired elements, it was necessary to perform RTG examinations of each repaired casting. Further investigation of the mechanical properties is also advisable for repaired castings, especially in the field of creep resistance. Such requirements should be included in the qualification procedure for Inconel 713C precision-casting repair technology.

The high-temperature brittleness range (HTBR) determined in the transvarestraint test was understood as the difference between the longest crack temperature and the NST. The range had a width of 246 °C, and extended from 1053 °C to 1299 °C (Table 4). It was found that the HTBR under remelting conditions was nearly 5 times wider than the HTBR determined for the base material [38]. This indicated that the material was much more susceptible to cracking in a remelting process involving concentrated arc energy than under conditions of even heat distribution involved in Gleeble 3500 simulations [38].

The level of plastic strain at which no cracking occurred in a casting under remelting conditions was 0.3%. This was the high-temperature brittleness threshold, or the so-called reserve of plasticity described in Prokhorov's theory [20].

The results of the transvarestraint tests also enabled the determination of crystallisation cracking criteria. The critical strain speed (CSS), which for IN713C was 1.71 (1/s) (Table 4), was used as the criterion for the strain rate during remelting. If this CSS value was exceeded, crystallisation cracks would appear in the material during remelting (Figure 14). Another indicator describing the cracking susceptibility of a casting during remelting is the critical strain rate for temperature drop (CST), which for IN713C was 0.0055 1/°C. If this value was exceeded, cracking occurred. The schematic value of the CST, defined as the tangent of the angle between the tangent to the ductility curve and the temperature axis, is shown in Figure 15. If angle α was wider than the critical angle, the material cracked.

Examinations of the surfaces of the crystallisation cracks that appeared during the deformation of the remelted specimens in the transvarestraint tests indicated a similar cracking mechanism to the case of specimens deformed using a Gleeble simulator. An area was observed on the crack surface where parallel dendrites developed. It was found that the fracture surface changed within the area where columnar dendrites were present (i.e., within the melted metal area) (Figure 17c). An area typical of cracking (close to the NST) was also identified, where ruptured bridges and dendrites in the liquid–solid state were present (Figure 17b). Below the solidus lines, brittle fracture surfaces were observed in which voids had formed in the liquid–solid state due to the partial melting of dendrite edges and carbides (Figure 17d). A schematic change in the fracture surface structure is shown in Figure 18.

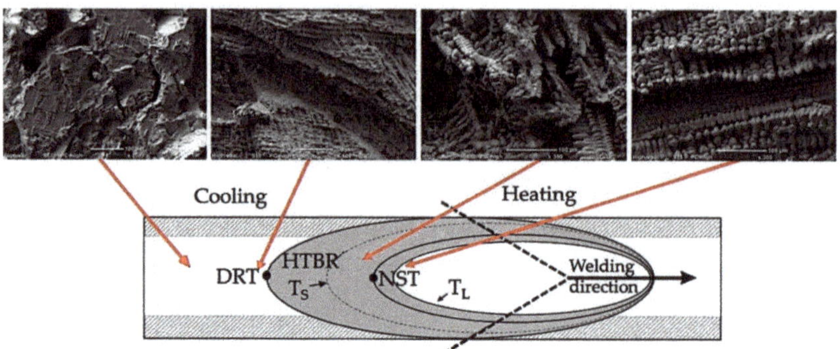

Figure 18. Change in surface morphology on the surface of a crystallisation crack that developed during the remelting of a casting under forced deformation conditions. T_{Lmax}—temperature at the end of the longest crack.

The obtained results of the technological and structural tests, including the description of the hot-cracking mechanism in HTBR and the determination of numerical fracture criteria in the form of indicators (ε_p, CST, and CSS), constituted a unique contribution to the understanding of the weldability of the Inconel 713C alloy. They are also a background to the evaluation of the possibility of using welding techniques for repair, regeneration, or surface modification of precision castings composed of Inconel 713C alloy.

The hot-cracking criteria and mechanisms described were used to devise technological tests for remelting and pad welding of Inconel 713C precision castings. Based on the results obtained and the requirements set by manufacturers and users, a number of welding technologies were selected that had the greatest potential for use in the repair of aircraft-engine components.

5. Conclusions

The test results presented confirmed the hypothesis that the possibility of repairing Inconel 713C precision castings is decided by hot-cracking susceptibility, which is the effect of structural phenomena occurring during padding-weld crystallisation. Based on their analysis, the following conclusions were formulated:

1. The critical strain speed (CSS) of 1.71 1/s and the critical strain rate for temperature drop (CST), in this case having the value of 0.0055 1/°C, should be adopted as the criteria for assessing the hot-cracking susceptibility of Inconel 713C within the high-temperature brittleness range.
2. Hot cracks appearing when the alloy was being remelted under forced deformation conditions developed within the high-temperature brittleness range. This was caused by voids, the formation of which was related to the loss of cohesion by the interdendritic liquid and the rupture of the solid body lattice formed of columnar dendrites. Areas with carbides in the Chinese script morphology favoured the development of hot cracks.
3. Hot cracks in the HAZ and the partially melted zone resulted from the critical strain being exceeded during the crystallisation of remelted areas or padding welds. The Inconel 713C alloy was susceptible to cracking during plastic deformation in the HAZ at temperatures above 1050 °C. The critical circumferential strain for this temperature was 0.48%.
4. The main difficulty in repairing Inconel 713C castings, as identified during the technological TIG tests, was due to microcracks initiating in the partially melted zone and propagating into the HAZ. Due to their size and location, such cracks were very difficult to detect by nondestructive testing methods.

5. Under variable-strain conditions characteristic of the remelting and pad-welding processes, the high-temperature brittleness range widened nearly 5-fold (the HTBR width was 246 °C), and extended from 1053 °C to 1299 °C. The strain below which the material was resistant to hot cracking was 0.3%.

Author Contributions: Conceptualization, K.Ł. and J.A.; methodology K.Ł. and J.A.; formal analysis, K.Ł.; writing—original draft preparation, K.Ł.; writing — review and editing, J.A.; visualization, K.Ł. and J.A.; supervision, J.A.; funding acquisition, K.Ł. All authors have read and agreed to the published version of the manuscript.

Funding: This research was funded by a Silesian University of Technology subsidy for statutory business activities (no. 11/030/BKM21/1058).

Institutional Review Board Statement: Not applicable.

Informed Consent Statement: Not applicable.

Data Availability Statement: The data supporting reported results are not stored in any publicly archived datasets. Readers can contact the corresponding author for any further clarification of the results obtained.

Conflicts of Interest: The authors declare no conflict of interest.

References

1. Du Pont, J.N.; Lippold, J.C.; Kiser, S.D. *Welding Metallurgy and Weldability of Nickel-Base Alloys*; John Wiley & Sons: Hoboken, NJ, USA, 2009. [CrossRef]
2. Zhang, S.; Zhao, D. *Aerospace Materials Handbook*, 1st ed.; CRC Press: Boca Raton, FL, USA, 2013.
3. Ma, D. Novel casting processes for single-crystal turbine blades of superalloys. *Front. Mech. Eng.* **2018**, *13*, 3–16. [CrossRef]
4. Andersson, J. Weldability of Precipitation Hardening Superalloys–Influence of Microstructure. Ph.D. Thesis, Chalmers University of Technology, Göteborg, Sweden, 1 January 2011.
5. Matysiak, H.; Zagorska, M.; Balkowiec, A.; Adamczyk-Cieslak, B.; Cygan, R.; Cwajna, J.; Nawrocki, J.; Kurzydłowski, K.J. The Microstructure Degradation of the IN 713C Nickel-Based Superalloy After the Stress Rupture Tests. *J. Mater. Eng. Perform.* **2014**, *23*, 3305–3313. [CrossRef]
6. Liu, S.; Yu, H.; Wang, Y.; Zhang, X.; Li, J.; Chen, S.; Liu, C. Cracking, Microstructure and Tribological Properties of Laser Formed and Remelted K417G Ni-Based Superalloy. *Coatings* **2019**, *9*, 71. [CrossRef]
7. Singh, S.; Andersson, J. Hot cracking in cast alloy 718. *Sci. Technol. Weld. Join.* **2018**, *23*, 568–574. [CrossRef]
8. Xu, J.; Lin, X.; Zhao, Y.; Guo, P.; Wen, X.; Li, Q.; Yang, H.; Dong, H.; Xue, L.; Huang, W. HAZ Liquation Cracking Mechanism of IN-738LC Superalloy Prepared by Laser Solid Forming. *Met. Mater. Trans. A* **2018**, *49*, 5118–5136. [CrossRef]
9. Kang, M.; Jiang, M.; Sridar, S.; Xiong, W.; Xie, Z.; Wang, J. Effect of Multiple Repair Welding on Crack Susceptibility and Mechanical Properties of Inconel 718 Alloy Casting. *J. Mater. Eng. Perform.* **2021**, *31*, 254–261. [CrossRef]
10. Trębacz, L.; Kuziak, R.; Pietrzyk, M. Simulation and interpretation of SICO test. *Mechanika* **2005**, *207*, 91–96.
11. Young, G.A.; Capobianco, T.E.; Penik, M.A.; Morris, B.W.; Mcgee, J.J. The Mechanism of Ductility Dip Cracking in Nickel-Chromium Alloys: Subsolidus cracking results from global stresses produced during fusion welding and local stresses generated when coherent or partially coherent second phases form. *Weld. J.* **2008**, *87*, 31–43.
12. Donachie, M.J.; Donachie, S.J. *Superalloys. A Technical Guide*; ASM International: Russell Township, OH, USA, 2002; ISBN 0–87170–749–7.
13. Yushchenko, K.; Savchenko, V. Classification and Mechanisms of Cracking in Welding High-Alloy Steels and Nickel Alloys in Brittle Temperature Ranges. In *Hot Cracking Phenomena in Welds II*; Böllinghaus, T., Herold, H., Cross, C.E., Lippold, J.C., Eds.; Springer: Berlin, Germany, 2008; pp. 95–114. [CrossRef]
14. TUZ, L. Badania skłonności do pęknięć gorących wybranych stopów niklu. *HUTNIK-WIADOMOŚCI HUTNICZE* **2016**, *1*, 35–42. [CrossRef]
15. Eilers, A.; Nellesen, J.; Zielke, R.; Tillmann, W. Analysis of the ductility dip cracking in the nickel-base alloy 617mod. *IOP Conf. Ser. Mater. Sci. Eng.* **2017**, *181*, 12020. [CrossRef]
16. Adamiec, J. Weldability of magnesium casting alloys. Habilitation Thesis, Silesian University of Technology, Gliwice, Poland, 7 July 2010.
17. Caron, J.; Sowards, J. Weldability of Nickel-Base Alloys. In *Comprehensive Materials Processing*; Hashmi, S., Tyne, C.J.V.T., Batalha, G.F., Yilbas, B., Eds.; Elsevier Ltd.: Amsterdam, The Netherladns, 2014; pp. 151–179. [CrossRef]
18. Nissley, N.E. Intermediate temperature grain boundary embrittlement in Ni-base weld metals. PhD Thesis, Ohio State University, Columbus, OH, USA, 2006.
19. Lancaster, J.F. *Metallurgy of Welding*, 6th ed.; Woodhead Publishing: Russell Township, OH, USA, 1999; ISBN 9781855734289.
20. Prokhorov, N.N. The problem of the strength of metals while solidifying during welding. *Svar. Proizv.* **1956**, *6*, 5–11.

21. DuPont, J.N. Application of Solidification Models for Controlling the Microstructure and Hot Cracking Response of Engineering Alloys. In *Hot Cracking Phenomena in Welds III*; Böllinghaus, T., Herold, H., Cross, C.E., Lippold, J.C., Eds.; Springer: Berlin, Germany, 2011; pp. 265–293.

22. Alexandrov, B.; Lippold, J.; Nissley, N. Evaluation of Weld Solidification Cracking in Ni-Base Superalloys Using the Cast Pin Tear Test. In *Hot Cracking Phenomena in Welds II*; Böllinghaus, T., Herold, H., Cross, C.E., Lippold, J.C., Eds.; Springer: Berlin, Germany, 2008; pp. 193–213.

23. Collins, M.G.; Lippold, J.C. An investigation of ductility dip cracking in nickel-based filler materials-Part I. *Weld. J.* **2003**, *82*, 288–295.

24. Nishimoto, K.; Saida, K.; Okauchi, H. Microcracking in multipass weld metal of alloy 690 Part 1—Microcracking susceptibility in reheated weld metal. *Sci. Technol. Weld. Join.* **2006**, *11*, 455–461. [CrossRef]

25. Nishimoto, K.; Saida, K.; Okauchi, H.; Ohta, K. Microcracking in multipass weld metal of alloy 690 Part 2—Microcracking mechanism in reheated weld metal. *Sci. Technol. Weld. Join.* **2006**, *11*, 462–470. [CrossRef]

26. Kou, S. *Welding Metallurgy*, 2nd ed.; John Wiley & Sons, Inc.: Hoboken, NJ, USA, 2003; ISBN 0-471-43491-4.

27. Nishimoto, K.; Saida, K.; Sakamoto, M.; Kono, W. Effect of Filler Metal La Additions on Micro-Cracking in Multi-Pass Laser Overlay Weld Metal of Alloy 690. In *Hot Cracking Phenomena in Welds II*; Böllinghaus, T., Herold, H., Cross, C.E., Lippold, J.C., Eds.; Springer: Berlin, Germany, 2008; pp. 389–408. ISBN 978-3-540-78627-6.

28. Ramirez, A.J.; Lippold, J.C. New Insight into the Mechanism of Ductility-Dip Cracking in Ni-Base Weld Metals. In *Hot Cracking Phenomena. in Welds*; Springer: Berlin, Germany, 2005; pp. 59–70. ISBN 978-3-540-78627-6.

29. Sidhu, R.; Ojo, O.; Chaturvedi, M. Weld Cracking in Directionally Solidified Inconel 738 Superalloy. *Can. Met. Q.* **2007**, *46*, 415–424. [CrossRef]

30. David, S.A.; Vitek, J.M.; Babu, S.S.; Boatner, L.A.; Reed, R.W. Welding of nickel base superalloy single crystals. *Sci. Technol. Weld. Join.* **1997**, *2*, 79–88. [CrossRef]

31. Davis, J.R. *ASM Specialty Handbook: Nickel, Cobalt, and Their Alloys*; ASM International: Russell Township, OH, USA, 2001; Volume 38, ISBN 978-0-87170-685-0.

32. Durand-Charre, M. *The Microstructure of Superalloys*, 1st ed.; CRS Press: Amsterdam, The Netherlands, 1997; ISBN 10 9056990977.

33. Andersson, J.; Sjöberg, G.; Brederholm, A.; Hänninen, H. Solidification Cracking of Alloy Allvac 718Plus and Alloy 718 at Transvarestraint Testing. In Zinc and Lead Metallurgy. In Proceedings of the 47th Annual Conference of Metallurgists, Winnipeg, MB, Canada, 24–27 August 2008; pp. 157–169.

34. Tasak, E.; Ziwewiec, A. *Spawalność Materiałów Konstrukcyjnych*; JAK Press: Krakow, Poland, 2009; ISBN 978-83-923191-9-1.

35. *ISO/TR 17641-3:2005*; Destructive tests on welds in metallic materials—Hot cracking tests for weldments—Arc welding processes—Part 3. The International Organization for Standardization: Geneva, Switzerland, 2005.

36. Łyczkowska, K.; Adamiec, J. Repair of Precision Castings Made of the Inconel 713C Alloy. *Arch. Foundry Eng.* **2017**, *17*, 210–216. [CrossRef]

37. Adamiec, J.; Łyczkowska, K. Structure of precision castings made of the Inconel 713C alloy. *Arch. Foundry Eng.* **2018**, *18*, 19–24. [CrossRef]

38. Łyczkowska, K.; Adamiec, J.; Jachym, R.; Kwieciński, K. Properties of the Inconel 713 Alloy Within the High Temperature Brittleness Range. *Arch. Foundry Eng.* **2017**, *17*, 103–108. [CrossRef]

materials

Article

Regeneration of Aluminum Matrix Composite Reinforced by SiC$_p$ and GC$_{sf}$ Using Gas Tungsten Arc Welding Technology

Katarzyna Łyczkowska *, Janusz Adamiec, Anna Janina Dolata, Maciej Dyzia and Jakub Wieczorek

Faculty of Materials Engineering, Silesian University of Technology, ul. Krasińskiego 8, 40-019 Katowice, Poland; janusz.adamiec@polsl.pl (J.A.); anna.dolata@polsl.pl (A.J.D.); maciej.dyzia@polsl.pl (M.D.); jakub.wieczorek@polsl.pl (J.W.)
* Correspondence: katarzyna.lyczkowska@polsl.pl

Abstract: The main motivation behind the presented research was the regeneration of the damaged surface of composite materials. The testing of melting and pad welding of the composite surface by Gas Tungsten Arc Welding (GTAW) with alternating current (AC) were carried out. The material of investigation was an AlSi12/SiCp + GCsf hybrid composite made by a centrifugal casting process. The composite was reinforced with 5 wt.% of silicon carbide particles and 5 wt.% of glassy carbon spheres. The composites were investigated in tribological tests. It was found that there was a possibility for modification or regeneration of the surface with pad welding technology. Recommended for the repairs was the pad welding method with filler metal with a chemical composition similar to the aluminum matrix composite (ISO 18273 S Al4047A (AlSi12 [A])). The surface of the pad welding was characterized by the correct structure with visible SiCp. No gases or pores were observed in the pad welding; this was due to a better homogeneity of the silicon carbide (SiCp) distribution in the composite and better filling spaces between liquid metal particles in comparison to the base material. Based on the tribological tests, it was found that the lowest wear was observed for the composite surface after pad welding. This was related to the small number of reinforcing particles and their agreeable bonding with the matrix. The plastic deformation of the Al matrix and scratching by worn particles were a dominant wear mechanism of the surface.

Keywords: aluminum matrix composites; silicon carbide; glassy carbon; GTAW welding; tribological test

Citation: Łyczkowska, K.; Adamiec, J.; Dolata, A.J.; Dyzia, M.; Wieczorek, J. Regeneration of Aluminum Matrix Composite Reinforced by SiC$_p$ and GC$_{sf}$ Using Gas Tungsten Arc Welding Technology. *Materials* **2021**, *14*, 6410. https://doi.org/10.3390/ma14216410

Academic Editor: Daolun Chen

Received: 24 September 2021
Accepted: 23 October 2021
Published: 26 October 2021

Publisher's Note: MDPI stays neutral with regard to jurisdictional claims in published maps and institutional affiliations.

1. Introduction

Aluminum matrix composites (AMCs) are common materials that are applied in many fields of industry, such as automotive, aerospace, electricity, chemical, etc., due to their unique properties [1–3]. The most used reinforcement phases are oxides (Al_2O_3, ZrO_2, SiO_2), carbides (SiC, TiC, B_4C, ZrC), and nitrides (Si_3N_4) [4]. The presence of these phases in strict proportion (normally between 5 and 30 wt.%) leads to an increase in material properties [1,4–7].

Aluminum matrix composites reinforced by silicon carbide (SiC) are a material solution successfully applied in the automotive field [8–10]. The popularity of these composites is due to a number of their properties, such as wear resistance, stiffness, compressive strength, low density, low coefficient of thermal expansion, good casting properties, and low production costs [11,12]. This kind of composite is widely used in tribological conditions where material is constantly subjected to variable loads and high-temperature conditions. For this reason, AMCs reinforced with SiC particles need to show high stability of tribological properties regardless of changing working conditions. These special types of materials are composites used in conditions where the surface is constantly subjected to wear, such as cylinder liners, brake disk, or pistons of engines or compressors [13–17].

Generally, in the case of wear, the whole element has to be replaced even if only a small part of the material surface is damaged by external conditions (e.g., pitting or scuffing mechanisms). The replacement of the whole element causes negative economic and environmental impacts because there remains a lack of effective methods of recycling aluminum matrix composites reinforced by silicon carbide [18–21].

There are various reports on the methods of welding aluminum composite castings. Based on the literature review, it can be stated that the most popular and promising method of Al/SiCp joining is Friction Stir Welding (FSW). Kurtyka et al. [22–24] described that one of the results of FSW process implementation can be a significant improvement in the distribution of the reinforcing phase particles. This process influences the mechanical properties of the composite and, compared to the starting material, it allows for an increase of approximately 40% of compressive strength and 30% of hardness. All the results confirm the effectiveness of the FSW method for joining aluminum matrix composites reinforced by SiC particles or other types of ceramic phases. This method is also intended for the joining of AMCs with different types of ceramic reinforcement [25].

Due to its technological solution, the FSW method is recommended for welding elements with specific dimensions, which significantly limits the regeneration possibilities of damaged areas.

The next part of the literature analysis focuses on bonding and remelting through other advanced welding technologies—Electron Beam Welding (EBW) and Laser Beam Welding (LBW). Wang at al. [26] described welding of $101Al/SiC_p$ composites by EBW technology. According to these studies, a small quantity of brittle Al_4C_3 compound and a single Si phase were generated in the welded joint. However, the authors proved that the interfacial reaction between SiC particles and the Al matrix could be greatly suppressed by high welding speed and low heat input. Based on the research, it was confirmed that LBW welding and hardfacing enables the production of welding joints with the required properties.

Studies conducted by Dahotre et al. [27] showed that the alloy matrix composites reinforced with 10 and 20 vol.% of SiC particulates were more readily welded by LBW. The opposite effect was observed in the composite materials where the fusion zone contained the fully melted matrix and the fully reacted SiC reinforcement, and where the heat-affected zone contained the partially melted matrix and the nearly unreacted SiC particles. Moreover, the authors showed that increasing the SiC content from 0 to 20% caused a decrease in the reflection of the laser beam and an increase in melt viscosity. This was potentially caused by an increasing amount of Al_4C_3 compound [28].

In turn, Wang et al. [29] described the results of an investigation with the use of micro-nano (Al–Si–Cu)–Ti foils as filler metal. The high-performance joints of aluminum matrix composites with high SiC particle content (Al-MMC$_s$/60% SiC$_p$) were observed. Moreover, the beneficial effect of adding Ti into the filler metal on improving wettability between SiC particles and the metallic brazed seam was confirmed.

Equally satisfactory results of Al/SiC$_p$ joining were obtained using the plasma spray process [30], welding by oxy-acetylene [31], and soldering and gluing [32].

Based on the literature review, it was found that more and more technologies are helping to obtain a permanent joint of Al/SiC$_p$ composites. However, it should be mentioned that the main problem during welding is the appearance of the Al_4C_3 phase, which may lead to a reduction in the strength properties of materials [22,27,29]. It is important to use strictly defined parameters of welding and pad welding that will limit the formation of the unfavorable Al_4C_3 phase.

Unfortunately, many technologies are too expensive and complicated to use, hence the need to develop a technology for the surfacing and regeneration of the damaged composite surface that will be relatively easy and available. One of the promising and still-developing technologies used for the regeneration of different kinds of materials is the Gas Tungsten Arc Welding (GTAW) method that helps to join metal–ceramic composites by GTAW DC welding, although SiC reinforcing particles have a much lower thermal conductivity than a metal matrix. It has also been found that the conditions favor the precipitation processes in the solidified mixture of the fused composite matrix and the additive material, which allows them to be effectively joined. These obtained results [33,34] are the main motivation for future work on the development of the GTAW method, which will allow for the regeneration of the composite at low cost.

The available research shows that there are some articles on Al/SiC_p composites, however there is no information on the surface regeneration of the Al/SiC_p composites with the addition of glassy carbon. It is necessary to conduct research that will contribute to the development an effective technology for the regeneration of the damaged piston surface of aluminum matrix composites reinforced by SiC and glassy carbon.

2. Materials and Methods

The material of investigation was a composite based on EN AC-48000 alloy (AlSi12CuNiMg) made by a centrifugal casting process. The material was reinforced with 5 wt.% of silicon carbide particles (SiC_p) of average size in the range of 30–70 μm, and 5 wt.% of spherical glassy carbon (GC_{sf}) of average size in the range of 5–15 μm.

In the first stage, composite suspensions were prepared by stir casting in an autoclave furnace with a moving graphite stirrer system, according to the procedure described in [7,14]. In the second stage, heated composite suspension was cast into the rotating mold (d = 60 mm; ω = 3000 rpm), according to the method described in previous own works [12,13]. An example of a cast of composite sleeves and its microstructure in the outer area is shown in Figure 1. Due to the gradient structure in this type of casting [12], remelting and pad welding tests were carried out on the outer surface of the composite sleeve. The welding test was carried out by the GTAW method using filler material AlSi12 with a diameter of 2 mm, and argon as the inert shielding gas, at a flow rate of 10 l/min. The parameters of welding technology are shown in Table 1. The remelting process was carried out with 120 A alternating current, and the pad welding with 140 A alternating current. The results of remelting and pad welding are shown in Figure 2. The material was not preheated prior to the pad welding process.

Figure 1. AlSi12/SiC_p + GC_{sf} composite sleeves: (**a**) view of a representative cast of the composite sleeve; (**b**) microstructure of the SiC_p particle-rich region.

Table 1. The parameters of remelting and pad welding of AlSi12/SiC_p + GC_{sf} composite.

Composite Material	Process	Welding Current (A)	Voltage (V)	Welding Speed (cm/min)	Gas Flow Rate (L/min)
AlSi12/SiC_p + GC_{sf}	Remelting	120	14	20	10
	Pad welding	140	16	20	10

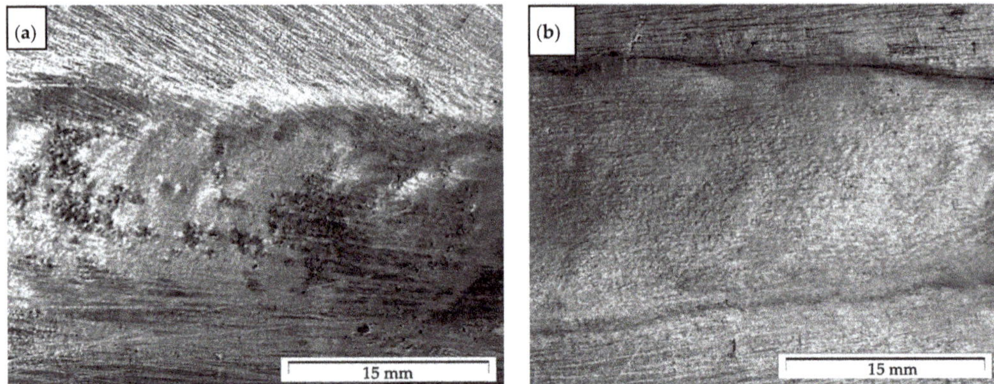

Figure 2. Face of the AlSi12/SiC$_p$ + GC$_{sf}$ composite weld: (**a**) remelting; (**b**) pad welding.

The tribological investigations were performed by reciprocating movement in ambient condition and without lubrication. A speed of 4 m/min, a load of 1.5 kg, and a distance of 500 m were applied as the main tribological parameters. The GJL300 iron was used as a typical material for a tribological partner. The counterpart was in the form of a pin with 6 mm diameter. The surfaces before testing were grinded by using sandpaper with 1000 gradation. The scheme of the device is shown in Figure 3.

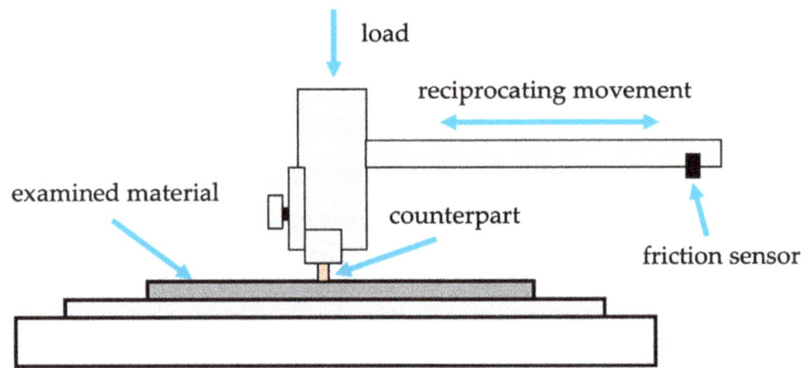

Figure 3. Scheme of a device for tribological testing in the reciprocating friction system.

The wear surfaces were analysed on a MicroProf 3000, FRT optical profilometer, FRT GmbH, Bergisch Gladbach (Germany). Based on these results, roughness values (Ra), root mean square (Rq), and average maximum height of the surface (Rz) of wear were achieved.

The metallographic examinations were conducted using an Olympus GX71, Warsaw (Poland) light microscope (LM) at magnifications of up to 500×. The structure of the surface after the welding and tribological tests were examined under the scanning electron microscopes (SEM) JEOL JCM-6000 Neoscope II, Tokyo (Japan), and Hitachi S-4200, Krefeld (Germany). Images were recorded in the secondary electron mode at a magnification of 1000× and at a voltage accelerating the electron beam to 15 keV.

3. Results and Discussion

Aluminum matrix composites made by centrifugal casting are characterized by a complex and heterogeneous structure. The differences depend on the casting structure of the aluminum alloy and the inhomogeneous distribution of reinforcing particles as a result of the centrifugal force during the casting process (Figure 1b).

The study materials are characterized by AlSi12/SiC$_p$ + GC$_{sf}$, made of silicon eutectic (α + Si), solid solution α, primary silicon crystals, and intermetallic crystal phases. Spherical glassy carbon particles were distributed in the matrix, mainly in the middle zone, and SiC particles were distributed in the surface zone (Figure 4). This particle distribution is characteristic of centrifugal casting, and it is a result of the density of the individual phases.

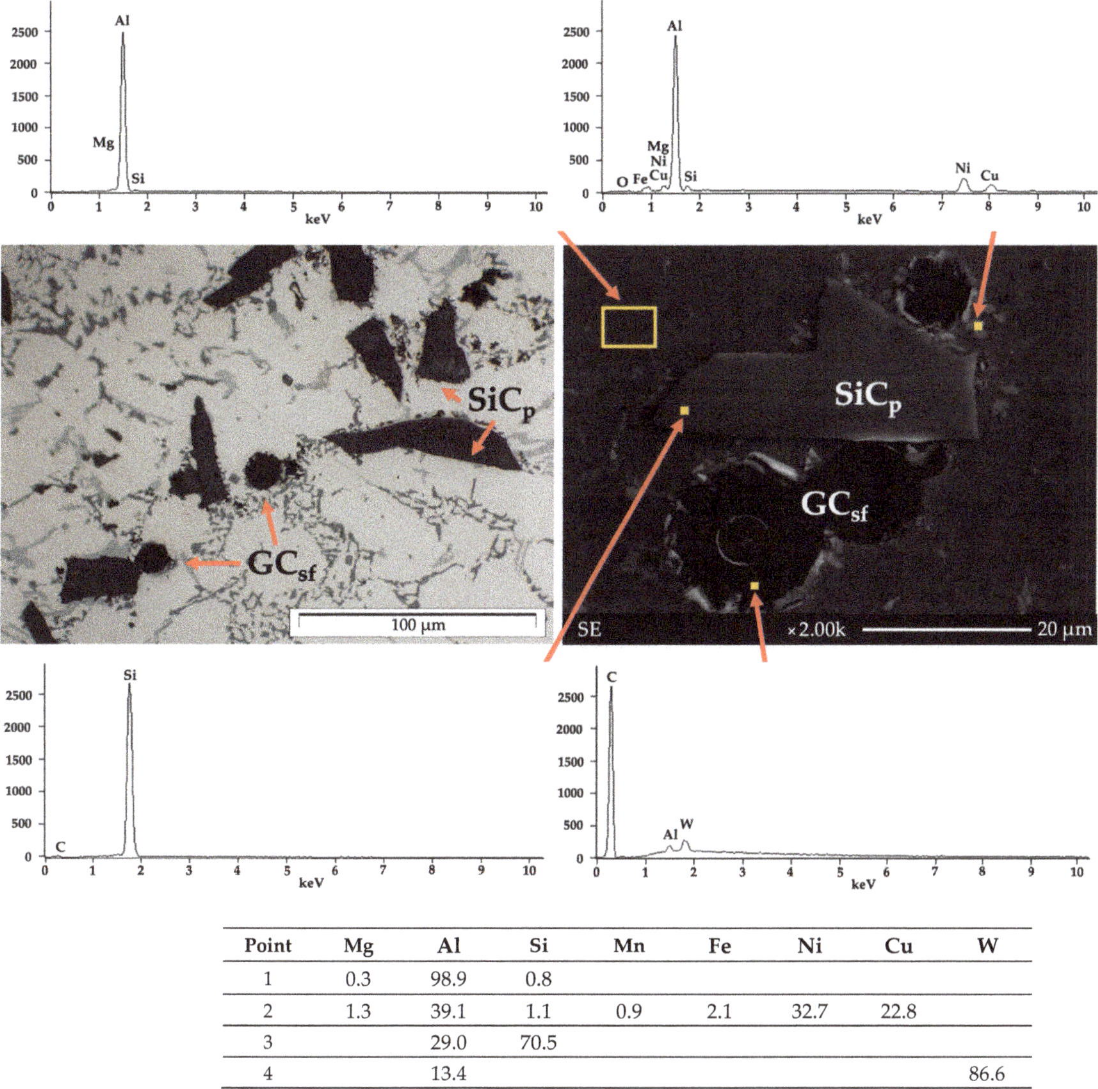

Point	Mg	Al	Si	Mn	Fe	Ni	Cu	W
1	0.3	98.9	0.8					
2	1.3	39.1	1.1	0.9	2.1	32.7	22.8	
3		29.0	70.5					
4		13.4						86.6

Figure 4. The microstructure and results of EDS analysis of AlSi12/SiC$_p$ + GC$_{sf}$ composite.

A significant problem was the existence of many pores and gas bubbles, especially in the concentration of the reinforcement phase, which is related to high surface tension and the inability to fully penetrate space mainly in the liquid phase during crystallization

(Figure 5). The second phenomenon determining the formation of voids is excessively found in fluid of low pressure and caused by the lack of supply of liquid metal to fill the space between the reinforcing particles. This phenomenon was described, among others, in works [35,36]. Based on the results, remelting and pad welding are one of the solutions to reduce the porosity of the subsurface zone, which works in the wear systems described.

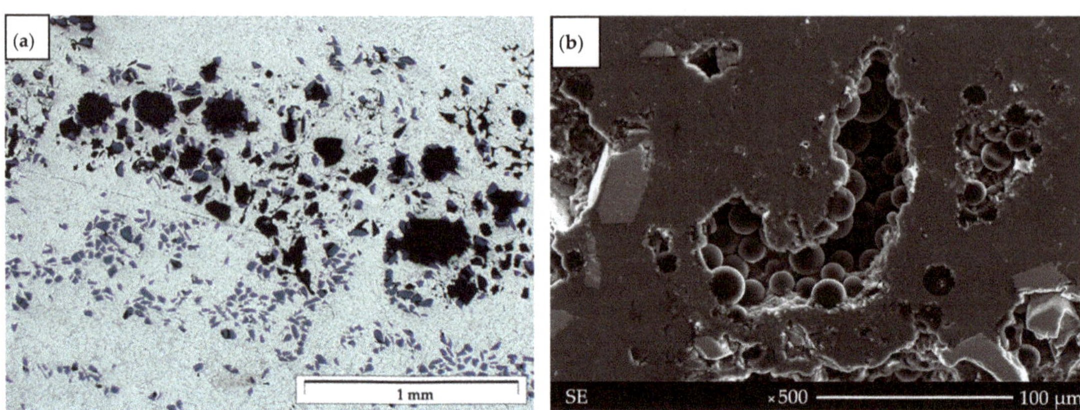

Figure 5. The microstructure of AlSi12/SiC$_p$ + GC$_{sf}$ composites: (**a**) gas porosity in SiC$_p$ area, (**b**) pores and voids with GC$_{sf}$ in areas of inhomogeneous distribution of particles.

The results of the visual examination of remelting revealed a uniform weld face with some small discontinuities. This is related to the presence of pores and the inhomogeneous distribution of the reinforcement phases, mainly for SiC$_p$ (Figure 2a). A similar weld face was observed in the surface of the pad welding (Figure 2b). The weld face was smooth and uniform, which indicated the correct selection of the pad welding parameters. No pores and bladders were observed on the surface. On this basis, it was concluded that the welding filler wire for pad welding (AlSi12) increased the area of the liquid metal pool and filled the space between the reinforcing particles and the aluminum matrix better. It is also important to reduce the number of reinforcing particle units of volume of the liquid metal pool.

The examination of the chemical composition of the AlSi12/SiC$_p$ + GC$_{sf}$ composites (pt. 1) by the EDS method in individual areas revealed SiC (pt. 3) and spherical glassy carbon (pt. 4). Fe-containing phases and Ni- and Cu-containing phases (pt. 2) were also disclosed.

The microstructure of the cross section with an orientation perpendicular to the remelting and pad welding direction indicated that pad welding had a positive impact on the quality of the subsurface layer. The reason for this is the homogenization of the SiC$_p$ reinforcement phases and the reduction in the number of pores in the composite (Figure 6a). In this area, the GC$_{sf}$ was not observed, providing information about the segregation of GC$_{sf}$ particles in the inner zone of the sleeve. Single reinforcing particles were revealed in the pad welding, however no pores and voids were found (Figure 6b). This indicates a properly selected pad welding technology.

In order to determine the operating conditions of the repaired composite elements, (e.g., low-loaded pistons engines [34]), the resistance tests of tribological wear in a reciprocating system were carried out. This is a typical friction system found in compressors and reciprocating internal combustion engines. The analysis of the coefficient of friction for materials without any modification indicated that the coefficient was constant over the entire range of the experiment (over a distance of 500 m) and was on average equal to 0.33 (Figure 7). The analysis of the remelting surface showed that in the initial period (approximately 150 m), the friction coefficient increased significantly to a value of up to 0.4.

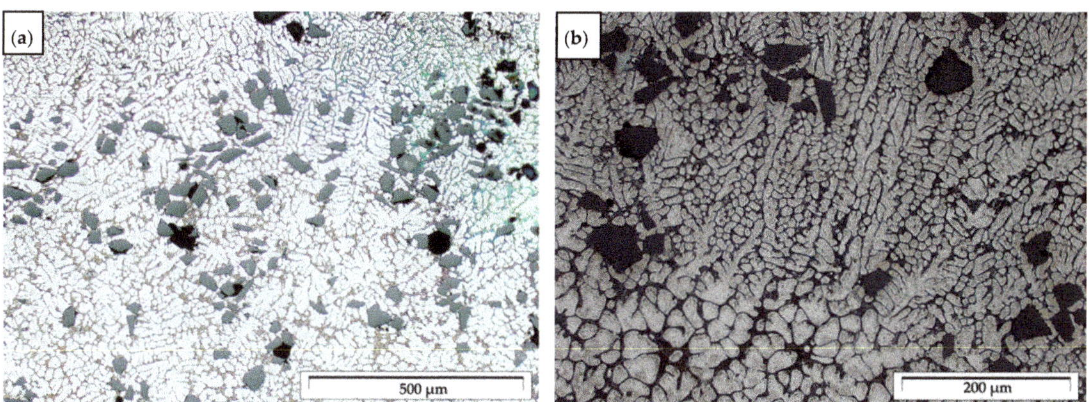

Figure 6. The microstructure of tested material after modification using welding techniques: (**a**) remelting structure with a visible reinforcement phase (SiC$_p$) and small number of pores; (**b**) pad welding with visible fusion zone and SiC$_p$.

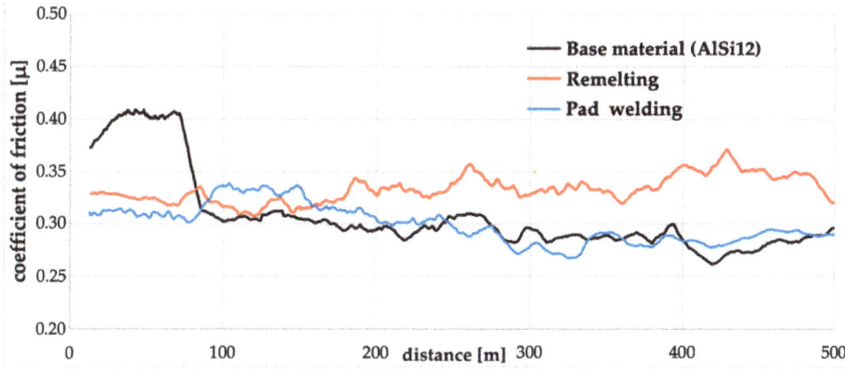

Figure 7. The variation of friction coefficient tested for base material (AlSi12/SiC$_p$ + GC$_{sf}$) and composites repaired by remelting and pad welding processes.

For the remelted material, slightly higher values of surface roughness parameters were obtained, Ra = 8.8 μm, Rq = 10.7 μm, and Rz = 37.0 μm, and the wear depth was approximately 130 μm (Figure 8b). This is a surface wear mechanism similar to that previously described, however, the cross-sectional profile indicates the scratching and ridging mechanisms. The profile of the worn surface depends on the behavior of individual SiC particles (Figure 8b). In the remelting process, the friction effect is reduced. It should be stated that the remelting of the composite surface does not yield a positive result for research.

This was related to the lapping process of the surface as a result of pulling out single SiC$_p$ particles (Figure 9b). After this stage, the coefficient of friction value stabilized at 0.3, which is lower than for the base material. Similar values were obtained for the surface of the pad welding, which indicates that after the abrasion step, the main friction surface becomes the matrix. The coefficient of friction value for pad welding was 0.3. Single SiC$_p$ particles were observed on the wear surface of the pad welding (Figures 9c and 10).

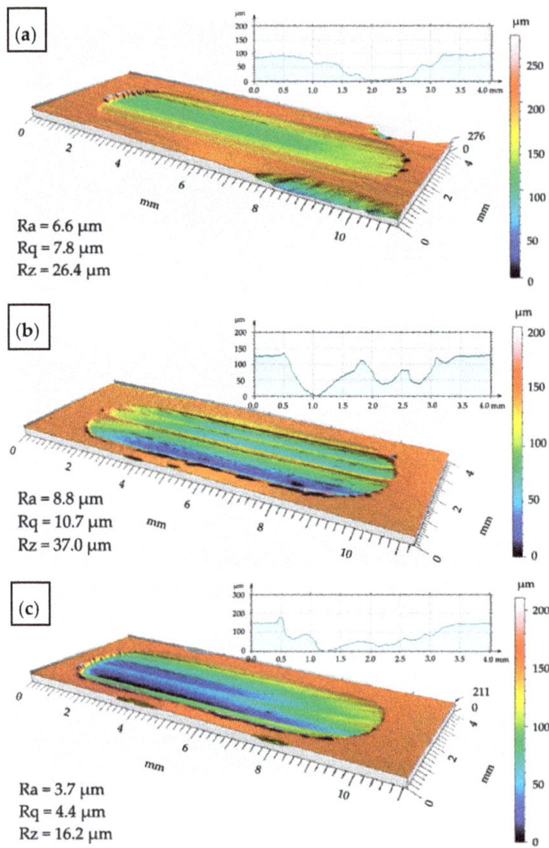

Figure 8. View of the wear surface after tribological tests: (**a**) surface of the base material (AlSi12/ SiC$_p$ + GC$_{sf}$); (**b**) remelting area; (**c**) pad welding area.

The profilometric observations of the wear process were supplemented by metallographic analysis (Figures 9 and 10). In the case of the base material, the surface development indices Ra, Rq, and Rz were the highest, with Ra = 6.6 μm, Rq = 7.8 μm, and Rz = 26.4 μm, respectively. The profile valley depth of the wear was approximately 150 μm (Figure 8a). These results were confirmed by the wear mechanism observed in metallographic tests (Figures 9 and 10). The wear of the surface was determined by reinforcing SiC particles, which furrowed after being pulled out of the matrix. In the next stage, the SiC particles were pushed into the matrix and the furrow was obliterated as a result of the plastic deformation of the Al matrix (Figures 9a, 10a and 11a).

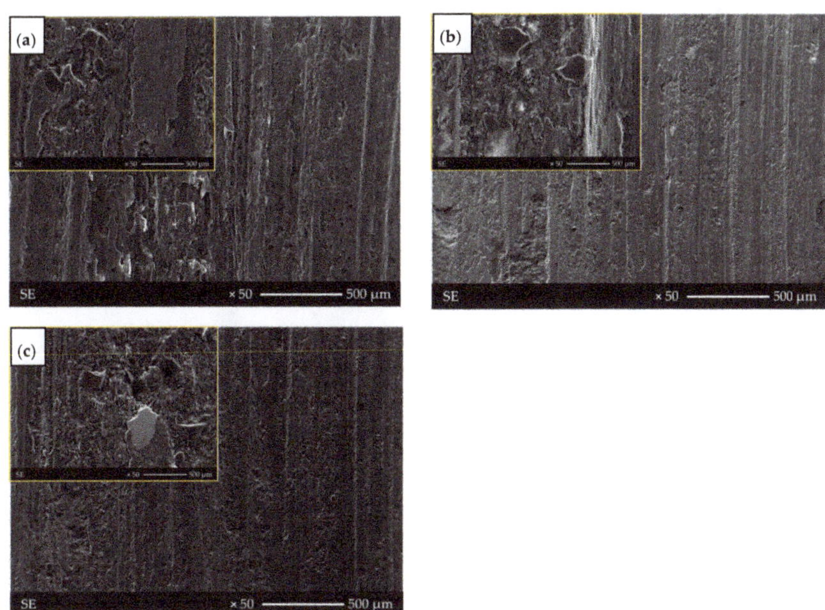

Figure 9. Surfaces after tribological test: (**a**) base material (AlSi12/SiCp + GCsf); (**b**) remelting area with visible SiCp; (**c**) pad welding area with single reinforcing particles.

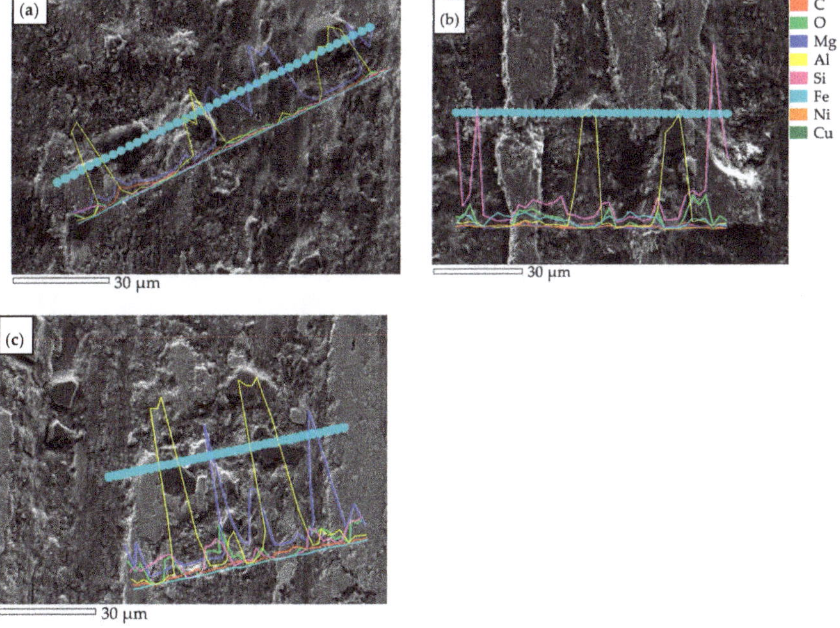

Figure 10. SEM–EDS line analysis after tribological test: (**a**) base material (AlSi12/SiC$_p$ + GC$_{sf}$); (**b**) remelting area; (**c**) pad welding area.

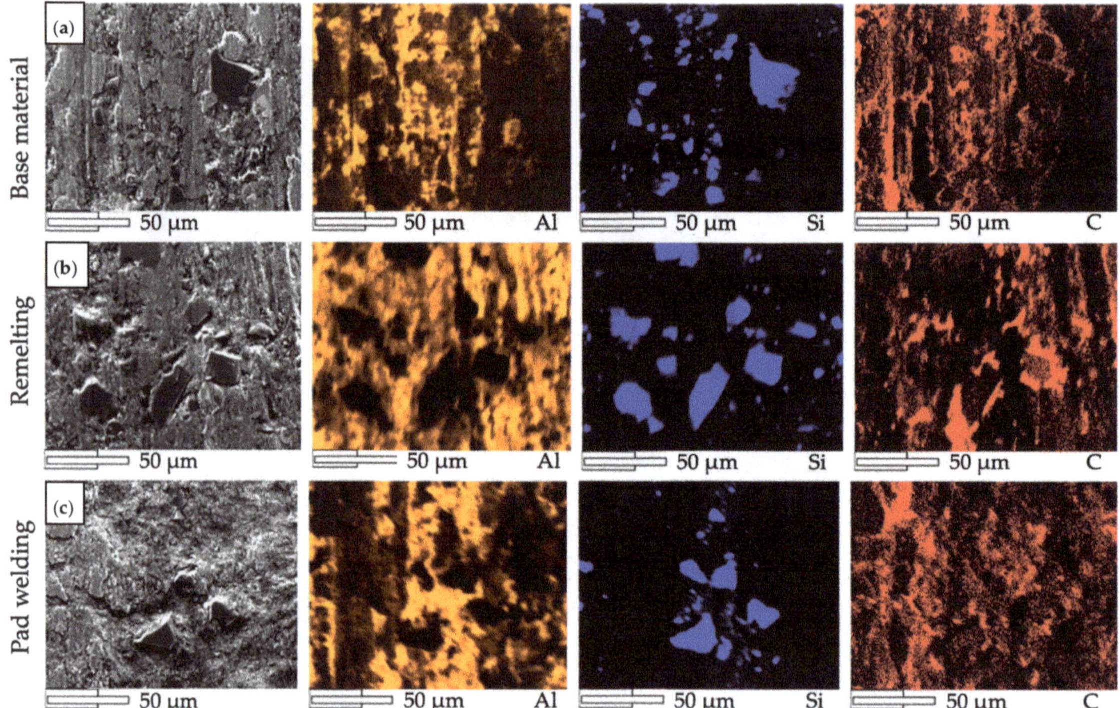

Figure 11. X-ray spectroscopy (EDS) mapping analysis of the wear track: (**a**) base material; (**b**) remelting area; (**c**) pad welding area.

The best results were obtained for the composite surface after pad welding (Figures 10c and 11c). This was also confirmed by the surface parameters index, which were Ra = 3.7 μm, Rq = 4.4 μm, and Rz = 16.2 μm, and the valley depth of the wear was approximately 100 μm (Figure 8c). It was found that, due to the small number of reinforcing particles (Figures 6 and 11c) and their agreeable bonding with the matrix (Figure 9c), no ridging or particle pull-out phenomena occur. There was even wear as a result of the plastic deformation of the matrix (Figure 11c) and surface scratching (Figure 9c).

On this basis, it can be assumed that a good technology for layer modification and for the repair or regeneration of elements working in reciprocating friction wear conditions is pad welding by GTAW method, with filler metal with a chemical composition suitable to the matrix.

4. Conclusions

Based on the research and the analysis of the obtained results, the following conclusions were drawn:

1. It is possible to modify or repair the surface of $AlSi12/SiC_p + GC_{sf}$ aluminum matrix composites reinforced by SiC_p/GC_{sf} made by centrifugal casting. The application of GTAW method with filler metal characterized by a chemical composition similar to the aluminum metal matrix composite is a confirmed method to achieve pad welding with the required properties. The process should be carried out in argon gas, at a flow rate of 10 l/min, and with an alternating current from 120 to 140 A.
2. The surface of the composite sleeve after remelting is characterized by the correct structure, in which the SiC_p/GC_{sf} reinforcing particles are observed. A much lower porosity of the remelted zone was found. This is due to a better homogeneity of

the SiC_p distribution in the composite and better filling spaces between liquid metal particles in comparison to the base material.

3. Single SiC_p particles were observed in the area of pad welding made with AlSi12 filler metal, this results from the major volume of the matrix in the liquid metal pool. No pores or gases were observed in the pad welding, which confirms the correct repair process.

4. The surface of the composite after the pad welding process is characterized by similar tribological properties as the base material, while the pad welding under the same conditions shows a lower degree of wear. This is due to a smaller number of reinforcing particles that cause the surface to be furrowed and the plastic to deformation.

Author Contributions: Conceptualization, K.Ł. and J.A.; methodology, K.Ł., J.A. and J.W.; formal analysis, K.Ł., J.A. and J.W.; writing—original draft preparation, K.Ł.; writing—review and editing, J.A. and A.J.D.; visualization, K.Ł.; supervision, J.A., A.J.D. and M.D.; funding acquisition; K.Ł. and J.A. All authors have read and agreed to the published version of the manuscript.

Funding: This research was funded by Silesian University of Technology subsidy for statutory business activities, No. 11/990/BK/21/0080 and subsidy for the maintenance and development of research potential, project No. 11/030/BKM20/1012.

Institutional Review Board Statement: Not applicable.

Informed Consent Statement: Not applicable.

Data Availability Statement: The data supporting reported results are not stored in any publicly archived datasets. The readers can contact the corresponding author for any further clarification of the obtained results.

Conflicts of Interest: The authors declare no conflict of interest.

References

1. Bodunrin, M.O.; Alaneme, K.K.; Chown, L.H. Aluminium matrix hybrid composites: A review of reinforcement philosophies; mechanical, corrosion and tribological characteristics. *J. Mater. Res. Technol.* **2015**, *4*, 434–445. [CrossRef]
2. *Comosite Materials Handbook*; MIL-HDBK-17; Department of Defense of USA: Philadelphia, PA, USA, 2002.
3. Nishchev, K.N.; Novopoltsev, M.I.; Mishkin, V.P.; Shchetanov, B.V. Studying the microstructure of AlSiC metal matrix composite material by scanning electron microscopy. *Bull. Russ. Acad. Sci. Phys.* **2013**, *77*, 981–985. [CrossRef]
4. Hekner, B.; Myalski, J.; Valle, N.; Botor-Probierz, A.; Sopicka-Lizer, M.; Wieczorek, J. Friction and wear behavior of Al-SiC(n) hybrid composites with carbon addition. *Compos. Part B Eng.* **2017**, *108*, 291–300. [CrossRef]
5. Mussatto, A.; Inam, U.A.; Reza, T.M.; Yan, D.; Dermot, B. Advanced Production Routes for Metal Matrix Composites. *Eng. Rep.* **2021**, *3*, 1–25. [CrossRef]
6. Wieczorek, J. Tribological Properties and a Wear Model of Aluminium Matrix Composites—SiC Particles Designed for Metal Forming. *Arch. Met. Mater.* **2015**, *60*, 111–115. [CrossRef]
7. Dolata, A.J. Hybrid Comopsites Shaped by Casting Methods, in Light Metal and their Alloys III—Technology, Microstructure and Properties. *Solid State Phenom.* **2013**, *211*, 47–52. [CrossRef]
8. Gao, Z.; Ba, X.; Yang, H.; Yin, C.; Liu, S.; Niu, J.; Brnic, J. Joining of Silicon Particle-Reinforced Aluminum Matrix Composites to Kovar Alloys Using Active Melt-Spun Ribbons in Vacuum Conditions. *Materials* **2020**, *13*, 2965. [CrossRef]
9. Panchenko, O.V.; Zhabrev, L.A.; Kurushkin, D.; Popovich, A.A. Macrostructure and Mechanical Properties of Al-Si, Al-Mg-Si, and Al-Mg-Mn Aluminum Alloys Produced by Electric Arc Additive Growth. *Met. Sci. Heat Treat.* **2019**, *60*, 749–754. [CrossRef]
10. Sobczak, J.; Slawinski, Z.; Darlak, P.; Asthana, R.; Rohatgi, P.; Sobczak, N. Thermal Fatigue Resistance of Discontinuously Reinforced Cast Aluminum-Matrix Composites. *J. Mater. Eng. Perform.* **2002**, *11*, 595–602. [CrossRef]
11. Hekner, B.; Myalski, J.; Pawlik, T.; Sopicka-Lizer, M. Effect of Carbon in Fabrication Al-SiC Nanocomposites for Tribological Application. *Materials* **2017**, *10*, 679. [CrossRef]
12. Dolata, A.J.; Dyzia, M.; Wieczorek, J. Tribological Properties of Single (AlSi7/SiCp, AlSi7/GCsf) and Hybrid (AlSi7/SiCp + GCsf) Composite Layers Formed in Sleeves via Centrifugal Casting. *Materials* **2019**, *12*, 2803. [CrossRef]
13. Dolata, A.J.; Golak, S.; Ciepliński, P. The Eulerian multiphase model of centrifugal casting process of particle reinforced Al matrix composites. *Compos. Theory Pract.* **2017**, *17*, 200–205.
14. Dyzia, M. Aluminum Matrix Composite (AlSi7Mg2Sr0.03/SiCp) Pistons Obtained by Mechanical Mixing Method. *Materials* **2017**, *11*, 42. [CrossRef]
15. Kumar, D.; Ottarackal, D.J.; Acharya, U.; Medhi, T.; Roy, B.S.; Saha, S.C. A parametric study of friction stir welded AA6061/SiC AMC and its effect on microstructure and mechanical properties. *Mater. Today Proc.* **2021**, *46*, 9378–9386. [CrossRef]

16. Kurganova, Y.A.; Chernyshova, T.A.; Kobeleva, L.I.; Kurganov, S.V. Service properties of aluminum-matrix precipitation-hardened composite materials and the prospects of their use on the modern structural material market. *Russ. Met. (Met.)* **2011**, *2011*, 663–666. [CrossRef]
17. Golak, S.; Dolata, A.J. Controlling the distribution of reinforcement in metal composite using a low frequency homogenised alternating electromagnetic field. *J. Compos. Mater.* **2016**, *50*, 1751–1760. [CrossRef]
18. Ciappa, M. Selected failure mechanisms of modern power modules. *Microelectron. Reliab.* **2002**, *42*, 653–667. [CrossRef]
19. Khan, S.; Manjunatha, K. Comparison of Pitting Corrosive Behaviour of as Casted and Heat Treated Al6061-SiC Metal Matrix Composite in Various Medium by Weight Loss Method. *Mater. Today Proc.* **2018**, *5*, 22517–22525. [CrossRef]
20. Xia, J.; Lewandowski, J.J.; Willard, M.A. Tension and fatigue behavior of Al-2124A/SiC-particulate metal matrix composites. *Mater. Sci. Eng. A* **2020**, *770*, 138518. [CrossRef]
21. Li, W.; Liang, H.; Chen, J.; Zhu, S.; Chen, Y. Effect of SiC Particles on Fatigue Crack Growth Behavior of SiC Particulate-reinforced Al-Si Alloy Composites Produced by Spray Forming. *Procedia Mater. Sci.* **2014**, *3*, 1694–1699. [CrossRef]
22. Kurtyka, P.; Sulima, I.; Wójcicka, A.; Ryłko, N.; Pietras, A. The influence of friction stir welding process on structure and mechanical properties of the AlSiCu / SiC composites. *J. Achiev. Mater. Manuf. Eng.* **2012**, *55*, 339–344.
23. Kurtyka, P.; Rylko, N.; Tokarski, T.; Wójcicka, A.; Pietras, A. Cast aluminium matrix composites modified with using FSP process—Changing of the structure and mechanical properties. *Compos. Struct.* **2015**, *133*, 959–967. [CrossRef]
24. Stawiarz, M.; Kurtyka, M.; Rylko, N.; Gluzman, S. Influence of FSP process modification on selected properties of Al-Si-Cu/SiCp composite surface layer. *Compos. Theory Pract.* **2019**, *19*, 161–168.
25. Salih, O.S.; Ou, H.; Sun, W.; McCartney, G. A review of friction stir welding of aluminium matrix composites. *Mater. Des.* **2015**, *86*, 61–71. [CrossRef]
26. Wang, S.G.; Ji, X.H.; Zhao, X.Q.; Dong, N.N. Interfacial characteristics of electron beam welding joints of SiCp/Al composites. *Mater. Sci. Technol.* **2011**, *27*, 60–64. [CrossRef]
27. Dahotre, N.B.; McCay, M.H.; McCay, T.D.; Gopinathan, S.; Allard, L.F. Pulse laser processing of a SiC/Al-alloy metal matrix composite. *J. Mater. Res.* **1991**, *6*, 514–529. [CrossRef]
28. Banerjee, A.J.; Biswal, M.K.; Lohar, A.K.; Chattopadhyay, H.; Hanumaiah, N. Review on experimental study of Nd:YAG laser beam welding, with a focus on aluminium metal matrix composites. *Int. J. Eng. Technol.* **2016**, *5*, 92. [CrossRef]
29. Wang, P.; Gao, Z.; Niu, J. Micro–nano filler metal foil on vacuum brazing of SiCp/Al composites. *Appl. Phys. A Mater. Sci. Process.* **2016**, *122*, 592. [CrossRef]
30. Harshavardhan, K.; Nagendran, S.; Shanmugasundaram, A.; Sankar, S.P.; Kowshik, K.S. Investigating the effect of reinforcing SiC and graphite on aluminium alloy brake rotor using plasma spray process. *Mater. Today Proc.* **2021**, *38*, 2706–2712. [CrossRef]
31. Tjahjanti, P.H.; Hermansyah, F.; Prasetya, L.H.; Sulistyanto, M.P.T. Casting and Welding of Aluminium Matrix Composite Materials Reinforced by SiC Particles. *IOP Conf. Series: Mater. Sci. Eng.* **2018**, *434*, 012222. [CrossRef]
32. Nowacki, J.; Sajek, A. Trends of Joining Composite AlSi-SiC Foams. *Adv. Mater. Sci.* **2019**, *19*, 70–82. [CrossRef]
33. Ogonowski, K.; Wysocki, J.; Gawdzińska, K.; Przetakiewicz, W. Direct current TIG welding of metal matrix composites. *Weld. Technol. Rev.* **2018**, *90*, 51–54. [CrossRef]
34. Pichumani, S.; Srinivasan, R.; Ramamoorthi, V. Mechanical properties, thermal profiles, and microstructural characteristics of Al-8 %SiC composite welded using pulsed current TIG welding. *J. Mech. Sci. Technol.* **2018**, *32*, 1713–1723. [CrossRef]
35. Molenaar, J.M.M.; Katgerman, L.; Kool, W.H.; Smeulders, R.J. On the formation of the stircast structure. *J. Mater. Sci.* **1986**, *21*, 389–394. [CrossRef]
36. Zhang, L.; Eskin, D.G.; Miroux, A.; Katgerman, L. Formation of Microstructure in Al-Si Alloys under Ultrasonic Melt Treatment. *Light Met.* **2012**, *2012*, 999–1004. [CrossRef]

materials

MDPI

Article

Hybrid Welding (Laser–Electric Arc MAG) of High Yield Point Steel S960QL

Michał Urbańczyk [1,*] and Janusz Adamiec [2]

1 Łukasiewicz Resarch Network—Institute of Welding, Błogosławionego Czesława 16-18, 44-100 Gliwice, Poland
2 Division of Engineering Materials, Faculty of Materials Engineering, Silesian University of Technology, Krasińskiego 8, 40-019 Katowice, Poland; Janusz.adamiec@polsl.pl
* Correspondence: michal.urbanczyk@is.lukasiewicz.gov.pl

Abstract: The article discusses the effect of the hybrid-welding process (laser–electric arc MAG *Metal Active Gas*) on the structure and properties of butt joints (having various thicknesses, i.e., 5 mm and 7 mm) made of steel S960QL. Welding tests were performed in the flat position (PA) and in the horizontal position (PC). Joints made of steel S960QL in the above-presented configuration are present in elements of crane structures (e.g., telescopic crane jibs). The welding tests involved the use of the G Mn4Ni1.5CrMo solid electrode wire and the Ar+18% CO_2 shielding gas mixture (M21) (used in the MAG method). Non-destructive visual and radiographic tests did not reveal the presence of any welding imperfections in the joints. The welded joints obtained in the tests represented quality level B in accordance with the requirements of the ISO 12932 standard. Microscopic metallographic tests revealed that the heat-affected zone (HAZ) contained the coarse-grained martensitic structure resulting from the effect of the complex welding thermal cycle on the microstructure of the joints. Destructive tests revealed that the joints were characterised by tensile strength similar to that of the base material. The hybrid welding (laser–MAG) of steel S960QL enabled the obtainment of joints characterised by favourable plastic properties and impact energy exceeding 27 J. The tests revealed the possibility of making hybrid-welded joints satisfying the quality-related requirements specified in the ISO 15614-14 standard.

Keywords: hybrid welding; steel S960QL; HLAW; laser beam; MAG metal active gas

check for **updates**

Citation: Urbańczyk, M.; Adamiec, J. Hybrid Welding (Laser–Electric Arc MAG) of High Yield Point Steel S960QL. *Materials* **2021**, *14*, 5447. https://doi.org/10.3390/ma14185447

Academic Editors: Artur Czupryński and Claudio Mele

Received: 31 July 2021
Accepted: 16 September 2021
Published: 20 September 2021

Publisher's Note: MDPI stays neutral with regard to jurisdictional claims in published maps and institutional affiliations.

1. Introduction

Presently, the implementation of advanced welding technologies in various industries is considered to be one of the most important trends enabling the modernisation of technological processes [1]. Laser welding is an advanced, continuously improved, and increasingly common welding process applied in numerous industries (Figure 1). The development of laser radiation sources has led to a situation where the market offer includes lasers having power exceeding 100 kW [2]. The laser-welding process and its variants (remote welding, hybrid welding, etc.) have become primary joining processes used in many industrial sectors [3–5].

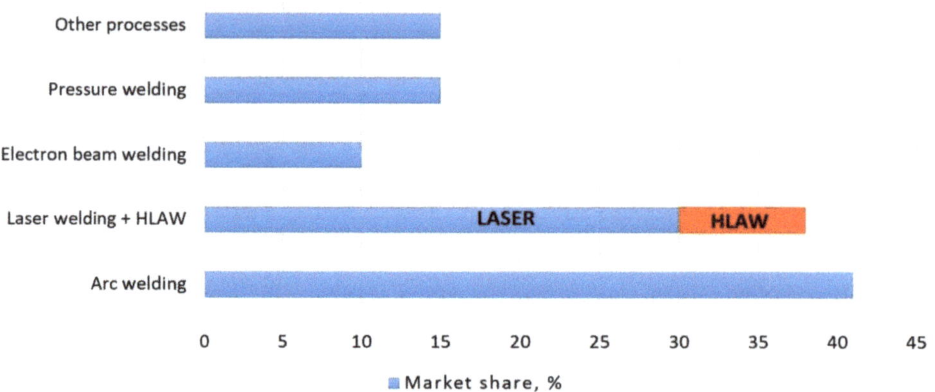

Figure 1. Various welding methods used in joining processes [6].

One of the variants of the welding process using laser radiation as the heat source is hybrid laser arc welding (HLAW). The process involves the simultaneous use of two heat sources, i.e., laser radiation and electric arc. During the welding process, the use of the two above-named heat sources leads to the formation of the common weld pool (Figure 2a). The hybrid laser welding process involving the use of the combined heat source (laser beam and electric arc) is characterised by a number of advantages in comparison with the advantages characteristic of each of the aforementioned processes when used separately (Figure 2b) [7].

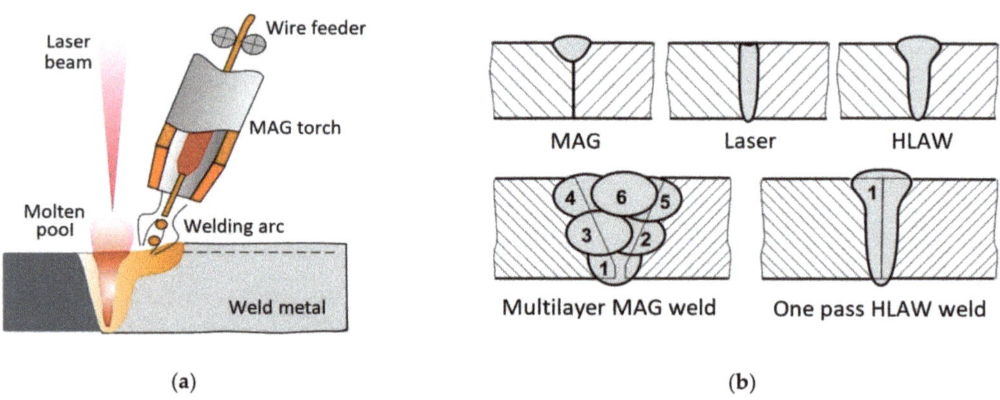

Figure 2. Hybrid laser arc welding (HLAW): (**a**) principle of the hybrid-welding process (laser–MAG *Metal Active Gas*) and (**b**) differences between the shape and geometry of welds obtained using the MAG welding, laser welding, and laser–MAG welding process [7].

The first tests, aimed to combine two heat sources (i.e., laser radiation and electric arc), were first performed by Steen and Eboo in the 1970s [8,9]. The two scientists demonstrated that the simultaneous use of laser and electric arc enabled the obtainment of higher welding rates and greater penetration depth than those obtainable using laser welding and arc welding procedures separately.

Obtainable results, the development of lasers, and the advantages resulting from the simultaneous use of two heat sources encouraged research centres worldwide to perform tests of hybrid technologies.

Silva et al. (2020) [10] discussed the hybrid laser welding process performed using various values of laser radiation power. The authors demonstrated that the application of the hybrid-welding process enabled the obtainment of joints containing fewer welding imperfections (gas pores in the weld) in comparison with joints made using laser radiation only. In addition, hybrid-welded joints obtained by the authors were characterised by weld face hardness lower by 100 HV than that of the weld root, which, in turn, translated into reduced steel brittleness.

One of the crucial aspects of the hybrid-welding process is the manner of joint preparation. Kah et al. (2011) [11] described the effect of a gap located between elements being joined on the shape and geometry of the weld formed as a result of the hybrid-welding process. The authors demonstrated that an excessively large gap between the aforesaid elements (>0.8 mm) resulted in the reduced height of the weld face and the increased height of the weld root.

Because of their advantages (significant penetration depth, higher welding rates, and reduced filler metal consumption), hybrid laser technologies are investigated by numerous research centres around the world. Researchers use HLAW to join various structural materials including steel [12,13], titanium alloys [14,15], aluminium alloys [16,17], or dissimilar materials [18–20].

Increased demand for high yield point structural steels (>900 MPa) in the crane-building industry has led to the intensification of tests concerning the applicability of hybrid-welding technologies in the joining of various steel grades [21,22].

Toughened steels having a yield point of 960 MPa are characterised by the fine-grained martensitic or martensitic-bainitic structure obtained through the toughening (i.e., hardening and quenching) of steel [23].

The joining of steels having the carbon equivalent value (CEV) $\leq 0.8\%$ must be performed using processes characterised by a low heat input to the joint. According to Siltanen et al. (2011) [24], the hybrid-welding method (laser + electric arc) is rated among the aforesaid processes due to the reduction of beads needed to make a joint with full penetration (regarding the welding of elements having thickness >4 mm). The researchers conducted hybrid butt welding tests involving 6 mm thick plates made of steel S960QL. The use of the hybrid method resulted in lower heat input to the joint (only one bead was made) in comparison with that accompanying the use of the MAG method (where the obtainment of the same thickness required three beads). The mechanical properties of the joint were similar to those of the base material.

The article discusses the effect of the hybrid-welding process (laser–electric arc MAG) on the structure and properties of butt joints having various thicknesses (i.e., 5 mm and 7 mm) and made of steel S960QL. Related welding tests were performed in the flat position (PA) and in the horizontal position (PC). Joints made of steel S960QL in the above-presented configuration are used in elements of crane structures (e.g., telescopic crane jibs) [25].

The hybrid-welding of plates having various thicknesses (e.g., 5 mm and 7 mm) made of high yield point steel S960QL and used in structural elements of cranes has been seldom discussed in scientific publications. Information on the subject is rudimentary and often subject to strict know-how-related confidentiality policy of individual manufacturers. The development of the crane industry and the growing demand for cranes have forced entrepreneurs to search for new high-performance welding technologies making it possible to increase production efficiency.

2. Materials and Methods

2.1. Materials

The tests discussed in the article involved the use of plates made of steel S960QL and having thicknesses of 5 mm and 7 mm; the remaining dimensions were 150 mm × 350 mm. The chemical composition of the plates was subjected to check analysis performed using a Q4 TASMAN 170 spark emission spectrometer (BRUKER; Billerica, MA, USA). The test results were compared with the requirements specified in the EN 10025-6+A1:2009 standard [26]. The results of the tests are presented in Table 1.

Table 1. Chemical composition and mechanical properties of the test plates made of steel S960QL.

	Chemical Composition, (%)										
	C	**Si**	**Mn**	**P**	**S**	**Cr**	**Cu**	**Ni**	**Mo**	**V**	**CEV**
EN 10025-6	max 0.20	max 0.80	max 1.70	max 0.02	max 0.01	max 1.5	max 0.50	max 2.0	max 0.70	-	max 0.82
Check analysis	0.13	0.39	1.40	0.009	0.001	0.01	0.01	0.19	0.44	0.03	0.47

Mechanical properties		
R_m [MPa]	R_e [MPa]	A_5 [%]
980 ÷ 1150	960	10

The filler metal used in the welding of steel S960QL was a Union SG700 electrode wire (G Mn4Ni1.5CrMo: SG700/ID-No. 822000508, ISO 16834-A) having a diameter of 1.2 mm (Böhler Schweisstechnik). The shielding gas used in the MAG method was an M21 group gas mixture (Ar—82% and CO_2—18%) (Messer). The shielding gas flow rate amounted to 16 dm^3/min.

2.2. Welding Method and Equipment

The hybrid-welding (laser–MAG) tests were performed at Łukasiewicz Research Network—Institute of Welding using a robotic welding station (Figure 3a) consisting of a TruDisk 12002 disc laser (TRUMPF; Stuttgart, Germany) having a power of 12 kW, a KRC30HA welding robot (KUKA; Augsburg, Germany) equipped with a hybrid-welding head (Figure 3b), and a PHOENIX 452 RC PULS MIG/MAG welding machine generating a maximum welding current of 450A (EWM Hightec Welding GmbH; Mündersbach, Germany).

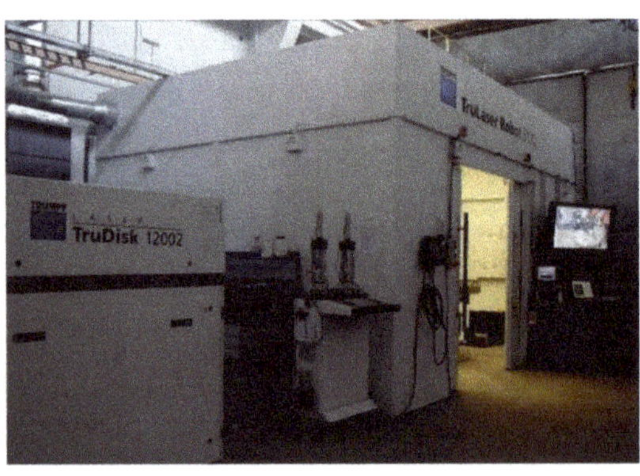

(a)

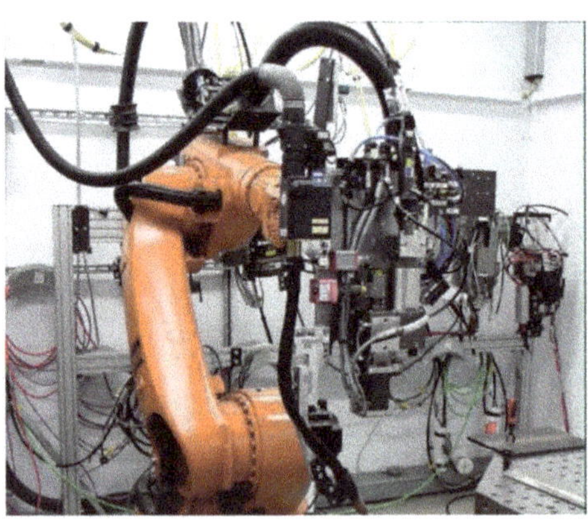

(b)

Figure 3. Robotic laser welding station (TruLaser Robot 5120) with the TruDisk 12002 disc laser: (**a**) main view, (**b**) D70 hybrid-welding head (Trumpf) (Łukasiewicz Research Network—Institute of Welding in Gliwice).

The welding tests were performed in two positions, i.e., in the flat position (PA) (Figure 4a) and in the horizontal position (PC) (Figure 4b). The edges of the plates to be joined were subjected to square butt weld preparation and set up without a gap (b = 0) at the interface.

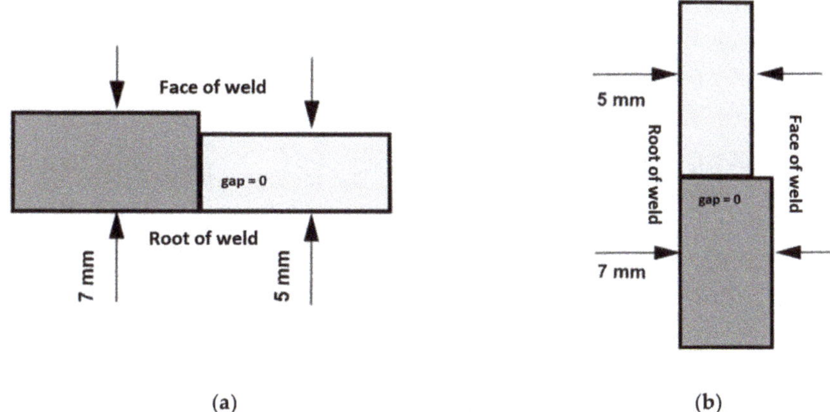

(a)   (b)

Figure 4. Setting up and plate thicknesses used in the tests: (**a**) flat position (PA) and (**b**) horizontal position (PC) (according to ISO standard positions).

During the hybrid-welding process, the laser radiation beam was transported using an optical fibre (dLLK) having a diameter of 400 μm and enabling the obtainment of laser beam focus diameter d_{og} = 0.8 mm (in relation to f_{col} = 200 mm—collimator lens focal length and f_{og} = 400 mm—focusing lens focal length; Figure 5a).

The laser radiation beam was positioned in the plane perpendicular to the surface of the plates, whereas the MAG welding torch was positioned at an angle of 65° (α1) in relation to the surface of the plates. The angle between the beam axis and the axis of the MAG welding torch (α2) amounted to 25°. The distance between the laser beam focus and the electrode tip (a) amounted to 2 mm (Figure 5b). During welding performed in the horizontal position (PC), the entire system was inclined at an angle of 90° (Figure 5c).

The welding tests were performed in the A–L (arc leading) configuration, i.e., with the arc power source leading the heat source in the process (Figure 5b).

A heat input was calculated using Equation (1) presented in the ISO 15614-14 standard [27].

$$Q = \frac{(P_{laser} \cdot U \cdot I)}{v_s} \cdot 10^{-3} [kJ/mm] \tag{1}$$

where Q—heat input (kJ/mm), P_{laser}—laser power (W), U—arc voltage (V), I—welding current (A), and v_s—welding rate (mm/s).

The setting up of the plates along with their thicknesses in the welding tests performed in various welding positions, are presented in Figure 4.

2.3. Tests of Welded Joints

The joints were subjected to visual tests (VT), radiographic tests (RT), and destructive tests performed in accordance with the requirements of the ISO 15614-14 standard (concerning hybrid-welding procedure qualification). The results of observations and measurements performed to identify the quality level (regarding the presence of welding imperfections) were assessed in accordance with the ISO 12932 standard [28].

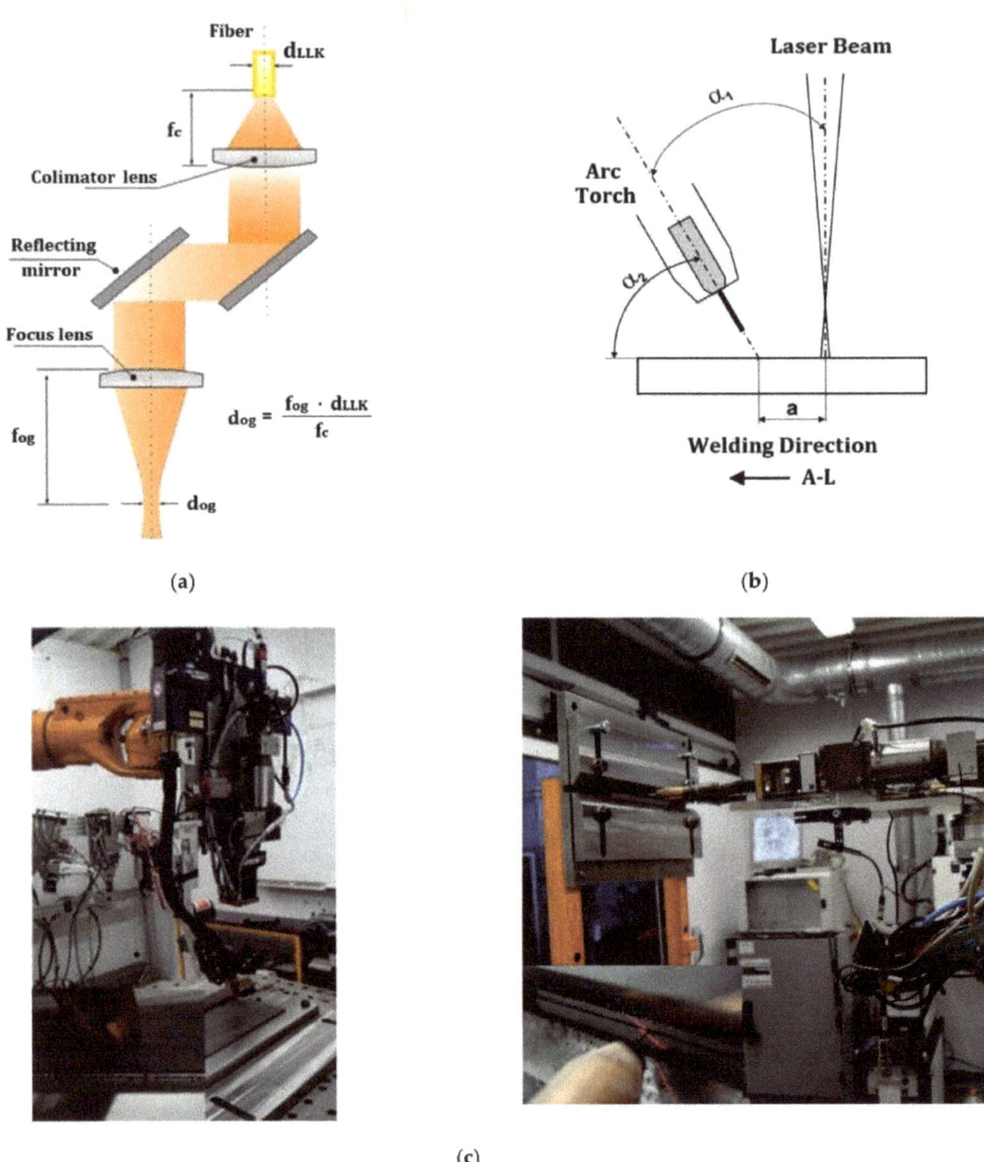

Figure 5. Schematic diagram of the laser optical system as well as the position of the laser beam and electric arc in the hybrid-welding technology: (**a**) laser optical system, (**b**) position of the laser beam and electric arc, and (**c**) processing head inclination during welding in the flat position (PA) and in the horizontal position (PC).

The research included the following tests:

- Visual tests performed in accordance with the requirements of the ISO 17637 standard;
- Radiographic tests of the welded joints, performed in accordance with the requirements of the ISO 17636-1 standard and involving the use of an Eresco 65 MF3 X-ray unit (GE Sensing&Inspection Technologies; Ahrensburg, Germany);

- Macroscopic metallographic tests, performed using an Olympus SZX9 stereoscopic microscope (Olympus, Tokyo, Japan). To identify their structure, the specimens were subjected to etching in Adler's reagent (Chmes, Poznań, Poland);
- Microscopic metallographic tests, performed in accordance with the requirements of the ISO 17639 standard and involving the use of a Nikon Eclipse MA200 light microscope (Leuven, Belgium). The specimens were subjected to grinding with abrasive paper having a granularity of 800 and 1000, polishing performed using a powerpro 4000 grinder/polisher (Buehler; Germany) and metaldi Monocrystalline Diamond Suspension (3 μm), as well as etching in 5% Nital (5% HNO_3 in ethanol);
- Tests performed using a scanning transmission electron microscope (STEM) involving the use of thin foils; the specimens were subjected to two-sided grinding (with abrasive paper) to reach a thickness of 0.5 mm. The process of electrochemical thinning was performed using a Struers tenupol-5 machine, with a voltage of 45 V and a temperature of 5 °C. The process was carried out in electrolyte composed of 70% $CH20H$, 20% glycerine, and 10% $hclo4$. The cooling agent was liquid nitrogen. The tests were performed using a Hitachi 2300A scanning-transmission electron microscope (STEM) (Japan), illuminating thin foils. The microscope was equipped with an FEG-type gun provided with the Schottky emitter. The accelerating voltage during the tests amounted to 200 kv;
- Hardness distribution tests were performed in accordance with the requirements of the ISO 9015-1 standard and involved the use of a GNEHM DIGITAL BRICKERS 220 hardness tester. Vickers hardness tests (HV) were performed along two measurement lines located 2 mm away from the upper and lower edge of the specimen. The imprints were made in the base material, heat-affected zone (HAZ) and in the weld;
- Static tensile tests involved 2 specimens cut out perpendicularly to the weld and prepared in accordance with the requirements of the ISO 6892-1 standard. The preparation of the specimens involved the removal of excessive root and face reinforcement as well as the mechanical reduction of specimen thickness from 7 mm to 5 mm (performed to obtain the even surface of the plates across the entire specimen). The dimensions of the specimens were 300 mm × 25 mm × 5.0 mm. The tension rate amounted to 10 mm/min. The tests were performed using an MTS 810 TEST SYSTEMS testing machine (Eden Prairie, MN, USA);
- Face bend test (FBB) and root bend test (RBB) of the butt weld were performed in accordance with the requirements of the ISO 5173 standard. The tests involved 4 specimens—two specimens on each side. The thickness of the plate was mechanically reduced from 7 mm to 5 mm (in order to obtain the even surface of the plates across the entire specimen). The dimensions of the specimens were 300 mm × 20 mm × 5.0 mm). The tests were performed using a LOS12126 testing machine (Losenhausenwerk AG; Düsseldorf, Germany);
- Impact strength tests, performed in accordance with the requirements of the ISO 9016 standard, involved the use of 2 sets of specimens (3 specimens in each set) sampled from the weld area and from the heat-affected zone (HAZ). The cross-section of the specimens used in the test was reduced. The dimensions of the specimens were 2.5 mm × 8.0 mm × 55 mm. The depth of the V notch amounted to 2 mm. Before the tests, the specimens were cooled to a temperature of −40 °C. The cooling process was performed using an FP89 cooling circulator (Julabo). Impact energy was identified using an RKP 300 impact-testing machine (Amsler).

The parameters of the hybrid-welding process (laser + electric arc MAG) performed both in the flat position (PA; joint no. 1) and in the horizontal position (PC; joint no. 2) are presented in Table 2.

Table 2. Parameters of the hybrid-welding process used when making the joints (having thicknesses of 5 mm and 7 mm) in the flat position (joint no. 1 PA) and horizontal position (joint no. 2 PC).

Welding Parameters	Joint No. 1 (PA)	Joint No. 2 (PC)
Laser power (kW)	3.75	3.75
Welding rate (m/min)	1.3	1.3
Filler metal wire feed rate (m/min)	8.5	8.5
Welding current (A)	290	275
Arc voltage (U)	27	27
Interface gap (between the plates) (mm)	0	0
Heat input (kJ/mm)	0.57	0.56

3. Results and Discussion

3.1. Weld Formation

The weld face side and the weld root side of the joints after the hybrid-welding process are presented in Figure 6.

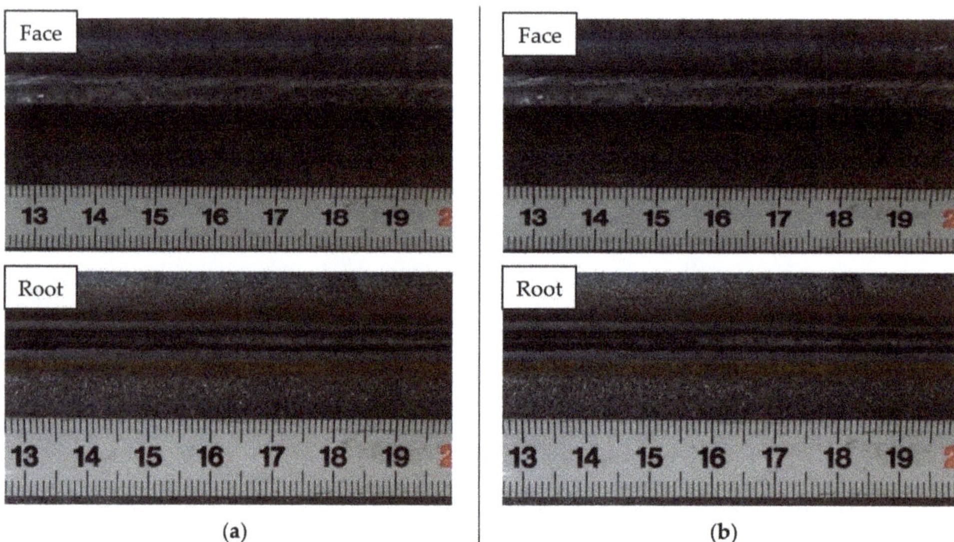

Figure 6. Joint (having thickness 5 mm and 7 mm) viewed from the face side and the root side after the hybrid-welding process: (**a**) joint no. 1 (PA) and (**b**) joint no. 2 (PC) (in accordance with Table 2).

The visual welding tests revealed that hybrid-welded joints no. 1 and 2 (Figure 6) made in the flat position (PA) and in the horizontal position (PC) were characterised by the smooth spatter-free weld face and the properly shaped weld root.

In accordance with the requirements specified in the ISO 12932 standard (concerning hybrid-welding procedure qualification), the joints made in the PA and PC positions satisfied related criteria and represented quality level B.

The subsequent stage included the performance of non-destructive radiographic (X-ray) tests aimed to detect (if any) internal welding imperfections. The X-ray tests involved 100% of the joint length. The X-ray photographs of joints no. 1 and 2 are presented in Figure 7. Joint no. 1 and joint no. 2 did not contain any internal welding imperfections.

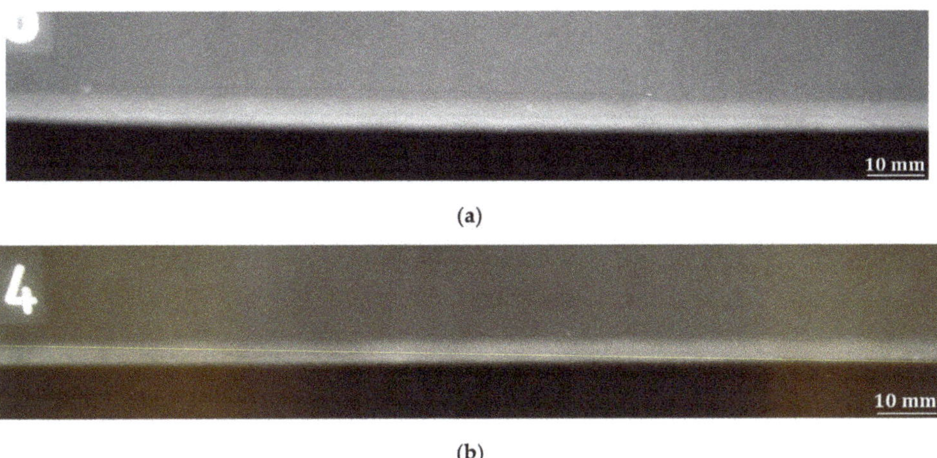

(a)

(b)

Figure 7. X-ray photograph of the hybrid-welded joints: (**a**) joint no. 1 (PA) and (**b**) joint no. 2 (PC).

The subsequent stage involved macrostructural tests of joints no. 1 and 2 (Figure 8). The macrostructural tests did not reveal any welding imperfections within the weld area and in the heat-affected zone (HAZ).

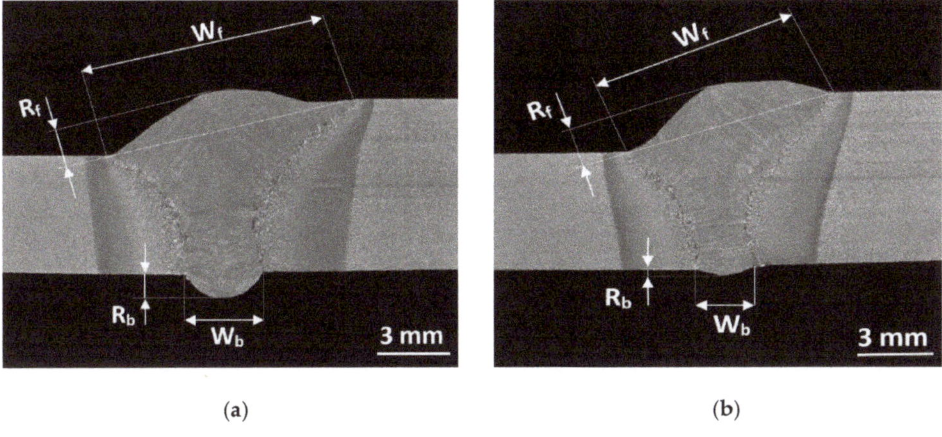

(a)   (b)

Figure 8. Macrostructure as well as the face and the root of the hybrid-welded joints (having thicknesses of 5 mm and 7 mm): (**a**) joint no. 1 (PA) and (**b**) joint no. 2 (PC).

The etched metallographic specimens revealed the clearly visible borders between the base material, HAZ, and the weld.

Geometrical dimensions of hybrid-welded joints no. 1 and 2 made in the flat position (PA) and in the horizontal position (PC) are presented in Table 3.

Table 3. Geometrical dimensions of hybrid-welded joints no. 1 and 2.

Dimensions	Joint No. 1	Joint No. 2
Weld face width (W_f/mm)	9.7	8.3
Weld face height (R_f/mm)	1.5	1.7
Weld root width (W_b/mm)	3.1	2.2
Weld root height (R_b/mm)	1	0.3

The width of the weld face (W_f/mm) of the joint made in the flat position (PA) amounted to 9.7 mm, whereas its height (R_f/mm) amounted to 1.5 mm. The width of the weld root (W_b/mm) amounted to 3.1 mm, whereas its height (R_b/mm) amounted to 1 mm. The width of the weld face (W_f/mm) of the joint made in the horizontal position (PC) amounted to 8.3 mm and was 1.4 millimetres lower in comparison with that of the joint made in the flat position (PA). The height of the weld face (R_f/mm) amounted to 1.7 mm. The width of the weld root (W_b/mm) of the joint made in the horizontal position (PC) amounted to 2.2 mm (0.9 mm less than that of the joint made in the flat position (PA)). The height of the weld root (R_b/mm) amounted to 0.3 mm (0.7 mm less than that of the joint made in the flat position (PA)) (Table 3).

The tests revealed that joints no. 1 and 2 represented quality level B in accordance with the requirements specified in the ISO 12932 standard (concerning hybrid-welding procedure qualification).

3.2. Microstructure Characteristics

Joint no. 2 (PC) was subjected to microscopic metallographic tests. The microscopic tests revealed the presence of three typical areas (Figure 9a), i.e., the base material (containing the fine-grained structure of tempered martensite (Figure 9b), the heat-affected zone having a width of approximately 1 mm (Figure 9c), and the weld area (Figure 9d).

The heat-affected zone (HAZ) contained the coarse-grained martensitic structure (having a thickness of 425 HV10) formed as a result of the welding thermal cycle effect (Figure 9c). In turn, the weld contained the homogenous acicular martensitic structure having a hardness of 350 HV10 (Figure 8d). The above-presented observation results were confirmed by tests performed using a scanning transmission electron microscope (Figure 10a–f).

In article [29], Guo et al. stated that the crucial aspects related to the welding of fine-grained, high-strength steel S960 MPa were the temperature of the process and the welding rate. Those two factors significantly affected the microstructure in the zones of the welded joint (HAZ, weld).

(a)   (b)

Figure 9. *Cont.*

(c) (d)

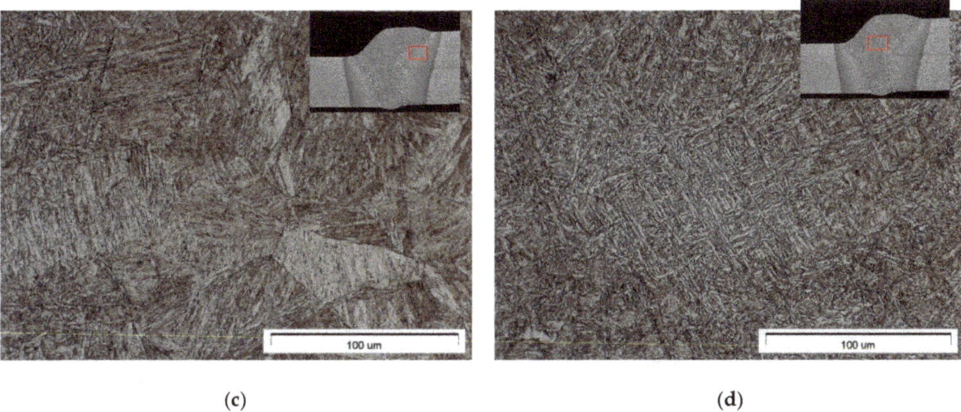

Figure 9. Structure of hybrid-welded joint no. 1 made of steel S960QL: (**a**) macrostructure, HAZ, and the weld; (**b**) fine-grained martensite in the base material; (**c**) HAZ containing coarse-grained martensit; (**d**) weld—martensitic structure.

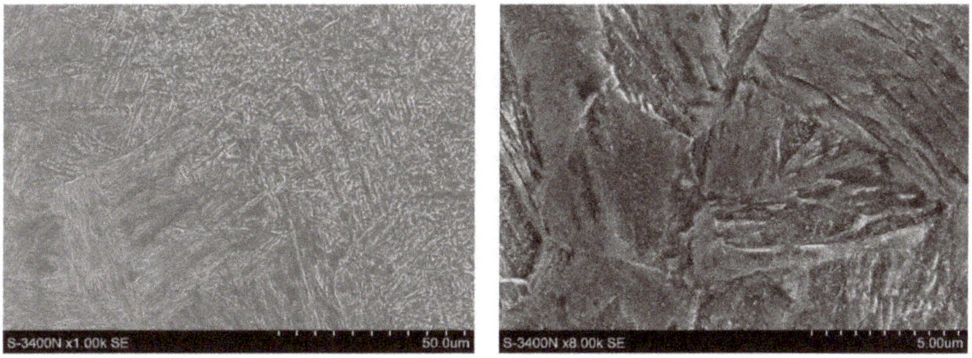

(a) (b)

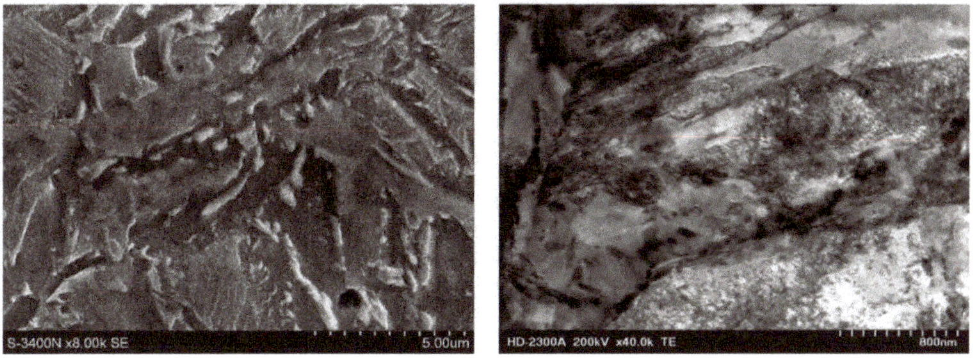

(c) (d)

Figure 10. *Cont.*

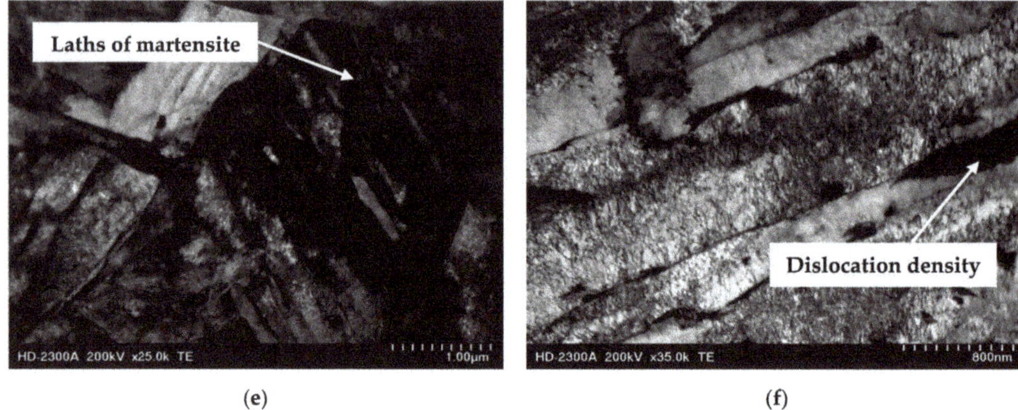

(**e**)   (**f**)

Figure 10. Microstructure of the welded joint made of steel S960QL: (**a**) fusion line (SEM), (**b**) HAZ (SEM), (**c**) weld (SEM), (**d**) base material—martensite with visible dislocations (STEM), (**e**) martensite observed in the HAZ (STEM), and (**f**) laths of martensite in the weld with a visible increase in dislocation density (STEM).

In the structure of the base material (in the martensitic laths), it was possible to observe significant dislocation density, which was connected with the manufacturing and hardening of high yield point steel S960QL. During welding, the thermal cycle effect triggered austenitisation and the growth of austenite grains in the HAZ. The cooling of the heat-affected zone led to the martensitic transformation, resulting in the formation of coarse-grained martensite (Figure 10b). The aforesaid area also contained fine carbides (Figure 10e). The weld contained fine-grained martensite with numerous fine carbides (Figure 10c). The martensite laths were characterised by significant dislocation density (indicating the hardening of the joint in the above-named area).

3.3. Hardness Distribution

Hardness measurements concerning joint no. 1 (PA) and joint no. 2 (PC) revealed that the highest hardness value was characteristic of the weld–HAZ interface. In relation to joint no. 1, the aforesaid hardness amounted to 436 HV10, whereas in relation to joint no. 2, the hardness amounted to 448 HV10. An increase in hardness in the aforesaid area could be ascribed to the rate of heat propagation and the hardening of the joint. A similar distribution of hardness in hybrid-welded joints made of fine-grained high-strength steel was observed by Lahdo et al. (2014) [30]. The authors stated that the weld–HAZ interface underwent hardening, which was connected with the rate at which the above-named area was cooled. The reduction of hardness at the weld–HAZ interface would require the application of preheating.

The difference in hardness between the base material and the HAZ in the joints made in the flat position and in the horizontal position amounted to 22%. Regarding joint no. 2, hardness in the central area of the weld was similar to that of the base material and amounted to 344 HV10 (Figure 11b). In turn, regarding joint no. 1, the difference between the above-named zones amounted to 20 HV10 (Figure 11a).

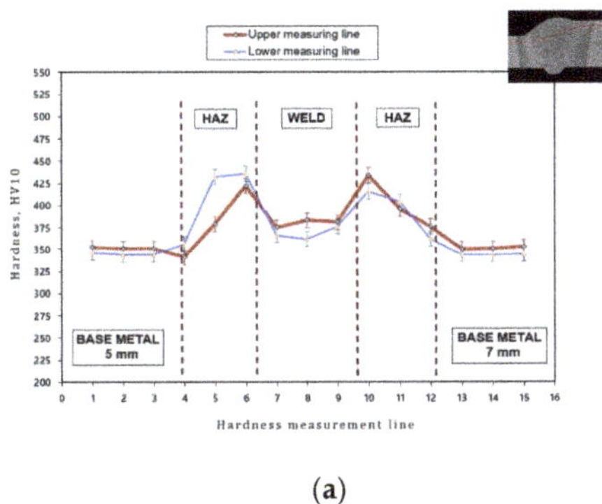

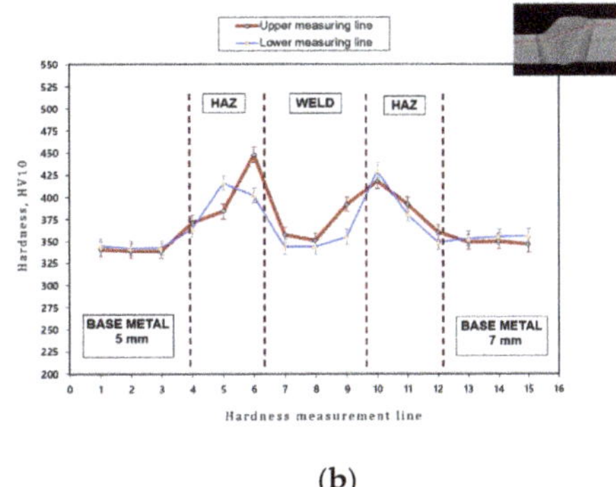

(a)   (b)

Figure 11. Measurement results and the distribution of hardness in the cross-section of the HLAW joint: (**a**) joint no. 1 (PA) and (**b**) joint no. 2 (PC).

Both in the joint made in the flat position (PA) and that made in the horizontal position (PC), the HAZ was characterised by Vickers hardness above 400 (448 HV10). The above-presented values are acceptable in accordance with the requirements of the ISO 15614-14 standard concerning hybrid-welding procedure qualification. The hardness increase in the aforesaid area did not affect other mechanical properties (tension, bending).

3.4. Mechanical Properties

The analysis of the destructive tests concerning the hybrid-welded joints (laser beam—MAG) revealed that both joint no. 1 and joint no. 2 satisfied the requirements of the ISO 15614-14 standard (Table 4). The hybrid-welding process did not trigger a decrease in the tensile strength of the joints (1053 MPa and 1068 MPa, respectively) in comparison with that of the base material (restricted within the range of 980 MPa to 1150 MPa; Table 1). The specimens ruptured in the HAZ (area characterised by a clearly visible grain growth). The decrease in tensile strength in the HAZ area resulted from the loss of mechanical properties obtained by the steel in the manufacturing process (as a result of hardening and tempering). Figure 12 presents the specimens after the static tensile test of the HLAW joint.

Table 4. Mechanical test results concerning the hybrid-welded joints made of steel S960QL.

Joint No.	Tensile Strength *,[1]		Bend Test *, Bend Angle, °		Impact Strength Test KCV **, Impact Energy J, (Testing Temperature: −40 °C)	
	Rm, MPa	Area of Rupture	Weld Face	Weld Root	HAZ	Weld
Joint no. 1	1053	HAZ	180	180	46	30
Joint no. 2	1068	HAZ	180	180	40	30

* Average result of two measurements; ** average result of three measurements, [1] standard deviation σ = 9.9.

The bend angle obtained in the bend test amounted to 180°—both during the face and root bend tests of the butt weld. The joints were characterised by high plastic properties.

Impact energy obtained at a temperature of −40 °C also indicated the favourable mechanical properties of the joints. In relation to the specimens sampled from the central area of the weld, both regarding joint no. 1 (PA) and joint no. 2 (PC), impact energy amounted to 30 J. In turn, regarding the specimens sampled from the HAZ area, impact energy amounted to 46 J in relation to the joint made in the flat position and 40 J in relation to the joint made in the horizontal position.

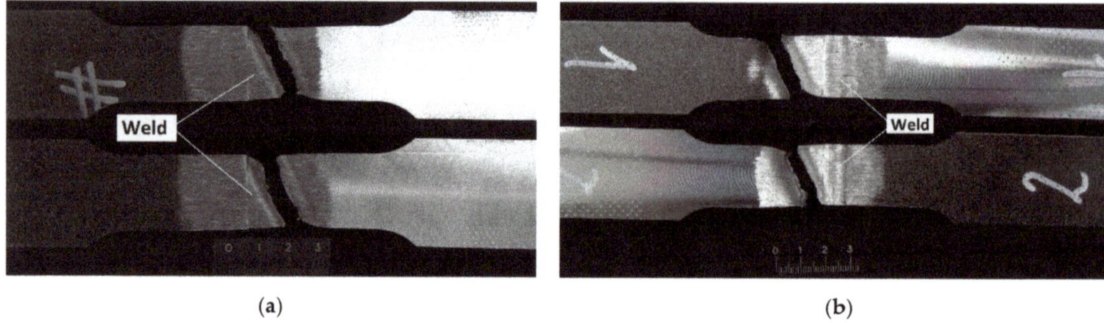

Figure 12. Specimens after the static tensile test of the HLAW joint: (**a**) joint no. 1 (PA) and (**b**) joint no. 2 (PC).

Siltanen et al. (2015) [31] obtained similar mechanical properties using another filler metal (in the hybrid-welding process). In relation to 6 mm thick hybrid-welded joints, the tensile strength amounted to 1000 MPa, whereas impact energy exceeded 27 J (34 J, 48 J) (at a testing temperature of −40 °C). Therefore, the joints satisfied the requirements specified in the EN 10025-6 standard [26].

4. Conclusions

The hybrid-welding tests (laser + MAG) involved the making and testing of butt joints with full penetration. The joints (having various thicknesses, i.e., 5 mm and 7 mm) were made of steel S960QL plates. The above-named steel is used, among other things, in the production of telescopic jibs of self-propelled cranes.

The tests revealed the possibility of obtaining welded joints satisfying the requirements of quality level B (i.e., top quality level) in accordance with the ISO 15614-14 standard.

Both the joint made in the flat position (PA—1G) and that made in the horizontal position (PC—2G) were characterised by the uniform, smooth, and spatter-free weld face as well as by the properly shaped weld root. The visual and radiographic tests did not reveal the presence of surface or internal welding imperfections.

The static tensile tests revealed that the joints were characterised by high strength (1053 MPa and 1068 MPa, respectively) and high plasticity (bend angle of 180°).

The specimens sampled from the weld area and from the heat-affected zone revealed that impact energy amounted to more than 27 J (30 J in relation to the weld as well as 40 J and 46 J in relation to the HAZ).

The microscopic metallographic tests revealed that the heat-affected zone (HAZ) contained the coarse-grained martensitic structure formed as a result of the complex welding thermal cycle effect. During welding, the thermal cycle effect triggered austenitisation, the growth of austenite grains and the precipitation of carbides in the heat-affected zone. In turn, the weld contained the homogenous structure of acicular martensite.

Author Contributions: M.U.: assumptions, practical tests, optimisation of welding parameters, destructive tests, and conclusions; J.A.: methodology, metallographic tests, and the analysis of test results. Both authors have read and agreed to the published version of the manuscript.

Funding: The research was funded by the National Centre for Research and Development, Project no. PBS3/B5/31/2015.

Conflicts of Interest: The authors declare no conflict of interest.

References

1. Acherjee, B. Hybrid laser arc welding: State-of-art review. *Opt. Laser Technol.* **2018**, *99*, 60–71. [CrossRef]
2. Kawahito, Y.; Wang, H.; Katayama, S.; Sumimori, D. Ultra high power (100 kW) fiber laser welding of steel. *Opt. Lett.* **2018**, *43*, 4667–4670. [CrossRef]

3. Nielsen, S.E. "High power laser hybrid welding—challenges and perspectives", 15th Nordic Laser Materials Processing Conference, Nolamp 15, 25–27 August 2015, Lappeenranta, Finland. *Phys. Procedia* **2015**, *78*, 24–34. [CrossRef]

4. Brian, M. Victor "Hybrid laser arc welding", Edison Welding Institute, ASM Handbook. *Weld. Fundam. Process.* **2011**, *6A*.

5. Shenghai, Z.; Yifu, S.; Huijuan, Q. The technology and welding joint properties of hybrid laser-TIG welding on thick plate. *Opt. Laser Technol.* **2013**, *48*, 381–388. [CrossRef]

6. Innovationen Für die Wirtschaft. Forschung in der Fügetechnik DVS. 2015. Available online: https://www.dvs-ev.de/fv/neu/GB2017.pdf (accessed on 30 April 2016).

7. Banasik, M.; Urbańczyk, M. Laser + MAG Hybrid Welding of Various Joints. *Biul. Inst. Spaw.* **2017**, *1*, 6–13.

8. Steen, W.M.; Eboo, M. Arc-augmented laser welding. *Met Constr.* **1979**, *11*, 332–335.

9. Steen, W.M. Arc-augmented laser processing of materials. *J. Appl. Phys.* **1980**, *51*, 5636–5641. [CrossRef]

10. Silva, R.; Mano, M.D.C.; Rodrigues, B.M.; Sousa, S.M.J.; Pereira, M.; Ramos, B.B.; Schwedersky, B.M.; Silva, G.H.R. A comparison between LBW and hybrid laser-GMAW processes based on microstructure and weld geometry for hardenable steels. *Int. J. Adv. Manuf. Technol.* **2020**, *110*, 2801–2814. [CrossRef]

11. Kah, P.; Salminen, A.; Martikainen, J. The influence of parameters on penetration, speed and bridging in laser hybrid welding. *Mechanika* **2011**, *17*, 324–333. [CrossRef]

12. Adamiec, J.; Więcek, M.; Gawrysiuk, W. Fibre laser usage in boiler elements' production for the power industry. *Weld. Int.* **2010**, *24*, 853–860. [CrossRef]

13. Górka, J.; Stano, S. Microstructure and properties of hybrid laser arc welded joints (laser beam-MAG) in thermo-mechanical control processed S700MC steel. *Metals* **2018**, *8*, 132. [CrossRef]

14. Li, C.; Muneharua, K.; Takao, S.; Kouji, H. Fiber laser-GMA hybrid welding of commercially pure titanium. *Mater Des.* **2009**, *30*, 109–114. [CrossRef]

15. Shi, J.; Zhou, Y.; Liu, L. Application of pulsed laser-TIG hybrid heat source in root welding of thick plates titanium alloys. *Appl Sci.* **2017**, *7*, 527. [CrossRef]

16. Casalino, G.; Campanelli, S.L.; DalMaso, U.; Ludovico, A.D. Arc leading versus laser leading in the hybrid welding of aluminium alloy using a fiber laser. *Procedia CIRP* **2013**, *12*, 151–156. [CrossRef]

17. Huang, L.; Wu, D.; Hua, X.; Liu, S.; Jiang, Z.; Li, F.; Wang, H.; Shi, S. Effect of the welding direction on the microstructure characterization in fiber laser-GMAW hybrid welding of 5083 aluminium alloy. *J Manuf. Process.* **2018**, *31*, 514–522. [CrossRef]

18. Han, X.; Yang, Z.; Ma, Y.; Shi, C.; Xin, Z. Comparative Study of Laser-Arc HybridWelding for AA6082-T6 Aluminum Alloy with Two Different Arc Modes. *Metals* **2020**, *10*, 407. [CrossRef]

19. Thomy, C.; Vollertsen, F. Laser-MIG hybrid welding of aluminium to steel—Effect of process parameters on joint properties. *Weld. World* **2012**, *56*, 124–132. [CrossRef]

20. Casalino, G.; Mortello, M.; Perulli, P.; Varone, A. Effects of laser offset and hybrid laser on microstructure and IMC in Fe-Al dissimilar welding. *Metals* **2017**, *7*, 282. [CrossRef]

21. Turichin, G.; Klimova-Korsmik, O.; Skylar, M.; Zhitenev, A.; Kurakin, A.; Pozdnyakov, A. Laser-arc hybrid welding perspective ultra-high strength steels: Influence of the chemical composition of weld metal on microstructure and mechanical properties. *Procedia CIRP* **2018**, *74*, 752–756. [CrossRef]

22. Atabaki, M.; Ma, J.; Yang, G.; Kovacevic, R. Hybrid laser/arc welding of advanced high strength steel in different butt joint configurations. *Mater Des.* **2014**, *64*, 573–587. [CrossRef]

23. Gogou, E. Use of High Strength Steel Grades for Economical Bridge Design. Master's Thesis, Delft University of Technology, Delft, The Netherlands, 2012.

24. Siltanen, J.; Kömi, J.; Laitinen, R.; Lehtinen, M.; Tihinen, S.; Jasnau, U.; Sumpf, A. Laser-GMA hybrid welding of 960 MPa steels. 30th International Congress on Laser Materials Processing. *Int. Congr. Appl. Lasers Electro-Opt.* **2011**, *2011*, 592–598.

25. Laser MSG Hybrid Welding of Mobile Crane Booms. Fraunhofer IWS Annual Report. 2014. Available online: https://www.iws.fraunhofer.de/content/dam/iws/en/documents/publications/annual_report_articles/JB_IWS_2014_en_S36-37.pdf (accessed on 25 June 2018).

26. EN. Hot rolled products of structural steels. In *Technical Delivery Conditions for Flat Products of High Yield Strength Structural Steels in the Quenched and Tempered Condition*; European Standard: Brussels, Belgium, 2019.

27. ISO. *Specification and Qualification of Welding Procedures for Metallic Materials. Welding Procedure Test. Part 14: Laser-Arc Hybrid Welding of Steels, Nickel and Nickel Alloys*; International Organization for Standardization: Geneva, Switzerland, 2013.

28. ISO. *Welding. Laser-Arc Hybrid Welding of Steels, Nickel and Nickel Alloys. Quality Levels for Imperfections*; International Organization for Standardization: Geneva, Switzerland, 2013.

29. Guo, W.; Crowther, D.; Francis, J.A.; Thompson, A.; Liu, Z.; Li, L. Microstructure and mechanical properties of laser welded S960 high strength steel. *Mater Des.* **2015**, *85*, 534–548. [CrossRef]

30. Lahdo, R.; Seffer, O.; Springer, A.; Kaierle, S.; Overmeyer, L. GMA-laser hybrid welding of high-strength fine-grain structural steel with an inductive preheating. *Phys. Procedia* **2014**, *56*, 637–645. [CrossRef]

31. Siltanen, J.; Tihinen, S.; Kömi, J. Laser and laser gas-metal-arc hybrid welding of 960 MPa direct-quenched structural steel in a butt joint configuration. *J. Laser Appl.* **2015**, *27*, S29007. [CrossRef]

Article

Microstructure Investigation of WC-Based Coatings Prepared by HVOF onto AZ31 Substrate

Ewa Jonda [1], Leszek Łatka [2,*], Anna Tomiczek [3], Marcin Godzierz [4,*], Wojciech Pakieła [1] and Paweł Nuckowski [5]

[1] Department of Engineering Materials and Biomaterials, Silesian University of Technology, 18a Konarskiego Str., 44100 Gliwice, Poland; ewa.jonda@polsl.pl (E.J.); wojciech.pakiela@polsl.pl (W.P.)

[2] Department of Metal Forming, Welding and Metrology, Faculty of Mechanical Engineering, Wroclaw University of Science and Technology, 5 Łukasiewicza Str., 50371 Wroclaw, Poland

[3] Scientific and Didactic Laboratory of Nanotechnology and Material Technologies, Faculty of Mechanical Engineering, Silesian University of Technology, 7a Towarowa Str., 44100 Gliwice, Poland; anna.tomiczek@polsl.pl

[4] Centre of Polymer and Carbon Materials Polish Academy of Sciences, 34 M. Curie-Skłodowskiej Str., 41819 Zabrze, Poland

[5] Materials Research Laboratory, Faculty of Mechanical Engineering, Silesian University of Technology, 18a Konarskiego Str., 44100 Gliwice, Poland; pawel.nuckowski@polsl.pl

* Correspondence: leszek.latka@pwr.edu.pl (L.Ł.); mgodzierz@cmpw-pan.edu.pl (M.G.); Tel.: +48-713-202-138 (L.Ł.); +48-322-716-077 (M.G.)

Abstract: In this paper, three commercial cermet powders, WC-Co-Cr, WC-Co and WC-Cr_3C_2-Ni, were sprayed by the High Velocity Oxy Fuel (HVOF) method onto magnesium alloy AZ31 substrate. The coatings were investigated in terms of their microstructure, phase analysis and residual stress. The manufactured coatings were analyzed extensively using optical microscopy (OM), X-ray diffraction (XRD), scanning (SEM) and transmission electron microscopy (TEM). Based on microstructure studies, it was noted that the coatings show satisfactory homogeneity. XRD analysis shows that in WC-Co, WC-Co-Cr and WC-Cr_3C_2-Ni coatings, main peaks are related to WC. Weaker peaks such as W_2C, $Co_{0.9}W_{0.1}$, Co and W for WC-Co and W_2C, Cr_3C_2 and Cr_7C_3 for WC-Cr_3C_2-Ni also occur. In all cermet coatings, linear stress showed compressive nature. In WC-Co and WC-Cr_3C_2-Ni, residual stress had a similar value, while in WC-Co-Cr, linear stress was lower. It was also proved that spraying onto magnesium substrate causes shear stress in the WC phase, most likely due to the low elastic modulus of magnesium alloy substrate.

Keywords: High Velocity Oxy Fuel; AZ31 magnesium alloy; microstructure; X-ray diffraction; residual stress analysis—$\sin^2\psi$ method

Citation: Jonda, E.; Łatka, L.; Tomiczek, A.; Godzierz, M.; Pakieła, W.; Nuckowski, P. Microstructure Investigation of WC-Based Coatings Prepared by HVOF onto AZ31 Substrate. *Materials* **2022**, *15*, 40. https://doi.org/10.3390/ma15010040

Academic Editors: Ahalapitiya H Jayatissa, Andrew Ruys and Guillermo Requena

Received: 15 October 2021
Accepted: 18 December 2021
Published: 22 December 2021

Publisher's Note: MDPI stays neutral with regard to jurisdictional claims in published maps and institutional affiliations.

1. Introduction

The elements of machines and equipment operated in the conditions of abrasive, erosion or corrosion wear and tear are exposed to damage. New element production costs significantly exceed its recovery enabling the restoration of its usable values and increased durability. One of the numerous methods allowing for a combination of the beneficial properties of the core with resistance to abrasive wear, heat resistance and increased hardness is the deposition of coating [1,2]. This enables lifetime increase and improvement of reliability and operating durability of the machinery elements. One of the most frequently applied technologies of protective coating deposition is thermal spraying [3–5]. Generally, it produces metallic, carbide, ceramic and composite coatings of any chemical and phase composition on an appropriately prepared base [6,7]. Among the most commonly used methods mentioned above is the HVOF (High Velocity Oxy Fuel) spraying method, which enables the production of dense coatings with compact structure and high adhesion to the substrate. In the HVOF method, the flammable gas (or liquid fuel) is fed to the combustion chamber together with the oxygen, and the stream of gases produced during combustion is formed in the nozzle. The HVOF gun works continuously while the liquid fuel is sprayed. Among the most frequently used flammable gases, ethylene, propylene and acetylene should be mentioned, while kerosene is a popular liquid fuel. In the stream of argon or nitrogen, the feedstock material is fed (mainly in the form of powder) axially or radially along the gun axis [1,8]. The most significant applications of the HVOF method include spraying cermet coatings based on tungsten carbide (WC), where, due to the low temperature obtained by the particles, carbide transformation takes place to a lesser extent. The cermet materials are a combination of metallic and metal materials in which the substrate is usually formed from Co, Ni, Al, Ti, Mo or their alloys. Because the tungsten carbide (WC) could be well wetted, without limitation, by cobalt (Co), nickel (Ni), iron (Fe) and cobalt–chromium (CoCr), the cermet materials based on WC are some of the most frequently used cermet materials [9–12]. The advantage of these materials is their high resistance to abrasive, erosion and cavitation wear and the ceramic coatings produced from them are characterized, without limitation, with higher hardness, low thermal conductivity coefficient, high corrosion and oxidation resistance and high resistance to abrasion and erosion [13–15].

In the literature, there is a small gap concerning the deposition of hard and wear-resistance coatings on soft and low melting temperature substrate. It is much more important when combined with mass reduction, e.g., in the automotive and aviation industry. This study's novelty is the purpose of using a magnesium alloy as a substrate. This group of materials has not yet been investigated in depth. Magnesium alloys are lighter than aluminum ones, which is a significant advantage. However, the main disadvantages are poor mechanical properties and resistance against wear, erosion, corrosion, etc. In recent years, some investigations have mainly concentrated on HVOF coating materials, such as amorphous Fe-based, stainless steel and hydroxyapatite [16–18]. Only a few articles are dedicated to cermet HVOF coating on magnesium alloy substrate [16,17]. Cermet coatings produced by HVOF could substantially improve these properties on the top surface. Moreover, because of relatively soft flame and average temperatures (ensured by an appropriate selection of process parameters), HVOF spraying allows manufacturing such coatings onto magnesium substrate without damaging it [19,20].

This study's main aim was to investigate and compare the microstructure and residual stress in the cermet coatings manufactured by the HVOF method on the AZ31 magnesium alloy substrate. Such a solution could make it possible to use it in aircraft structures.

2. Materials and Methods

2.1. Powders

In this study, three commercially available powders were used as feedstock material. They are labelled as follows:

- P1-WC-Co-Cr (86-10-4, Höganäs, Amperit 558.074);
- P2-WC-Co (88-12, Höganäs, Amperit 518.074); and
- P3-WC-Cr_3C_2-Ni (73-20-7, Woka 3702-1).

Chemical compositions have been given in wt %. For all powders, delivery conditions were agglomerated and sintered. Moreover, the particle size range was −45 + 15 μm for each one. The main diameter d_{50} was around 30 μm for all powders.

2.2. Deposition Process

The magnesium alloy AZ31 with 5 mm thickness was used as a substrate. Before the spraying, the surfaces of the samples were sand-blasted with corundum and ultrasonic treated. The JP 5000 spray system TAFA (Indianapolis, IN, USA) by RESURS (Warszawa, Poland) was used to manufacture the coatings. Kerosene and oxygen were used as the fuel media, whereas nitrogen was used as the carrier gas. The schematic diagram of the HVOF coating process is presented in Figure 1, and the spraying parameters are listed in Table 1.

The coatings manufactured from P1, P2 and P3 powders are labelled in the text as C1, C2 and C3, respectively.

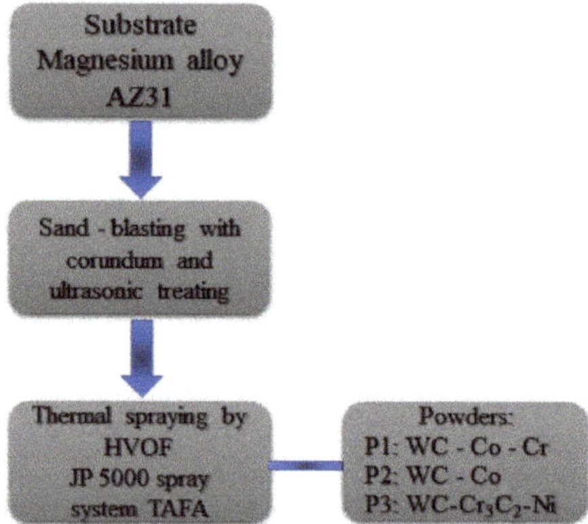

Figure 1. The schematic diagram of the HVOF coating manufacturing process.

Table 1. Spraying parameters of cermet coatings.

Oxygen Flow Rate, L/min	900
Kerosene flow rate, L/h	26.1
N_2 flow rate, L/min	12
Powder feed rate, g/min	70
Water flow rate, L/min	23
Spray distance, mm	360

2.3. Coatings' Characterization

Microscopic investigations and fracture morphology were carried out by scanning electron microscope (Supra 35, Zeiss, Oberkochen, Germany) with secondary electron and backscattered detectors. The chemical composition was analyzed by EDS (energy dispersive X-ray spectroscopy) (Supra 35, Zeiss, Oberkochen, Germany). The area EDS measurements were randomly distributed in the coating. It was carried out in one sample in 10 areas, and three coatings were tested. TEM investigations were undertaken with a field emission transmission electron microscope (S/TEM Titan 80-300 from FEI, Hillsboro, OR, USA) with a super twin-lens operated at 300 kV and equipped with an annular dark-field detector. A focused ion beam method (FIB) prepared thin foils for TEM analysis. The lamella extraction was performed on the SEM/Ga-FIB FEI Helios NanoLab 600i (FEI, Brno, Czech Republic) device, while the thinning and removal of the amorphous layer were performed on SEM/Xe-PFIB FEI Helios G4 PFIB CXe (FEI, Brno, Czech Republic). The coatings' cross-sections were observed by a Keyence VHX6000 (Keyence International, Mechelen, Belgium) microscope. Based on these images, at 2000× magnification, the porosity of sprayed coatings was estimated according to ASTM E2109-01 standards. Image J open-source software (1.50i version) was used to calculate porosity. At the same magnification, 10 measurements carried out at random locations along the coatings cross-sections were taken into account to calculate the average thickness value and standard deviation.

Microhardness of manufactured coatings were estimated with Vickers indenter under the load of 2.94 N (HV0.3) using the HV1000 hardness tester (Sinowon Innovation Metrology), according to the ISO 4516 standard. Ten imprints at the cross-sections of each coating were made to calculate the average value and standard deviation.

XRD studies were performed using the D8 Advance diffractometer (Bruker, Karlsruhe, Germany) with a Cu-Kα cathode (λ = 1.54 Å) operating at 40 kV voltage and 40 mA current. The scan rate was 0.60°/min with a scanning step of 0.02° in the range of 20° to 120° 2Θ.

Identification of fitted phases was performed using the DIFFRAC.EVA program using the ICDD PDF#2 database, while the exact lattice parameters of the fitted phase were calculated using Rietveld refinement in the TOPAS 6 program, based on the Williamson–Hall theory [21–23]. The pseudo-Voigt function described diffraction line profiles at the Rietveld refinement. The Rwp (weighted-pattern factor) and S (goodness-of-fit) parameters were used as numerical criteria of the quality of the fit of calculated to experimental diffraction data.

Residual stress analyses (RSA) were performed using the iso-inclination mode of the D8 Advance diffractometer (Bruker, Karlsruhe, Germany) with the use of the (211) peak of the WC phase, according to EN-15305 standards. RSA measurements were performed at six different φ angles ($0°$, $45°$, $90°$, $135°$, $180°$, $225°$) to obtain a reliable stress mode [24,25]. Results were evaluated using the DIFFRAC.LEPTOS program, and all peaks were fitted using standard fit, while the applied stress mode was established as biaxial [24,25] with consideration of shear stress contribution, due to low hardness and elastic modulus of substrate material, which is mainly omitted in literature. The following material parameters were used for residual stress analysis: Young's modulus 600 GPa and Poisson ration 0.20, which gives $S_1 = -3.333\ 10^{-7}$ MPa and $1/2S_2 = 2.000\ 10^{-6}$ MPa^{-1} and are in agreement with literature data [25]. The 45 MPa limit was used as a stress-free WC material, while a 22.5 MPa limit was used for shear stress contribution.

3. Results and Discussion

3.1. Feedstocks

The morphology of feedstock powders is given in Figure 2. All powders have similar particles size and spherical shapes. This is important from a technological point of view because it provides suitable flowability of the powder particles during spraying.

Figure 2. Morphology of: (**a**) P1, (**b**) P2, and (**c**) P3 powder (SEM).

3.2. Microstructure of the Coatings

The detailed examination at high magnification (Figure 3b,d,f) revealed a dense structure with fine pores (much lower than 1 μm) and a typical low porosity level. This dense structure is due to the inherent characteristic of the HVOF process (mainly the high kinetic energy of the particles). In Figure 3b,d, the hard particles are homogeneously distributed in the cobalt matrix, whereas for C3 (Figure 3f), there are some areas of nickel matrix islands without hard particles. This is similar to the phenomenon reported by [26–28].

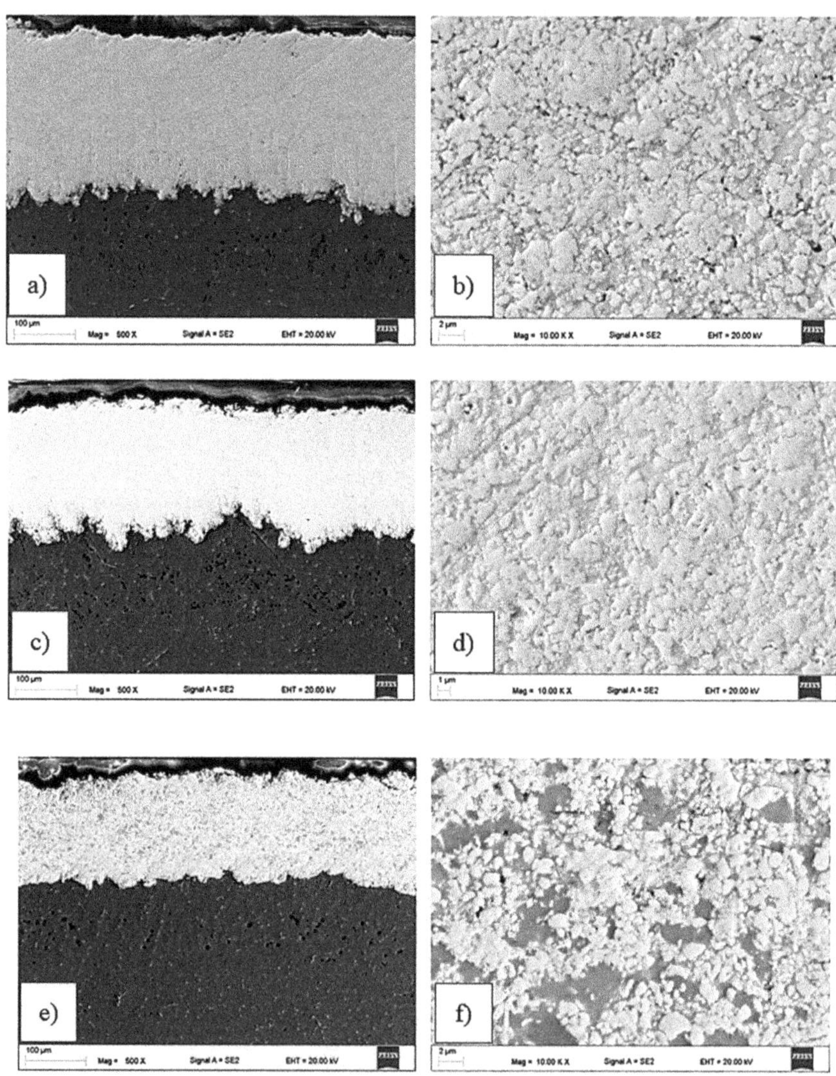

Figure 3. SEM images of cross-sections of HVOF-sprayed coatings: (**a,b**) C1, (**c,d**) C2, (**e,f**) C3 ((**a,c,e**)—mag. 500×; (**b,d,f**)—mag. 10,000×).

The examination of low-magnification polished cross-sections (Figure 3a,c,e) showed a relatively smooth, dense and homogeneous structure of HVOF-sprayed coatings. The microstructure is typical for thermal spraying coatings. The interface between cermet coating and AZ31 substrate was clear in all samples, and no evidence of delamination was observed.

The image analysis results in coatings' porosity determination are collected in Table 2, and the results are quite similar. The lowest porosity value for the C3 sample could be related to lower hardness and better porosity filling by nickel than cobalt. In their work, Yao et al. [13] reported that the coating porosity is related to the powder composition and oxygen flow rate, and decreased with the oxygen flow increase. In Table 2, the coatings' thickness values and microhardness (HV0.3) are presented.

Table 2. Average thickness, porosity and microhardness of deposited coatings.

	C1	C2	C3
Thickness, μm	279 ± 24	206 ± 8	177 ± 20
Porosity, vol %	2.9 ± 0.7	2.6 ± 0.5	1.9 ± 0.5
HV0.3	1198 ± 195	1269 ± 167	989 ± 124

The chemical composition of the C1, C2 and C3 sprayed coatings is presented in Figure 4, and the chemical element distributions in the micro areas are shown in Figures 5–7.

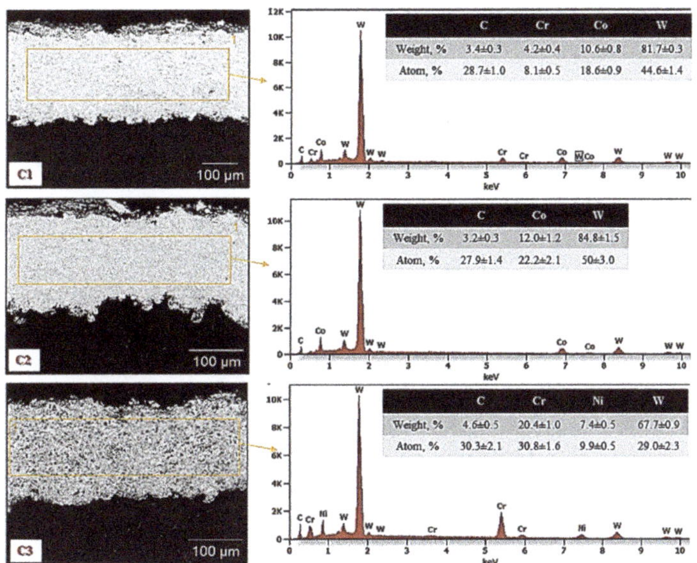

Figure 4. The chemical composition of the HVOF-sprayed coatings.

The map analysis revealed the areas with a higher concentration of individual chemical elements in the analyzed coatings. In the case of the sample C1, the highest concentration of tungsten (light area in Figure 5a and purple in Figure 5e) and Cr (black area in Figure 5a and yellow area in Figure 5c), as well as Co (a gray area in Figure 5a and yellow Figure 5d), was observed. The increased share of these elements corresponds to tungsten carbide and a metallic CoCr matrix, respectively. Analysis of the distribution of elements in the area of the C2 coating showed an even distribution of tungsten carbide (light area in Figure 6a and purple area in Figure 6d) in the Co matrix (a gray area in Figure 6a and green in Figure 6c). In sample C3, areas with a large mass fraction of chromium (black area in Figure 7a and yellow in Figure 7c) and tungsten (bright area in Figure 7a and purple in Figure 7d), as well as a nickel (a gray area in Figure 7a and blue in Figure 7e) were observed, which correspond to carbides and a metallic Ni matrix used during the process.

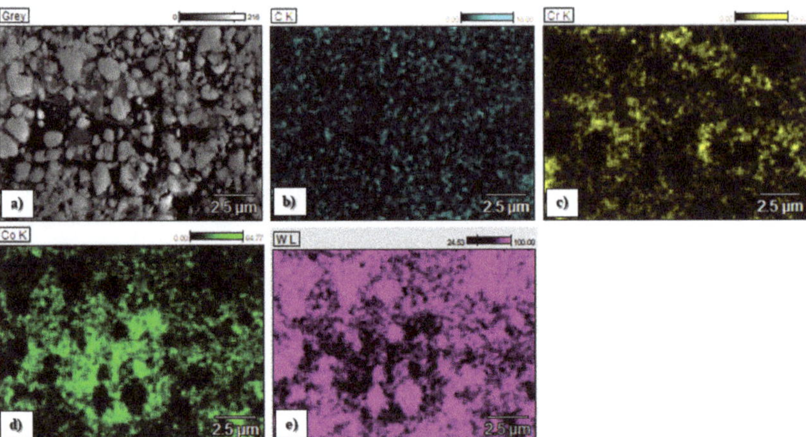

Figure 5. Elemental distribution maps of spraying elements in the analyzed area of the C1 coating obtained during thermal spraying: (**a**)—central part of the layer, (**b**)—map of the carbon, (**c**)—map of the chromium, (**d**)—map of the cobalt, (**e**)—map of the tungsten.

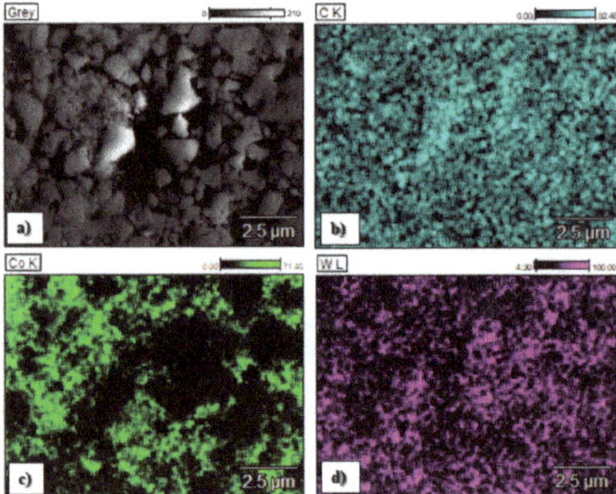

Figure 6. Elemental distribution maps of spraying elements in the analyzed area of the C2 coating obtained during thermal spraying: (**a**)—central part of the layer, (**b**)—map of the carbon, (**c**)—map of the cobalt, (**d**)—map of the tungsten.

The microhardness of the coatings depends on several factors, including porosity, carbide particle size and degree of decarburization. Process parameters (among others, spray distance) determines the temperature of the particles during spraying, which has a significant effect on hardness value. The coatings' hardness increases with increasing particle temperature. It could be explained that decarburization and dissolution of W, Cr and C in the metal matrix (CoCr) take place at a higher temperature. Consequently, this leads to the hardness increasing. In general, the matrix hardness is higher and also W_2C hard carbides are formed during spraying, which results in coatings' hardness increasing. A similar value of microhardness and porosity for the C3 sample was observed in other investigations [29] and C1 and C2 samples [30]. Yuan et al. reported that the physical features such as morphology and density of the WC-Co powders play a very important role in determining the microhardness of the coatings by affecting the coating porosity and extent of decarburization [31].

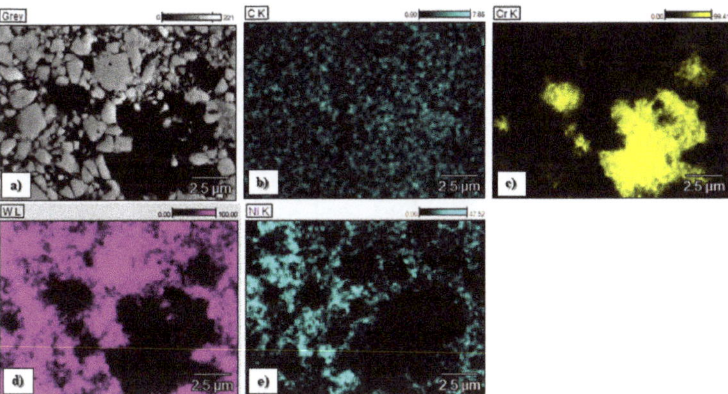

Figure 7. Elemental distribution maps of spraying elements in the analyzed area of the C3 coating obtained during thermal spraying: (**a**)—central part of the layer, (**b**)—map of the carbon, (**c**)—map of the chromium, (**d**)—map of the tungsten, (**e**)—map of the nickel.

Results of TEM analysis are divided into three parts, according to the type of coating material. The C1 sample analysis revealed that coating contains a matrix and two-particle types (Figure 8). The particles marked with red arrows and named with the letter A (Figure 8a) are larger (1–2 µm) than others and irregular shapes. The analysis of the chemical composition (Figure 8b) confirmed the presence of tungsten (100 at. %).

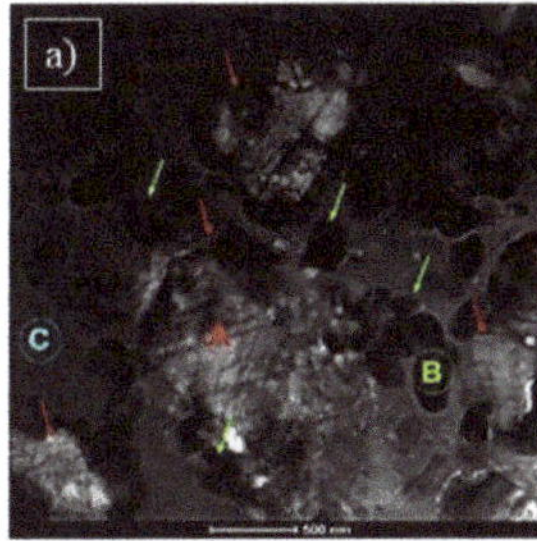
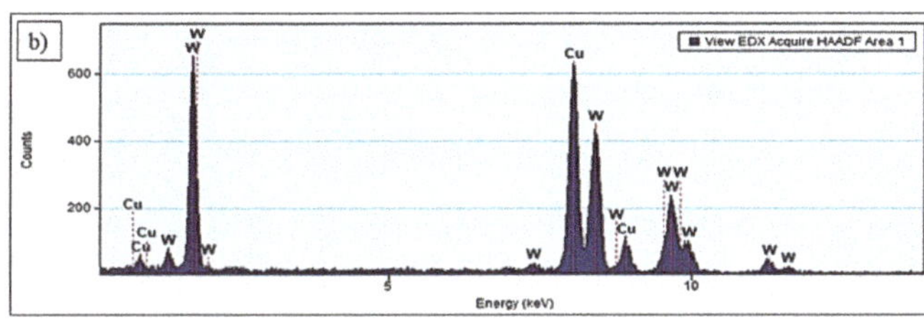

Figure 8. STEM micrograph in BF of sample C1 (**a**); the results of chemical analysis marked with letter A (**b**).

The spectrum of the energy-dispersive X-ray spectroscopy (EDS) also shows the signal from Cu, which was omitted in the analysis. It could results e.g., from holder and pole pieces. EDS technique has a limitation in the study of light elements (Z < 11). Based on the obtained spectrum, their presence (especially carbon) in the tested material cannot be excluded. Electron diffraction SAED (Figure 9a) identified the particles as WC, the hexagonal phase and the P-6m2 space group [32].

The second type of precipitation, marked with green arrows, has a more regular and spherical shape. It is occurring in the matrix or around the WC shown earlier. Diffraction investigation showed the W_2C phase (Figure 9b), where Co and Cr replace some W atoms by the structure. The W_2C phase belongs to the hexagonal system, space group P-3m1 [33]. Process parameters significantly influence the microstructure of the coating. Especially important is the point when high temperature affects particles of feedstock material. The 2000–3000 K W_2C phase is more stable than the WC one in the temperature range.

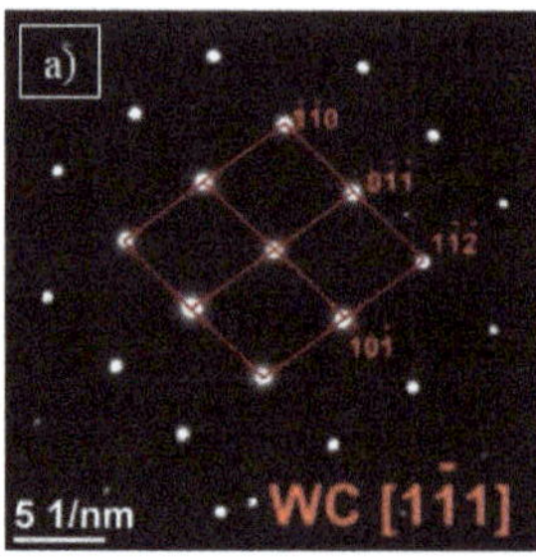
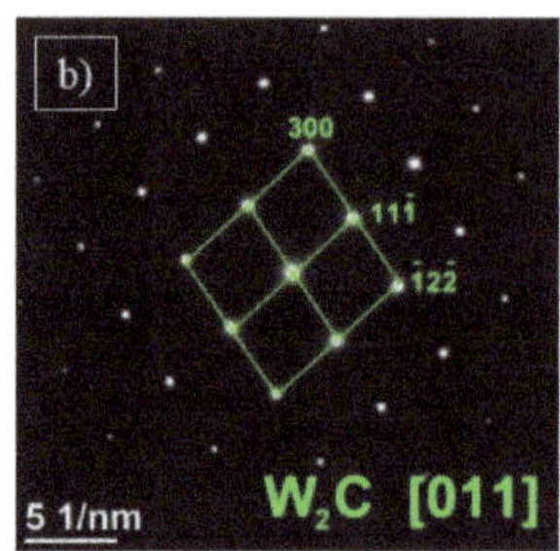
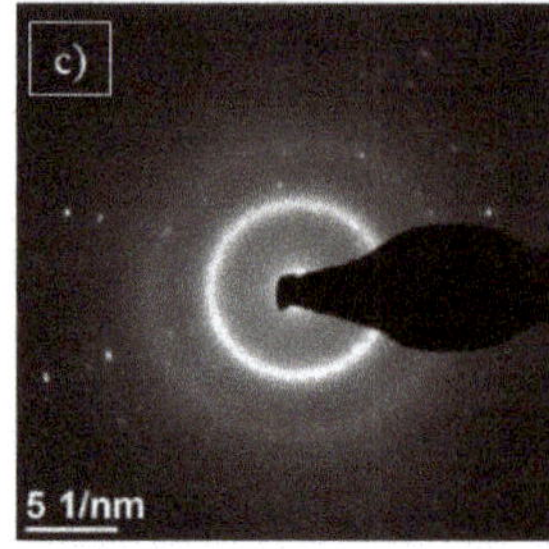

Figure 9. SAED diffraction analysis of sample C1; for the area marked on Figure 8a with letter A: WC [111] (**a**); with letter B: W2C [11] (**b**); with letter C as the matrix (**c**).

Moreover, the higher temperature of the particles causes decarburization of the WC phase and leads to W_2C precipitation, which is a harder and more brittle phase. Myalska et al., in their work, provided a detailed explanation of this phenomenon [34]. It was confirmed by SAED electron diffraction that the matrix is amorphous (Figure 9c).

In the C2 sample, the precipitates marked with the letter B (Figure 10a) occur around larger WC carbides. They are irregular in shape and composed of many smaller grains. EDS analysis (Figure 10b) showed a high proportion of W (72 at. %) and Co (28 at. %). The SAED electron diffraction, performed for the area marked as B, confirmed the polycrystalline structure of the precipitation (Figure 11a). SAED diffraction identified the cubic tungsten with space group Im-3m [35]. The matrix has an amorphous structure, which may result from the high cooling rate of the particles while striking the substrate surface. It was confirmed by SAED electron diffraction (Figure 11b). The analysis of the chemical composition of the matrix (Figure 10c) confirmed the content of Co (58 at. %) and W (42 at. %). A similar morphology has been observed by other researchers [32,36,37].

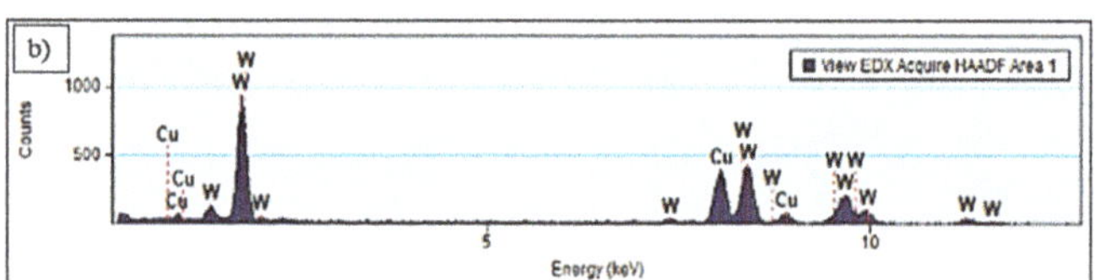
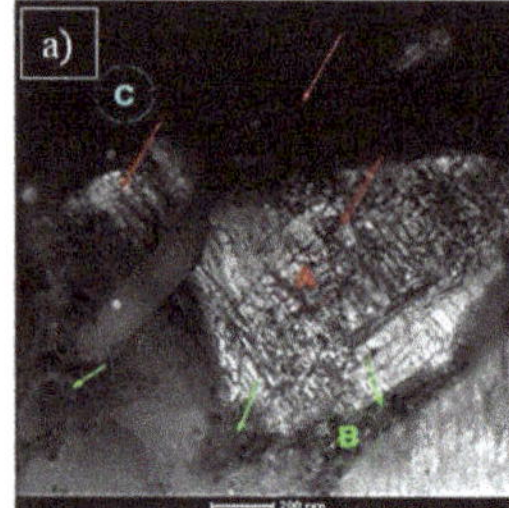
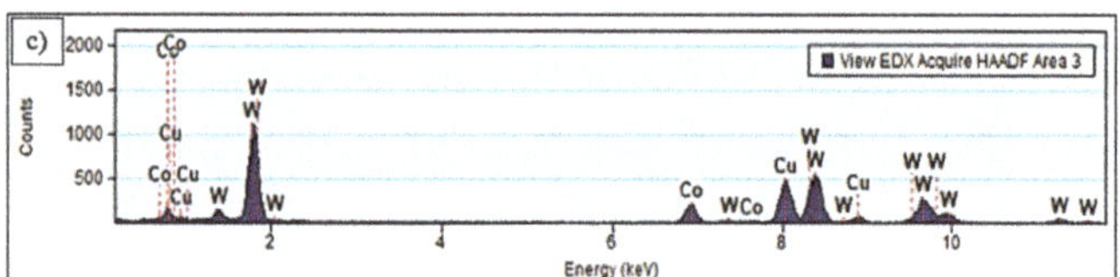
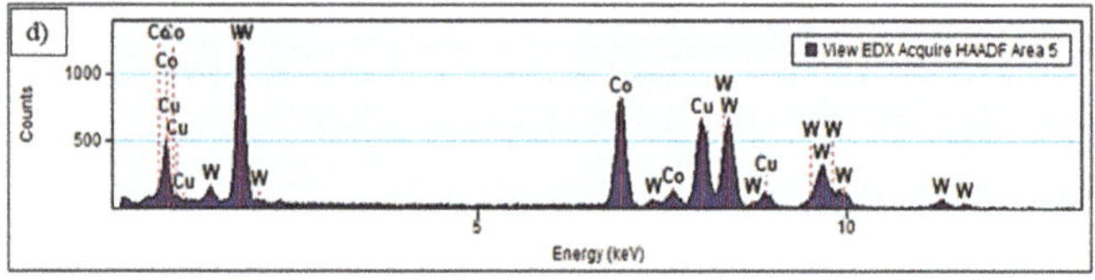

Figure 10. STEM micrograph in BF of sample C2 (**a**); (**b**) the results of chemical analysis for precipitation marked with letter A and (**c**) with letter B, (**d**) the results of chemical analysis for matrix marked with letter C.

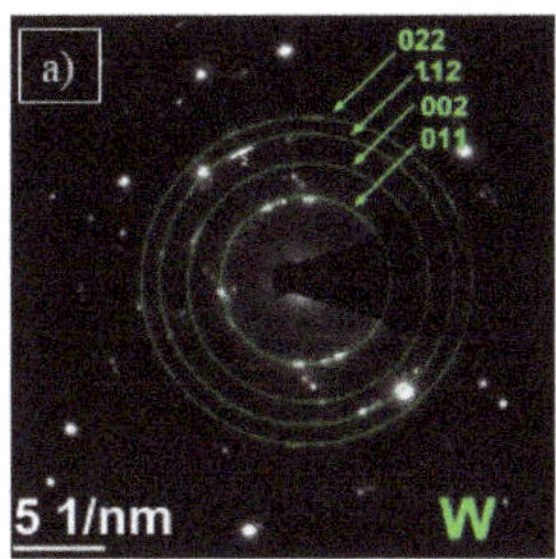
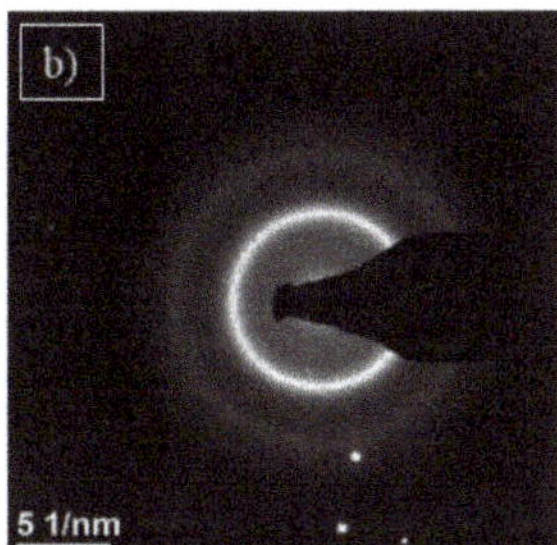

Figure 11. The SAED electron diffraction of sample C2, performed for the area marked as B (**a**) and amorphous matrix (**b**).

STEM analysis of C3 coating showed a matrix and two types of precipitations (Figure 12a). The ones marked by red arrows were identified as WC. They are characterized by irregular shapes and varied sizes (from 200 nm up to 1 μm). Chemical composition analysis (Figure 12b) confirmed the presence of W (100 at. %) inside these carbides. SAED pattern of WC phase and its solution with direction [210] is presented in Figure 13a. Separations marked by green arrows (Figure 12a) were identified as chromium carbide Cr_3C_2. They are rounded with size c.a. several hundred nanometers. Chemical composition analysis (Figure 12c) confirmed the presence of chromium (94 at. %) and tungsten (6 at. %). The SAED pattern of Cr_3C_2 is presented in Figure 13b. It is an orthorhombic space group Pnma [38]. SAED electron diffraction confirmed an amorphous structure of the matrix (Figure 13c). It could also be confirmed by uniform contrast in STEM bright-field images (Figure 12a).

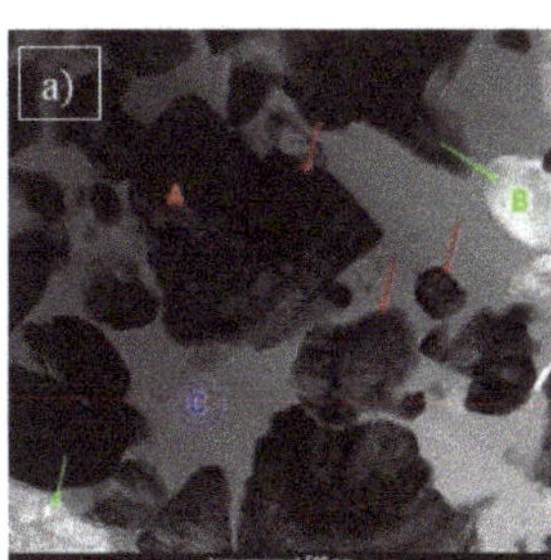
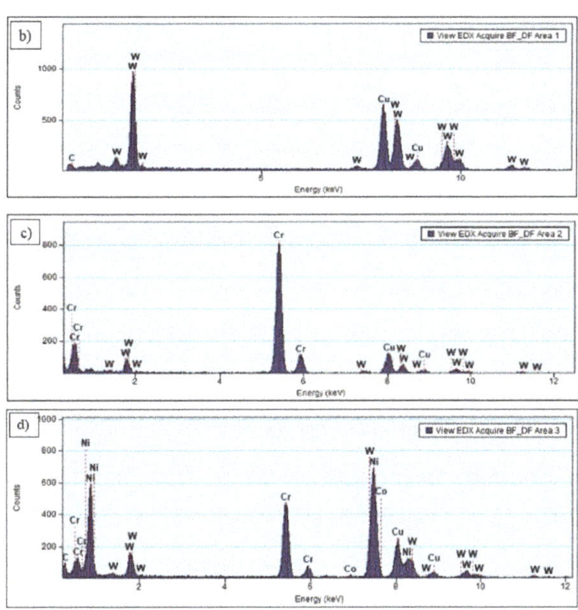

Figure 12. STEM micrograph in BF of sample C3 (**a**); the results of the chemical analysis performed for the area marked as A (**b**) and B (**c**) and C (**d**).

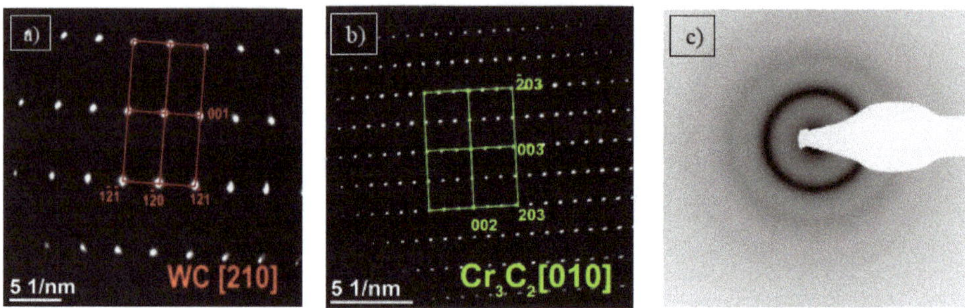

Figure 13. SAED diffraction analysis of sample C3; for the area marked on Figure 12a with letter A: WC with direction [210] (**a**); with the letter B: Cr_3C_2 with direction [10] (**b**); with letter C: amorphous matrix (**c**).

3.3. Phase Composition

Phase compositions of feedstock powders in the delivery conditions are shown in Figure 14. As expected, mainly the WC phase was detected. This phase composition was confirmed by other authors working with similar powders [34,36,39].

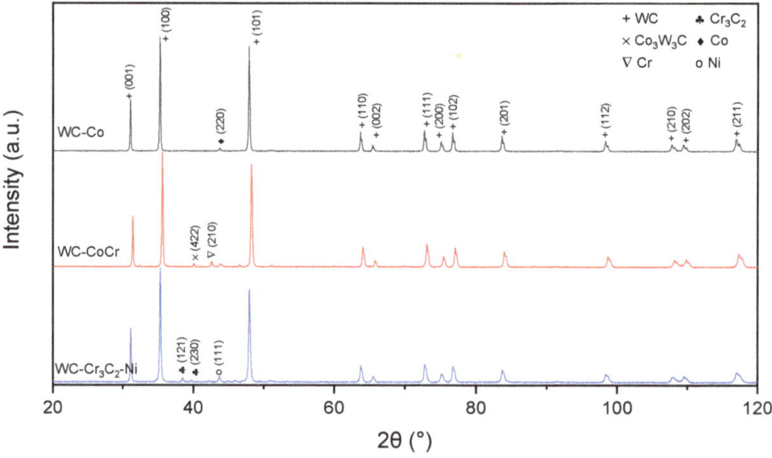

Figure 14. XRD patterns of feedstock powders.

HVOF spraying resulted in changes in the coatings' phase composition. The phase composition of cermet coatings consists of hexagonal WC (PDF#00-061-0244), hexagonal W_2C carbide (PDF#00-035-0776), hexagonal Co (PDF#03-065-9722) and a cubic solid solution of W in Co with composition $Co_{0.9}W_{0.1}$ (PDF#03-065-9928). Additionally, in the C3 coating, the Cr_3C_2 (PDF#00-035-0804) and Cr_7C_3 (PDF#00-036-1482) carbides have been identified. Moreover, crystallites were detected in the C2 coating in the presence of cubic W (PDF#00-001-1204), which is in agreement with literature data [40]. It should be noted that no peaks coming either from the WC_{1-x} phase or from the Co_3W_3C or Co_6W_6C phases were found in the coatings, but those phases were identified in other papers and feedstock powders (Figure 15) [24,41–43]. During the deposition process of cermet coating, high temperature and oxygen lead to the decarburization process of carbides; thus, the formation of new carbides was detected instead of metal oxidation [24,41–44].

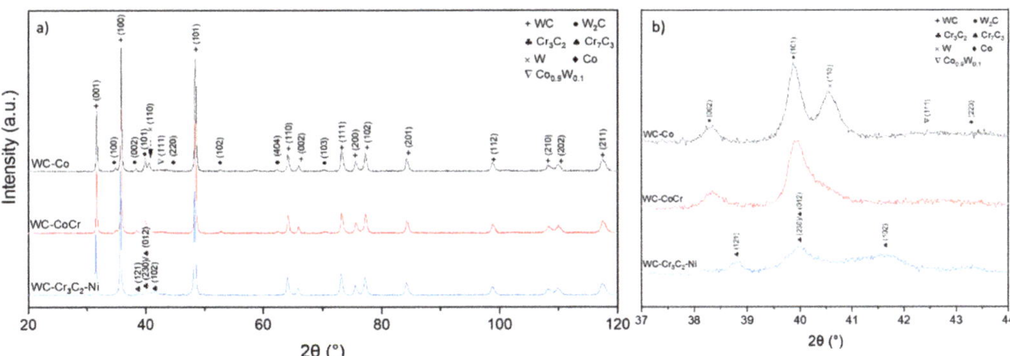

Figure 15. Typical XRD patterns obtained from cermet coatings sprayed onto magnesium substrate: (**a**) whole pattern, (**b**) magnification of chromium carbides region.

3.4. Residual Stress Analysis

The presence of residual stress in material might implicate unwanted effects during the exploitation of elements, such as cracks or coating delamination from a substrate. Thus, it is important to obtain a coating with low residual stress. Stress generation might have two natures:

(a) Thermal—during spraying, a high temperature is used, resulting in a change in phase composition and generating thermal stress in the main phase, which is used in WC coatings.

(b) Impact—in HVOF, a high speed of particles is achieved. When hot particles hit the substrate, additional stress is generated, which might have both linear and shear components.

Therefore, the generation of linear stress most likely has a thermal nature, related to thermal expansion of WC, while the generation of shear stress most likely has an impact nature. Such a phenomenon was not described earlier as an effect of relatively high hardness of used substrate materials (steel, cast iron, nickel alloys, etc.), resulting in cracking of WC particles during impact. Used magnesium alloy has the lowest Young's modulus and hardness of all engineering alloys and might deform during the HVOF process.

In all cermet coatings, linear stress shows a compressive nature (Figure 16, Table 3). In C2 and C3, residual stresses have a similar value, with a different part of shear stress contribution, while in C1, linear and shear stresses are almost even. However, shear stress contribution in cermet is very high, most likely as an effect of spraying onto magnesium substrate, which might deform during the HVOF process. In C2, shear stress is higher than in C1 (Figure 16), most likely due to the presence of chromium in C2 cermet, which might partially absorb energy during the HVOF process. The lowest shear stress contribution was detected in C3 coatings, most likely as an effect of the Cr_3C_2 carbide presence in powder, which may also absorb the impact energy (see Table 3).

Due to the complex nature of presence stress, it is nearly impossible to determine the order of the overall stress. However, compressive stress in HVOF coatings is unlikely to be eliminated; thus, only shear stress should be considered. Furthermore, it is a new aspect in HVOF-derived coatings. The C3 coating should be considered the best one because it is almost a shear stress-free material. On the other hand, in C2 coatings, high shear stress might have a negative impact on tribological properties, resulting in cracking of WC.

Observed results of residual stress (Figure 16) in the WC phase are in agreement with literature data describing residual stress in cermet coatings with similar thickness [24,25,45–47]. Oladijo et al. [25] observed residual stress of WC-Co coatings thermally sprayed onto different metal substrates, and residual stress was in the range of -130 MPa (an aluminum substrate) to -50 MPa (brass substrate). Książek et al. [46] calculated stress in a Cr_3C_2-NiCr coating in the range of −230 to −420 MPa, but Cr_3C_2 carbide has a much lower Young's modulus than WC carbide. On the other hand, Masoumi et al. [45] detected residual stress around −130 MPa in 400 μm thick WC-Co-Cr coating, but they used E = 316 GPa in their calculations. Santana et al. [24] show that in WC-Co coatings with thickness in the range of 300–450 μm, residual stresses are −180 to −220 MPa.

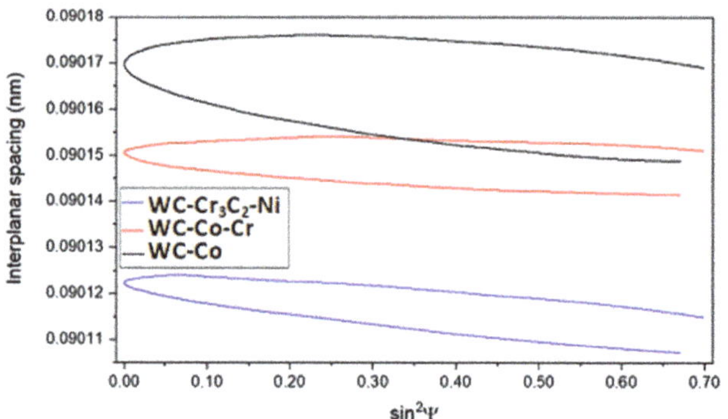

Figure 16. Residual stress diagrams were obtained for examined cermet coatings.

Table 3. Mean residual stress values in various cermet coatings sprayed onto AZ31 magnesium substrate.

Sample	Residual Stress, MPa	
	Linear Stress	**Shear Stress**
C1	−65.0 ± 28.8	56.5 ± 25.1
C2	−109.3 ± 29.3	86.1 ± 33.4
C3	−113.8 ± 3.7	27.6 ± 11.5

4. Conclusions

This work was focused on HVOF spraying with feedstock powders WC-Co-Cr, WC-Co and WC-Cr$_3$C$_2$-Ni. The coatings were studied in terms of the influence of feedstock powder content on the microstructure, phase composition and residual stress.

It can be summarized that:

1. All the coatings revealed relatively smooth, dense and homogeneous structure. In all samples, the interface between the coating and magnesium alloy substrate was clear, and no evidence of delamination was observed.
2. The porosity in all of the investigated coatings was quite similar (in vol %)—2.9 ± 0.7 for C1, 2.6 ± 0.5 for C2 and 1.9 ± 0.5 for C3—and the thickness was in the range of 177 ± 20 μm to 279 ± 24 μm. In addition, the lowest microhardness (HV0.3) was observed for the C3 sample (989 ± 124), while the highest was observed for the C2 (1269 ± 167).
3. Based on the results of the TEM analysis, the C1 coating contains an amorphous matrix and two types of precipitates: WC and W_2C. The C2 coating contains a matrix with an amorphous structure and precipitation of WC. Finally, analysis of the C3 coating showed a matrix with an amorphous structure and two types of precipitations: WC and Cr_3C_2.
4. XRD studies showed that phase composition of cermet coatings consists of hexagonal WC, hexagonal W_2C carbide, hexagonal cobalt and a cubic solid solution of tungsten in cobalt with composition $Co_{0.9}W_{0.1}$. Additionally, in the WC-Cr$_3$C$_2$-Ni coating, Cr_3C_2 and Cr_7C_3 carbides were identified.

5. In all cermet coatings, linear stress shows a compressive nature. However, in C2 and C3, residual stress has a similar value, with a different part of shear stress contribution, while in C1, both linear and shear stresses are almost even and lower than in other coatings.

Author Contributions: Conceptualization, E.J.; methodology, E.J., A.T. and P.N.; investigation, E.J., L.Ł., A.T., M.G., W.P. and P.N.; writing—original draft preparation, E.J. and L.Ł.; writing—review and editing, E.J., L.Ł., A.T., M.G. and W.P. All authors have read and agreed to the published version of the manuscript.

Funding: The research was financed in the framework of scientific activity No. DEC-2019/03/X/ST5/00830, funded by the Polish Ministry of Science and Higher Education.

Institutional Review Board Statement: Not applicable.

Informed Consent Statement: Not applicable.

Data Availability Statement: Not applicable.

Conflicts of Interest: The authors declare no conflict of interest.

References

1. Hejwowski, T. Characteristics of the Thermal Spraying Process: Modern Heat Applied Coatings Resistant to Abrasive and Erosive Wear. Lublin University of Technology, Centre for Publishing and Digital Library. 2013. Available online: http://bc.pollub.pl/dlibra/publication/5141/edition/4059/content?ref=desc (accessed on 16 September 2021).
2. Czupryński, A. Flame Spraying of Aluminum Coatings Reinforced with Particles of Carbonaceous Materials as an Alternative for Laser Cladding Technologies. *Materials* **2019**, *12*, 3467. [CrossRef]
3. Espallargas, N. 1—Introduction to thermal spray coatings. In *Future Development of Thermal Spray Coatings—Types, Designs, Manufacture and Applications*; Woodhead Publishing, Elsevier Ltd.: Oxford, UK, 2015. [CrossRef]
4. Mishra, T.K.; Kumar, A.; Sinha, S.K. Experimental investigation and study of HVOF sprayed WC-12Co, WC-10Co-4Cr and Cr_3C_2-25NiCr coating on its sliding wear behaviour. *Int. J. Refract. Met. Hard Mater.* **2021**, *94*, 105404. [CrossRef]
5. Barbezat, G. Application of thermal spraying in the automobile industry. *Surf. Coat. Technol.* **2006**, *201*, 2028–2031. [CrossRef]
6. Wielage, B.; Wank, A.; Pokhmurska, H.; Grund, T.; Rupprecht, C.; Reisel, G.; Friesen, E. Development and trends in HVOF spraying technology. *Surf. Coat. Technol.* **2006**, *201*, 2032–2037. [CrossRef]
7. Vuoristo, P. Thermal Spray Coating Processes. *Compr. Mater. Proc.* **2014**, *4*, 229–276. [CrossRef]
8. Sobolev, V.V.; Guilemany, J.M.; Nutting, J. *HVOF Spraying Systems in High Velocity Oxy-Fuel Spraying—Theory: Structure-Property Relationships and Applications*; Maney Materials Science IOM3; The Institute of Materials, Minerals and Mining: London, UK, 2004; ISBN 1-902653-72-6.
9. Pawłowski, L. *Thermal Spraying Techniques in the Science and Engineering of Thermal Spray Coatings*, 2nd ed.; John Wiley & Sons: Chichester, UK, 2008; ISBN 978-0-471-49049-4.
10. Qiao, L.; Wu, Y.; Hong, S.; Long, W.; Cheng, J. Wet abrasive wear behavior of WC—Based cermet coatings prepared by HVOF spraying. *Ceram. Int.* **2021**, *47*, 1829–1836. [CrossRef]
11. Zhang, F.G. Friction and wear behavior of WC/Ni cemented carbide tool material irradiated by high -intensity pulsed electron beam. *Ceram. Int.* **2019**, *45*, 15327–15333. [CrossRef]
12. Zhang, M.; Li, M.; Chi, J.; Wang, S.; Yang, S.; Yang, J.; Wei, Y. Effect of Ti on microstructure characteristics, carbide precipitation mechanism and tribological behavior of different WC types reinforced Ni-based gradient coating. *Surf. Coat. Technol.* **2019**, *374*, 645–655. [CrossRef]
13. Yao, H.-L.; Yang, C.; Yi, D.-L.; Zhang, M.-X.; Wang, H.-T.; Chen, Q.-Y.; Bai, X.-B.; Ji, G.-C. Microstructure and mechanical property of high velocity oxy-fuel sprayed WCCr$_3$C$_2$-Ni coatings. *Surf. Coat. Technol.* **2020**, *397*, 126010. [CrossRef]
14. Berger, L.M. Application of hardmetals as thermal spray coatings. *Int. J. Refract. Met. Hard Mater.* **2015**, *49*, 350–364. [CrossRef]
15. Oksa, M.; Turunen, E.; Suhonen, T.; Varis, T.; Hannula, S.-P. Optimization of high velocity oxy-fuel sprayed coatings: Techniques. Materials and Applications. *Coatings* **2011**, *1*, 17–52. [CrossRef]
16. Sun, Y.; Yang, R.; Xie, L.; Wang, W.; Li, Y.; Wang, S.; Li, H.; Zhang, J. Interfacial bonding mechanism and properties of HVOF-sprayed Fe-based amorphous coatings on LA141 magnesium alloy substrate. *Surf. Coat. Technol.* **2021**, *426*, 127801. [CrossRef]
17. García-Rodríguez, S.; Torres, B.; Pulido-González, N.; Otero, E.; Rams, J. Corrosion behavior of 316L stainless steel coatings on ZE41 magnesium alloy in chloride environments. *Surf. Coat. Technol.* **2019**, *378*, 124994. [CrossRef]
18. Mardali, M.; Salimijazi, H.; Karimzadeh, F.; Luthringer-Feyerabend, B.J.; Blawert, C.; Labbaf, S. Fabrication and characterization of nanostructured hydroxyapatite coating on Mg-based alloy by high-velocity oxygen fuel spraying. *Ceram. Int.* **2018**, *44*, 14667–14676. [CrossRef]

19. García-Rodríguez, S.; López, A.J.; Bonache, V.; Torres, B.; Rams, J. Fabrication, Wear, and corrosion resistance of HVOF sprayed WC-12Co on ZE41 magnesium Alloy. *Coatings* **2020**, *10*, 502. [CrossRef]

20. Aulakh, S.S.; Kaushal, G. Laser texturing as an alternative to grit blasting for improved coating adhesion on AZ91D magnesium alloy. *Trans. IMF* **2019**, *97*, 100–108. [CrossRef]

21. Rietveld, H.M. Line profiles of neutron powder-diffraction peaks for structure refinement. *Acta Crystallogr.* **1967**, *22*, 151–152. [CrossRef]

22. Rietveld, H.M. A profile refinement method for nuclear and magnetic structures. *J. Appl. Crystallogr.* **1969**, *2*, 65–71. [CrossRef]

23. Karolus, M.; Łągiewka, E. Crystallite size and lattice strain in nanocrystalline Ni-Mo alloys studied by Rietveld refinement. *J. Alloys Compd.* **2004**, *367*, 235–238. [CrossRef]

24. Santana, Y.Y.; Renault, P.O.; Sebastiani, M.; La Barbera, J.; Lesage, J.; Bemporad, E.; Le Bourhis, E.; Puchi-Cabrera, E.; Staia, M. Characterization and residual stresses of WC-Co thermally sprayed coatings. *Surf. Coat. Technol.* **2008**, *202*, 4560–4565. [CrossRef]

25. Oladijo, O.P.; Venter, A.M.; Cornish, L.A.; Sacks, N. X-ray diffraction measurements of residual stress in WC-Co thermally sprayed coatings onto metal substrates. *Surf. Coat. Technol.* **2012**, *206*, 4725–4729. [CrossRef]

26. Hong, S.; Wu, Y.; Wang, B.; Lin, J. Improvement in Tribological Properties of Cr12MoV Cold Work Die Steel by HVOF Sprayed WC-Co-Cr Cermet Coatings. *Coatings* **2019**, *9*, 825. [CrossRef]

27. Mayrhofer, E.; Janka, L.; Mayr, W.P.; Norpoth, J.; Ripoll, M.R.; Gröschl, M. Cracking resistance of Cr_3C_2-NiCr and WC-Cr_3C_2-Ni thermally sprayed coatings under tensile bending stress. *Surf. Coat. Technol.* **2015**, *281*, 169–175. [CrossRef]

28. Wang, H.; Qiu, Q.; Gee, M.; Hou, C.; Liu, X.; Song, X. Wear resistance enhancement of HVOF-sprayed WC-Co coating by complete densification of starting powder. *Mater. Des.* **2020**, *191*, 108586. [CrossRef]

29. Goyal, K.; Sapate, S.G.; Mehar, S.; Vashishtha, N.; Bagde, P.; Rathod, A.B.; A Bagde, P. Tribological properties of HVOF sprayed WC-Cr_3C_2-Ni coating. *Mater. Res. Express* **2019**, *6*, 106415. [CrossRef]

30. Richert, M.W. The wear resistance of thermal spray the tungsten and chromium carbides coatings. *J. Achiev. Mater. Manuf. Eng.* **2011**, *47*, 177–184.

31. Yuan, J.; Ma, C.; Yang, S.; Yu, Z.; Li, H. Improving the wear resistance of HVOF sprayed WC-Co coatings by adding submicron-sized WC particles at the splats' interfaces. *Surf. Coat. Technol.* **2016**, *285*, 17–23. [CrossRef]

32. Litasov, K.; Shatskiy, D.A.; Fei, Y.; Suzuki, A.; Ohtani, E.; Funakoshi, K. Pressure-volume-temperature equation of state of tungsten carbide to 32 GPa and 1673 K. *Jpn. J. Appl. Phys.* **2010**, *8*, 053513. [CrossRef]

33. Epicier, T.; Dubois, J.; Esnouf, C.; Fantozzi, G.; Convert, P. Neutron powder diffraction studies of transition metal hemicarbides M_2C_{1-x}—II. In situ high temperature study on W_2C_{1-x} and Mo_2C_{1-x}. *Acta Metall. Mater.* **1988**, *36*, 1903–1921. [CrossRef]

34. Myalska, H.; Swadźba, R.; Rozmus, R.; Moskal, G.; Wiedermann, J.; Szymański, K. STEM analysis of WC-Co coatings modified by nano-sized TiC and nano-sized WC addition. *Surf. Coat. Technol.* **2017**, *318*, 279–287. [CrossRef]

35. Dubrovinsky, L.S.; Saxena, S.K. Thermal expansion of periclase (MgO) and tungsten (W) to melting temperatures. *Phys. Chem. Miner.* **1997**, *24*, 547–550. [CrossRef]

36. Hong, S.; Wu, Y.P.; Gao, W.W.; Wang, B.; Guo, W.M.; Lin, J.R. Microstructural characterization and microhardness distribution of HVOF sprayed WC-10Co-4Cr coating. *Surf. Eng.* **2014**, *30*, 53–58. [CrossRef]

37. Żórawski, W. The microstructure and tribological properties of liquid-fuel HVOF sprayed nanostructured WC–12Co coatings. *Surf. Coat. Technol.* **2013**, *220*, 276–281. [CrossRef]

38. Rundqvist, S.; Runnjso, G. Crystal structure refinement of Cr_3C_2. *Acta. Chem. Scand.* **1969**, *23*, 1191–1199. [CrossRef]

39. Digvijay, G.B.; Walmik, S.R. Tribo-behavior of APS and HVOF sprayed WC-Cr_3C_2-Ni coatings for gears. *Surf. Eng.* **2021**, *37*, 80–90. [CrossRef]

40. Xu, L.; Song, J.; Zhang, X.; Deng, C.; Liu, M.; Zhou, K. Microstructure and Corrosion Resistance of WC-Based Cermet/Fe-Based Amorphous Alloy Composite Coatings. *Coatings* **2019**, *8*, 393. [CrossRef]

41. Myalska, H.; Lusvarghi, L.; Bolelli, G.; Sassatelli, P.; Moskal, G. Tribological behavior of WC-Co HVAF-sprayed composite coatings modified by nano-sized TiC addition. *Surf. Coat. Technol.* **2019**, *371*, 401–416. [CrossRef]

42. Verdon, C.; Karimi, A.; Martin, J.L. A study of high velocity oxy-fuel thermally sprayed tungsten carbide based coatings. Part 1: Microstructures. *Mater. Sci. Eng. A* **1998**, *246*, 11–24. [CrossRef]

43. Mateen, A.; Saha, G.C.; Khan, T.I.; Khalid, F. Tribological behavior of HVOF sprayed near-nanostructured and microstructured WC-17wt.%Co coatings. *Surf. Coat. Technol.* **2011**, *206*, 1077–1084. [CrossRef]

44. Myalska, H.; Moskal, G.; Szymański, K. Microstructure and properties of WC-Co coatings, modified by sub-microcrystalline carbides, obtained by different methods of high velocity spray processes. *Surf. Coat. Technol.* **2014**, *260*, 303–309. [CrossRef]

45. Masoumi, H.; Safavi, S.M.; Salehi, M.; Nahvi, S.M. Effect of grinding on the residual stress and adhesion strength of HVOF thermally sprayed WC-10Co-4Cr coating. *Mater. Manuf. Process.* **2014**, *29*, 1139–1151. [CrossRef]

46. Książek, M.; Boron, Ł.; Tchórz, A. Microstructure, mechanical properties and wear behaviour of high-velocity oxygen-fuel (HVOF) sprayed (Cr_3C_2-NiCr+Al) composite coating on ductile cast iron. *Coatings* **2019**, *9*, 840. [CrossRef]

47. Venter, A.M.; Pirling, T.; Buslaps, T.; Oladijo, O.P.; Steuwer, A.; Ntsoane, T.P.; Cornish, L.; Sacks, N. Systematic investigation of residual strains associated with WC-Co coatings thermal sprayed onto metal substrates. *Surf. Coat. Technol.* **2012**, *206*, 4011–4020. [CrossRef]

 *materials*

 MDPI

Article

Performance Prediction of Erosive Wear of Steel for Two-Phase Flow in an Inverse U-Bend

Saifur Rahman [1], Rehan Khan [2,*], Usama Muhammad Niazi [3], Stanislaw Legutko [4], Muhammad Ali Khan [2], Bilal Anjum Ahmed [2], Jana Petrů [5], Jiří Hajnyš [5] and Muhammad Irfan [1]

1 Electrical Engineering Department, College of Engineering, Najran University, Najran 61441, Saudi Arabia
2 Department of Mechanical Engineering, College of Electrical and Mechanical Engineering, National University of Sciences and Technology, Islamabad 44000, Pakistan
3 Department of Mechanical Engineering Technology, National Skills University, Islamabad 44000, Pakistan
4 Faculty of Mechanical Engineering, Poznan University of Technology, 60-965 Poznan, Poland
5 Department of Machining, Assembly and Engineering Metrology, Mechanical Engineering Faculty, VŠB-Technical University of Ostrava, 17. Listopadu 2172/15, 708 00 Ostrava, Czech Republic
* Correspondence: mrehan.khan@ceme.nust.edu.pk

Abstract: Erosion of the elbow due to non-Newtonian viscous slurry flows is often observed in hydrocarbon transportation pipelines. This paper intends to study the erosion behavior of double offset U-bends and 180° U-bends for two-phase (liquid-sand) flow. A numerical simulation was conducted using the Discrete Phase Model (DPM) on carbon steel pipe bends with a 40 mm diameter and an R/D ratio of 1.5. The validity of the erosion model has been established by comparing it with the results quantified in the literature by experiment. While the maximum erosive wear rates of all evaluated cases were found to be quite different, the maximum erosion locations have been identified between 150° and 180° downstream at the outer curvature. It was seen that with the increase in disperse phase diameter, the erosive wear rate and impact area increased. Moreover, with the change of configuration from a 180° U-bend to a double offset U-bend, the influence of turbulence on the transit of the disperse phase decreases as the flow approaches downstream and results in less erosive wear in a double offset U-bend. Furthermore, the simulation results manifest that the erosive wear increases with an increase in flow velocity, and the erosion rate of the double offset U-bend was nearly 8.58 times less than the 180° U-bend for a carrier fluid velocity of 2 m/s and 1.82 times less for 4 m/s carrier fluid velocity. The erosion rate of the double offset U-bend was reduced by 120% compared to the 180° U-bend for 6 m/s in liquid-solid flow.

Keywords: erosion; wear; U-bends; discrete phase model; sand; elbow

Citation: Rahman, S.; Khan, R.; Niazi, U.M.; Legutko, S.; Khan, M.A.; Ahmed, B.A.; Petrů, J.; Hajnyš, J.; Irfan, M. Performance Prediction of Erosive Wear of Steel for Two-Phase Flow in an Inverse U-Bend. *Materials* **2022**, *15*, 5558. https://doi.org/10.3390/ma15165558

Academic Editors: Claudio Mele and Artur Czupryński

Received: 15 July 2022
Accepted: 11 August 2022
Published: 12 August 2022

1. Introduction

Erosion of pipeline components can cause serious malfunction for the hydrocarbon extraction and processing industries, as the sand produced can impact the walls of the pipeline, resulting in wear damage [1,2]. When the dispersed phase has to be transported, flow-changing devices, i.e., elbows, are inclined to erosion damage because of the redirected flow at curvature. Most of the available experimental and numerical data focus on long radius 90° elbow erosion [3–5]. In erosive wear modeling, the understanding of flow physics around the erodent and the target surface is crucial. Many researchers investigated to understand the flow of physics during particle wall interaction. The particle trajectories and the erosion rate are strongly affected by the momentum exchange between flowing fluid and solid particles [6–10].

Elemuren et al. [11] investigated 90° elbow erosion by employing experimentation. It was noticed from the surface topographies of the elbow upstream, middle and downstream sections that ridges and valleys are discerned downstream of the elbows at higher particle loadings.

X. Zhao et al. [12] performed numerical prediction on the wear of bends installed in a series configuration in the two phase (gas-solid) flow. The result of the study shows a V-shaped erosion pattern due to the secondary impaction of the disperse phase at outer curvatures. Additionally, they found that the erosion of downstream bends is significantly influenced by the upstream pipe due to the considerable change in particle impaction velocity and angle.

Q. Wang et al. [13] simulated erosion in elbow geometry using large eddy simulation (LES). It was observed that the highest erodent impaction remains at the outer wall of the elbow outlet. Moreover, the maximum erosion rate decreases as the bend curvatures increase.

Wang et al. [14] utilized the CFD approach to predict the erosion of a 90° elbow. It was noticed that the erosion distribution is also influenced by the erodent size; the maximum impaction and erosion location will be located near the elbow outlet. The location of maximum impaction shifted close to the exit section with the increase in particle size.

Khan et al. [15] investigated multiphase erosion flow for bend angles (60° and 90°) with sand particles using a flow loop. They observed that as the elbow angle increases, the erosive wear increased for the 90-degree pipe bend, and the erosion rate decreases as bend angles decrease. It was also found that the maximum erosion location will be at the downstream section for both elbow configurations. Duarte et al. [16] decoupled the relationship between sand erodent concentrations and erosive wear for elbow pipes by employing CFD. It was observed that the influence of interparticle collisions on the erosion rate is significant for low mass loading conditions, and erosive wear decreases with an increase in mass loading due to particle–particle interactions. Karimi et al. [17] predicted sand fines erosion of a 90-degree elbow utilizing numerical simulations. They found that CFD inaccurately predicts erosion due to sand fines in the elbow configuration; however, for direct impact cases, CFD results show good agreement with experiments. In elbow configuration, the rebound models were not simulating the fine particle trajectories inside the pipe correctly.

Bilal et al. [4] studied the influence of pipe bend R/D ratio on the wear rate and concluded that the erosive wear rate in multiphase flow is larger than that of single-phase flow conditions. Cui et al. [18] conducted a computational fluid dynamics simulation to quantify particle erosion in the elbow pipe for the bubble flow regime.

Li et al. [19] calculated erosion for continuous elbows in different directions by utilizing CFD-based simulation. It was noticed that the erosion rates of 50-micron particles are larger than those of the 10-micron particle size for the identical gas flow rate. H. Zhu and Y. Qi [20] numerically investigated the flow erosion of multiphase flow in a U-bend. They found that flow velocity and particle size significantly influence erosion rate and lead to excessive erosion. Mazumder [21] performed a numerical and experimental investigation on S-bend geometry erosion in multiphase flow to identify the location of erosion inside the pipeline.

In erosive wear modeling, the understanding of flow physics around the erodent and the target surface is crucial. Many researchers investigated erosive wear to understand the flow of physics during particle wall interaction. The particle trajectories and the erosion rate are strongly affected by the momentum exchange between flowing fluid and solid particles [6–9]. In gas-solid flows due to the high inertia of erodent, they cross the flow streamlines, while in liquid-solid flow the drag forces on erodent are higher due to high viscosity. Numerical predictions of impact conditions require many assumptions to quantify erosion rate, but CFD has been a widely adopted method to predict the impact condition in the fluid flow [22–24]. The most important factor that strongly influences the accuracy of CFD erosion prediction in liquid-solid-gas flow is particle distribution and size in carrier phases. For erosion-induced wear, hydrodynamics plays an important role. Numerous flow dynamics parameters due to slurry (liquid-solid) transport through 90° elbow configurations were numerically simulated and experimentally investigated in previous studies, and erosion prediction models were developed to quantify the erosion-induced damage for elbow configurations [25,26].

Previous studies on the erosion of elbow pipelines have mainly focused on 90° elbows. To the best of our knowledge, studies related to double offset U-bend and 180° U-bend erosion in liquid-solid flow have not been reported in the literature. In the hydrocarbon production industry, there are cases where elbow pipes were suspected to have an erosional impact on production systems, especially in transient operations. Prediction of such an effect will be beneficial to material selection, wall thickness design as well as erosion mitigation methods.

In this paper, the erosion of inverse double offset U-bend and 180° U-bend is studied for different flow velocities and particle sizes using a CFD-DPM. To validate the simulation model and flow physics, the predicted erosive wear of an elbow was compared with the data obtained by Mazumder [21] and W. Peng, X. Cao [27].

2. Problem Description

2.1. Model Geometry

Figure 1a,b is a geometry of a double offset U-bend and a 180° U-bend used as a computational domain in this study. The double offset U-bend and 180° U-bend with an internal diameter (ID) of 40 mm and an R/D ratio of 1.5 are selected in the research. The entry and exit pipe of the bend geometry is 600 mm to ensure the fully developed flow in the upstream pipe. The elbows are made of carbon steel and the orientation is inverse (vertical upward–vertical downward).

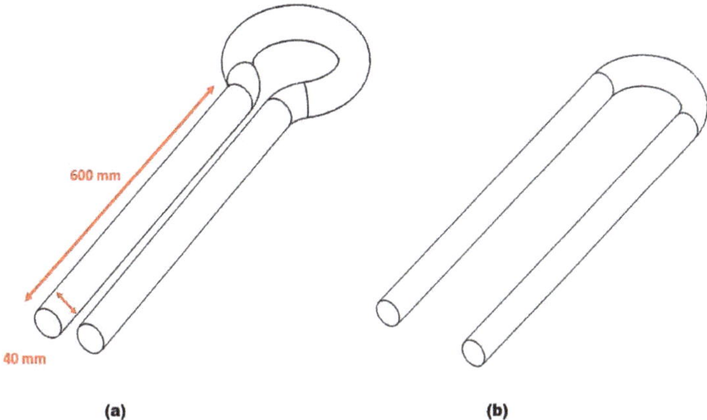

Figure 1. Elbow geometry: (**a**) double offset U-bend; (**b**) 180° U-bend.

2.2. Carrier and Solid Phase Model

In this study, the numerical equations of the FLUENT module of ANSYS used to solve the liquid-solid flow physics are represented as Equations (1) and (2):

$$\frac{\partial \rho}{\partial t} + \nabla \cdot \rho \vec{V} = 0 \tag{1}$$

$$\frac{\partial}{\partial t}(\rho \vec{V}) + \nabla \cdot \rho(\vec{V}\vec{V}) = -\nabla P + \nabla \cdot (\bar{\bar{\tau}}) + \rho \vec{g} + \vec{S_M} \tag{2}$$

$$m_e \frac{d\vec{u}_p}{dt} = \vec{F}_1 + \vec{F}_2 + \vec{F}_3 + \vec{F}_4 \tag{3}$$

In Equation (2), ρ = density, τ = stress tensor, ρg = body force, and S_M = momentum. The disperse phase model can be expressed in Equation (3), where m_p = erodent mass, F_1 = drag force, F_2 = pressure force, F_3 = particle force, and F_4 = buoyancy force.

2.3. Erosion and Turbulence Model

The erosion model defined by Oka is used in this study to quantify the erosion distribution of the double offset U-bend and the 180° U-bend, which is the reasonably accurate erosion prediction model of curve pipes [1] and is defined as:

$$ER = 10^9 \times \rho_t kF(\alpha)(H_v)^{k_a}\left(\frac{V_e}{V'}\right)^{k_b}\left(\frac{d_e}{d'}\right)^{k_c} \tag{4}$$

In Equation (4), ρ_t = density of wall, α = sand incidence angle, H_v = Vickers hardness of wall, V_e = incidence speed, V' = reference speed, d_e = sand size, and d' = reference sand size. In this study, the Grant and Tabakoff model [28] is selected to model particles and wall collision. The turbulence model (k–ε) was selected in this study since the Reynolds number calculated is larger than 4000.

2.4. Mesh and Model Validation

As shown in Figure 2, hexahedral structured meshing was generated on the U-bend configurations with approximately 478,000 cells. The boundary layer grids with 10-layer were implemented near the wall to capture the flow field structure near the wall. The flow solution chosen for this study is steady-state with 10^{-6} of the convergence criterion and the SIMPLEC numerical procedure is used to discretize the multiphase phase flow.

Figure 2. Computational mesh.

To further improve the simulation accuracy, the mesh independence study was performed. The validation study was performed by comparison with the maximum rate of erosion with the benchmark case of W. Peng and X. Cao [27]. The flow conditions are set the same in comparison with the benchmark study. It can be concluded that, for this geometry, most of the erosion is predicted at the outlet location irrespective of the type of mesh. The erosion rate was evaluated from three meshes of different sizes, as presented in Figure 3. It can be seen that there is less than a 2% difference in the results between mesh 2 and mesh 3, and the numerical simulations for the present investigation was conducted using mesh 2 with 478,000 cells. Fourteen cases with carrier fluid water (W) and Air (A) are scrutinized, as summarized in Table 1.

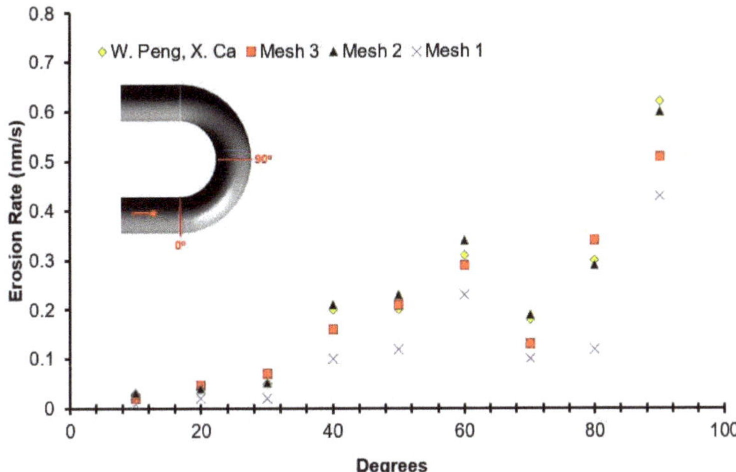

Figure 3. Comparison of the erosion rate obtained with three different mesh sizes and numerical results in W. Peng, X. Ca [27].

Table 1. Summary of simulation cases.

Case	Fluid	Orientation	Sand Size (μm)	Bend Type	Flow Velocity (m/s)	Erodent Flow Rate (kg/s)
1	A	H–V upward	300	180°	45.72	1
2	W	H–V upward	200	180°	10	0.2
3	W	V-V	450	180°	2	0.3
4	W	V-V	450	180°	4	0.3
5	W	V-V	450	180°	6	0.3
6	W	V-V	450	Double offset	2	0.3
7	W	V-V	450	Double offset	4	0.3
8	W	V-V	450	Double offset	6	0.3
9	W	V-V	75	180°	6	0.3
10	W	V-V	150	180°	6	0.3
11	W	V-V	250	180°	6	0.3
12	W	V-V	75	Double offset	6	0.3
13	W	V-V	150	Double offset	6	0.3
14	W	V-V	250	Double offset	6	0.3

A validation case was simulated to adjudge the effectiveness of the numerical model and particle tracking algorithm used in this study. The same parameters as the experiment of Mazumder [21] were set in the validation case. Air-solid flow enters a U-bend (ID = 12.7 mm, r/D = 1.5) at a velocity of 45.7 m/s, and 300 microns sand size. The comparison of the erosive wear location between the simulation and experiment at the outer wall is shown in Figure 4. The results obtained by CFD-DPM are consistent with the experiment with the first location of erosion observed at 19–69° and the second observed at 106–159°. As seen in Figure 4, the CFD identified accurately the erosion location that occurs at the outer wall as compared to the qualitative experiment.

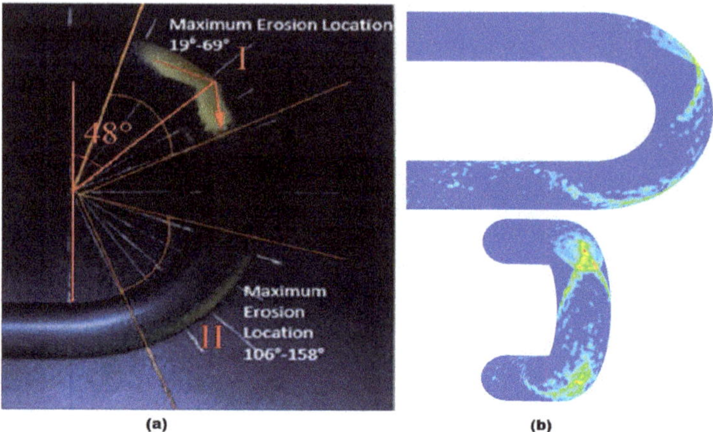

Figure 4. Comparison of the erosion location: (**a**) experimental results in Mazumder [21]; (**b**) numerical results (present study).

3. Results and Discussion

3.1. Effects of the Flow Velocity on Erosion

To decouple the influence of flow velocity on erosive wear distribution in the 180° U-bend and double offset U-bend, to avoid settling of erodent particles by maintaining the carrier flow velocities above certain levels in the pipelines, the cases with inlet velocity set to 2, 4 and 6 m/s are considered. The carrier fluid is water and the length of the pipe before and after the bend is set to 600 mm. Figures 5 and 6 show the erosive wear distribution contours on the 180° U-bend and double offset U-bend for different flow velocities. It can be discerned that the maximum erosive wear remains located in the outlet for all inlet velocities, and the maximum erosion rate of both the 180° U-bend and double offset U-bend increases drastically with the accretion in flow velocity. As presented in Figure 5, at the lowest flow velocity of 2 m/s, the highest erosion rate of the 180° U-bend locates adjacent to the exit of the bend. With the accretion of the fluid velocity, the exit of the bend at the outer curvature is significantly eroded and becomes the highest impaction location when the flow velocity reaches 6 m/s. The most severely eroded location of the double offset U-bend is shown in Figure 6, the simulation predicts a reduction of maximum erosion rate by a factor of 8.58 from Case 1 to Case 4 because more sand impaction occurs at the outer elbow curvature of Case 1 as opposed to Case 4. Furthermore, for Case 2 and Case 5, the decrease in the maximum erosion rate was about 1.81 times. As the redirection of flow is smoother for double offset U-bend, more sand follows the fluid stream, and less sand impaction occurs at the outer elbow curvature. For similar cognitions, Case 6 eroded less compared to Case 3, as shown in Figure 7.

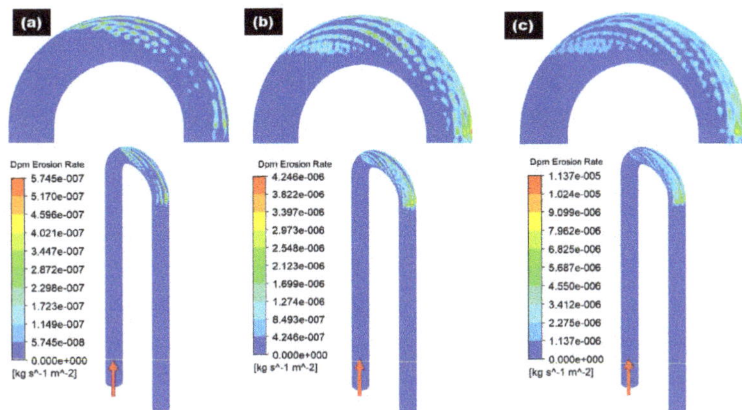

Figure 5. Erosion distribution in 180° U-Bend: (**a**) 2 m/s; (**b**) 4 m/s; (**c**) 6 m/s.

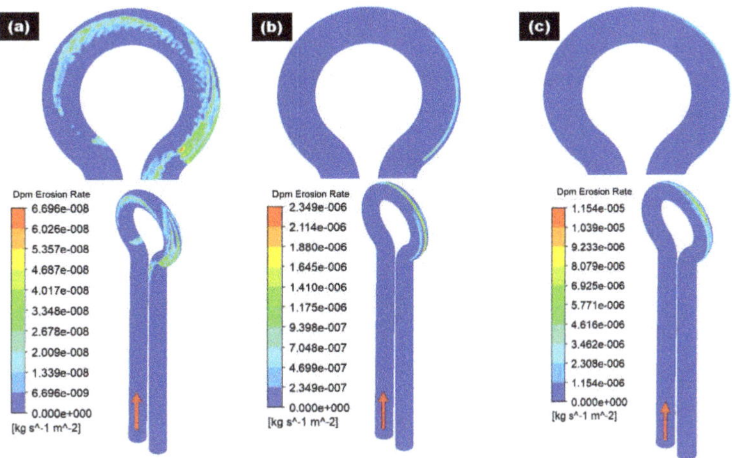

Figure 6. Erosion distribution in double offset U-bend: (**a**) 2 m/s; (**b**) 4 m/s; (**c**) 6 m/s.

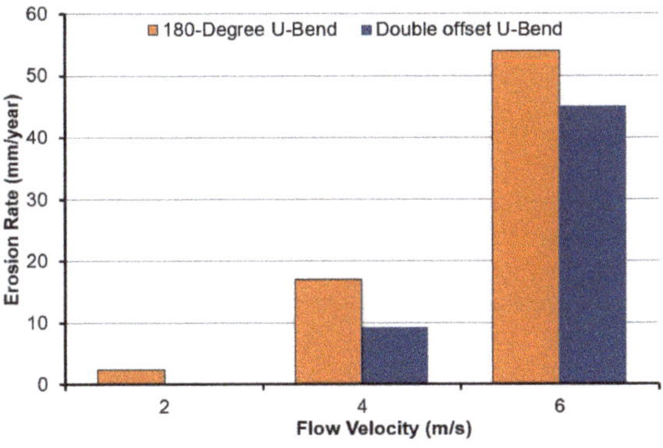

Figure 7. Maximum erosion rate in 180° U-bend and double offset U-bend for three flow velocities.

The erosion contour changes from symmetric to a more concentrated outer wall with the change from the 180° U-bend and double offset U-bend, and the zone affected by sand erosion is the minimum.

Figures 8 and 9 show the pressure distribution at selected planes inside the 180° U-bend and double offset U-bend. Simulations were undertaken with input velocities of 2, 4, and 6 m/s for the 180° U-bend and double offset U-bend. The centrifugal force acted on the fluid and was impelled to the outer curvature of the bend, which was exposed to higher pressure, while the inner curvature was subjected to lower pressure. However, as an example, only two extreme velocity distributions in the 180° U-bend and double offset U-bend are shown here for the case with the maximum erosion rate in Figure 10a,b. It is evident from these figures that the changing geometry from the 180° U-bend to the double offset U-bend can significantly influence the flow field. As the input velocity increased from 6 to 8.8 m/s in the 180° U-bend and from 6 to 22 m/s in the double offset U-bend, the flow pattern changed significantly, especially at the curvature of the 180° U-bend with a 1.46 times increase in velocities. In contrast, the input velocity accreted from 6 to 22 m/s in the double offset U-bend with a 3.66 times increase in velocities.

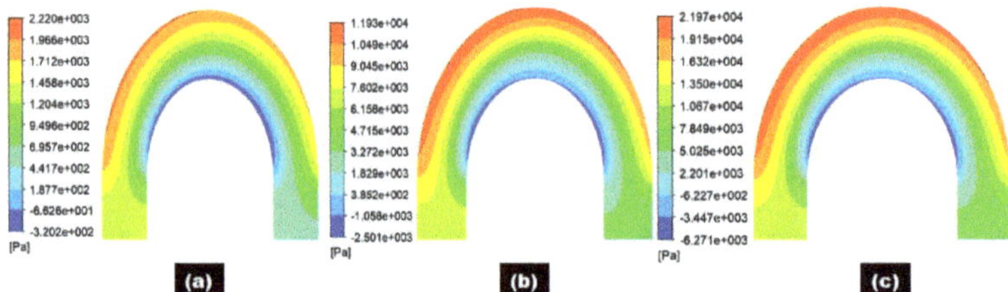

Figure 8. Pressure distribution contours of 180° U-bend for three different flow velocities: (**a**) 2 m/s; (**b**) 4 m/s; (**c**) 6 m/s.

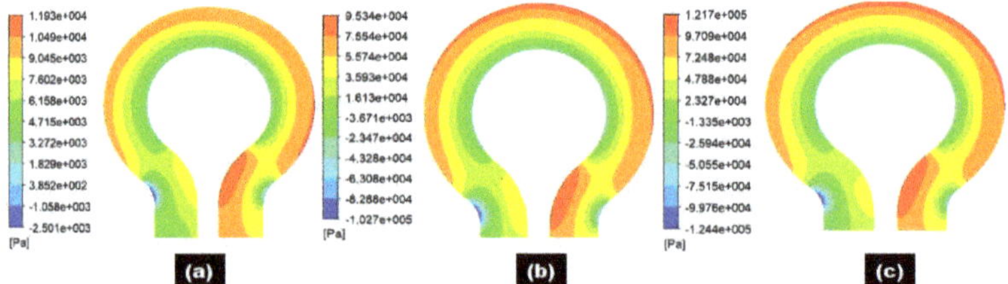

Figure 9. Pressure distribution contours of double offset U-bend for three different flow velocities: (**a**) 2 m/s; (**b**) 4 m/s; (**c**) 6 m/s.

Wall shear stresses inside the 180° U-bend and double offset U-bend under 6 m/s velocities of carrier fluid were simulated by CFD-DPM as shown in Figure 10c,d. The highest wall shear stress is observed at the inner curvature in the 180° U-bend and outer curvature in the upstream pipe of the double offset U-bend and then the second location of maximum wall shear located in the inner curvature of pipe in the double offset U-bend. When the carrier fluid is transported into the bend pipe, the fluid direction is altered and the pressure of the fluid on the outer wall of the bend pipe enhances under the action of centrifugal force. The carrier fluid kinetic energy on the outer curvature results in the pressure, and the carrier fluid velocity on the outer curvature of the bend reduces. Consequently, the reduction in wall shear stress at outer curvature is observed as shown in Figure 10c,d.

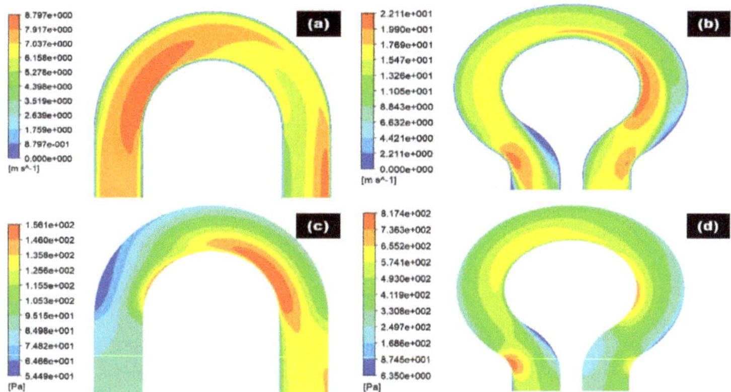

Figure 10. Contours of 180° U-bend and double offset U-bend for 6 m/s flow velocities: (**a,b**) velocity distribution; (**c,d**) wall shear stress.

3.2. Effect of Sand Size on Erosion

The maximum sand impaction zones will dynamically change with the change of particle diameter in the 180° U-bend and double offset U-bend in Figures 11 and 12. Zone A is inclined to maximum erosive wear when the particle diameter is 75 μm as shown in Figure 11. Consequently, as particle diameter increases, this maximum particle impaction will remain in Zone 1 with the addition of a medium erosion zone in the downstream section. When the particle size is 75 μm, the drag force is prepotent, and the transportation of sand is due to secondary flow. The flow pushes the sand in the circumferential path from the curvature to the exit of the bend and then moves the sand to the exit of the bend's outer curvature and, as a result, causes maximum erosion in Zone A. For 250 μm sand, the inertia force is a significant parameter in the transportation of the sand. The sand erodent has high momentum; thus, the flow velocity and direction have a slight effect on the sand. Since the sand particles divagate from the flow streamlines, impaction occurs at Zones A, B, and C. It is noticed that for all the evaluated case, the one maximum erosion zone appears along the curvature of the 180° U-bend. Nevertheless, the erosive wear becomes more serious as the particle size accrete. For 75 μm, erosion spots are located at Zone A at the outer curvature. As the sand size changes from 75 to 250 μm, the area of erosion location is significantly increased and hence leads to maximum erosion. Additionally, the severe erosion in the exit elbow section becomes visible for the sand size of 250 μm. The width of the maximum eroded region grows as the size of sand further increases.

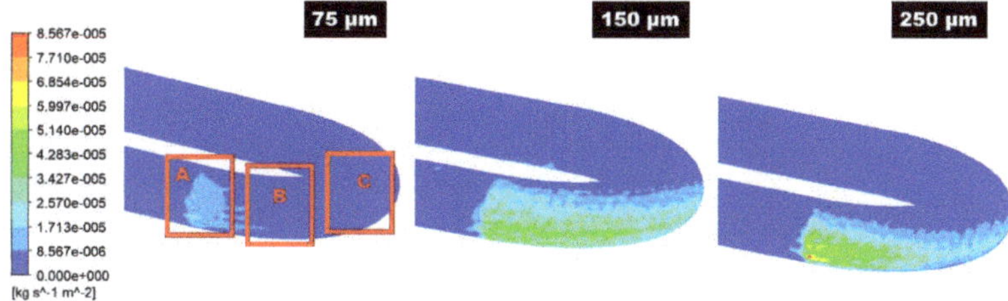

Figure 11. The contour of the erosion distribution in the 180° U-bend elbow under three different particle sizes.

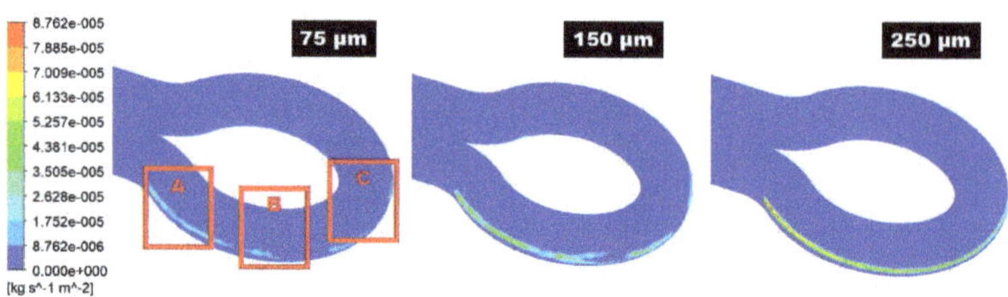

Figure 12. The contour of the erosion distribution in the double offset U-bend elbow under three different particle sizes.

Figure 13 shows that turbulence intensity increases sand impaction impact and erosive wear at the curvature near the outlet of the 180° U-bend and double offset U-bend. The turbulence in the fluid stream at the outlet of both bend configurations enhances the motion of sand in the radial directions and turns out to be a maximum erosive zone. This signifies that turbulence intensity causes the shift of particle trajectories inside the 180° U-bend and double offset U-bend. Based on analysis, turbulence weigh more in the 180° U-bend at the downstream as compared to the double offset U-bend. Figure 13 shows the contours of turbulent intensity in the entry, middle, and exit of the 180° U-bend and double offset U-bend. Figure 13 suggests that the reduction in the blue area and the enhancement in the red and yellow area signify the increase in turbulence. This means that maximum turbulence turns out in the 180° U-bend elbow exit sections.

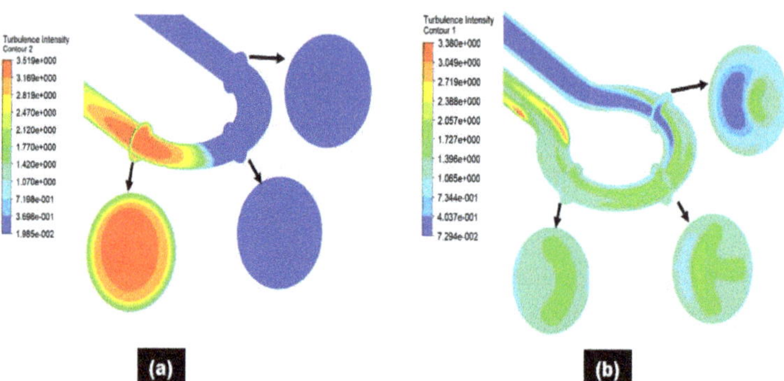

Figure 13. Turbulence intensity at the different locations inside the elbow pipe for 250 μm particle size: (**a**) 180° U-bend; (**b**) double offset U-bend.

Figure 14a,b show the trajectories of selected sand within the 180° U-bend and double offset U-bend. When the sand is 250 μm, the turbulence leads to the sand distribution and sand particles distribute evenly in the upstream pipe in a 180° U-bend. With the change of configuration from the 180° U-bend to the double offset U-bend, the effect of turbulence on the motion of sand particles maximizes in the downstream section and results in erosive wear in Zones A, B, and C. Moreover, the gravitation force direction is parallel to the flow, and the sand particles are more likely to alter the path followed and fall down. A typical sand trajectory is simulated, and no significant rebounding takes place. It can be observed from sand tracks that particles travel in a direction close to the outer wall in both elbow configuration and continuing downstream.

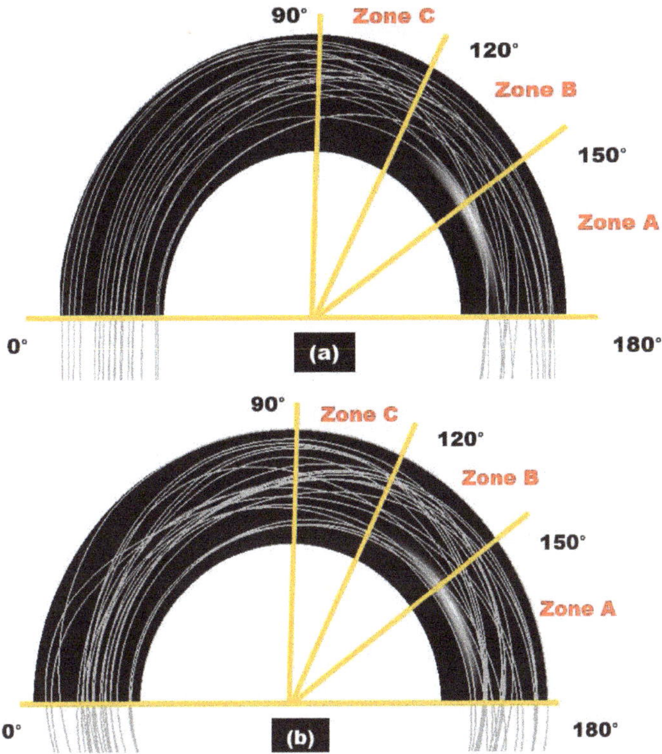

Figure 14. Trajectories of 250 µm sand particles under the particle velocity of 6 m/s: (**a**) 180° U-bend; (**b**) double offset U-bend.

The erosion rate is one of the most vital measures of flow-changing devices (i.e., elbow) lifetime in erosive flow conditions. Although the maximum erosion rate was seen to accrete with the increase in sand size for both the 180° U-bend and double offset U-bend for all evaluated cases, the double offset U-bend is less prone to erosive wear as shown in Figure 15. Notwithstanding this, the double offset U-bend proves its worth, as the highest rate of erosion is 1.23 times less than that of the 180° U-bend in the worst-case scenario.

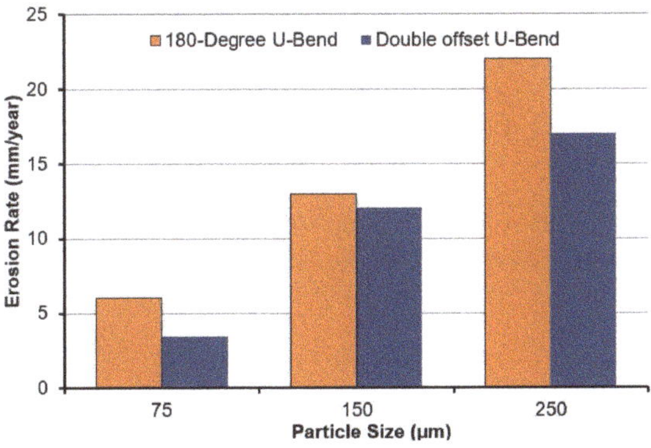

Figure 15. Maximum erosion rate in 180° U-bend and double offset U-bend for three different particle sizes.

4. Conclusions

A liquid-solid erosion simulation of the 180° U-bend and double offset U-bend has been employed to predict the erosive wear rate and distribution on both types of bend for sand-water flow conditions. Several simulations were performed to understand erosion patterns and trajectories of sand particles. Conforming to the numerical results, the following conclusions can be derived:

1. Erosive wear is the maximum at the outer curvature of the 180° U-bend and double offset U-bend for all evaluated cases. The maximum erosion area occurs between curvature angles of 150° and 180°. Additionally, the double offset U-bend proves its worth, as the highest wear rate is 1.23 times less than that of the 180° U-bend in the worst-case scenario.
2. The erosion rate of the double offset U-bend was nearly 8.58 times less than 180° U-bend for the fluid velocity of 2 m/s and 1.82 times less for 4 m/s fluid velocities. The maximum erosion rate of double offset U-bend was reduced by 120% compared to the 180° U-bend for 6 m/s in liquid-solid flow.
3. The 180° U-bend can be replaced with a double offset U-bend to slow down pipe erosion, especially for inverse orientation. Since many hydrocarbon and mineral processing plants require sand particle transportation, the double offset U-bend elbow appears to be a worthwhile alternative.
4. The formation of an erosive wear pattern at the double offset U-bend and 180° U-bend is explained through the sand particle tracking. With the change of configuration from the 180° U-bend to the double offset U-bend, the effect of turbulence on the motion of sand decreases as the flow approaches downstream and results in less erosive wear in the double offset U-bend.

Author Contributions: R.K. designed the methodology, performed the numerical analysis, and wrote the manuscript; M.A.K. and U.M.N. guided the data analysis and validation; B.A.A. and M.A.K. guided in investigation and project management; S.R., M.I., S.L., J.P. and J.H. performed the analysis, conceptualization, evaluation, visualization, editing, project management, and resources. All authors have read and agreed to the published version of the manuscript.

Funding: The APC was paid through the Department of Machining, Assembly and Engineering Metrology, Mechanical Engineering Faculty, VŠB-Technical University of Ostrava, 17. listopadu 2172/15, 708 00 Ostrava, Czech Republic.

Institutional Review Board Statement: Not applicable.

Informed Consent Statement: Not applicable.

Data Availability Statement: The data presented in this study are available upon request from the corresponding author.

Acknowledgments: The authors acknowledge the support from the Deanship of Scientific Research, Najran University. Kingdom of Saudi Arabia, for funding this work under the National Research Priorities funding program grant code number (NU/NRP/SERC/11/13).

Conflicts of Interest: The authors declare no conflict of interest.

References

1. Khan, R.; Ya, H.; Pao, W.; bin Abdullah, M.Z.; Dzubir, F.A. Influence of Sand Fines Transport Velocity on Erosion-Corrosion Phenomena of Carbon Steel 90-Degree Elbow. *Metals* **2020**, *10*, 626. [CrossRef]
2. Ma, G.; Ma, H.; Sun, Z. Simulation of Two-Phase Flow of Shotcrete in a Bent Pipe Based on a CFD–DEM Coupling Model. *Appl. Sci.* **2022**, *12*, 3530. [CrossRef]
3. Jia, W.; Zhang, Y.; Li, C.; Luo, P.; Song, X.; Wang, Y.; Hu, X. Experimental and numerical simulation of erosion-corrosion of 90° steel elbow in shale gas pipeline. *J. Nat. Gas Sci. Eng.* **2021**, *89*, 103871. [CrossRef]
4. Bilal, F.S.; Sedrez, T.A.; Shirazi, S.A. Experimental and CFD investigations of 45 and 90 degrees bends and various elbow curvature radii effects on solid particle erosion. *Wear* **2021**, *476*, 203646. [CrossRef]
5. Adedeji, O.E.; Duarte, C.A.R. Prediction of thickness loss in a standard 90° elbow using erosion-coupled dynamic mesh. *Wear* **2020**, *460–461*, 203400. [CrossRef]

6. Khan, R. Numerical Investigation of the Influence of Sand Particle Concentration on Long Radius Elbow Erosion for Liquid-Solid Flow. *Int. J. Eng.* **2019**, *32*, 1485–1490. [CrossRef]

7. Kannojiya, V.; Deshwal, M.; Deshwal, D. Numerical Analysis of Solid Particle Erosion in Pipe Elbow. *Mater. Today Proc.* **2018**, *5 Pt 1*, 5021–5030. [CrossRef]

8. Florio, L.A. Estimation of particle impact based erosion using a coupled direct particle—Compressible gas computational fluid dynamics model. *Powder Technol.* **2017**, *305*, 625–651. [CrossRef]

9. Duarte, C.A.R.; de Souza, F.J. Innovative pipe wall design to mitigate elbow erosion: A CFD analysis. *Wear* **2017**, *380–381*, 176–190. [CrossRef]

10. Khan, R.; Ya, H.H.; Pao, W.; Khan, A. Erosion–Corrosion of 30°, 60°, and 90° Carbon Steel Elbows in a Multiphase Flow Containing Sand Particles. *Materials* **2019**, *12*, 3898. [CrossRef]

11. Elemuren, R.; Tamsaki, A.; Evitts, R.; Oguocha, I.N.A.; Kennell, G.; Gerspacher, R.; Odeshi, A. Erosion-corrosion of 90° AISI 1018 steel elbows in potash slurry: Effect of particle concentration on surface roughness. *Wear* **2019**, *430–431*, 37–49. [CrossRef]

12. Zhao, X.; Cao, X.; Xie, Z.; Cao, H.; Wu, C.; Bian, J. Numerical study on the particle erosion of elbows mounted in series in the gas-solid flow. *J. Nat. Gas Sci. Eng.* **2022**, *99*, 104423. [CrossRef]

13. Wang, Q.; Ba, X.; Huang, Q.; Wang, N.; Wen, Y.; Zhang, Z.; Sun, X.; Yang, L.; Zhang, J. Modeling erosion process in elbows of petroleum pipelines using large eddy simulation. *J. Pet. Sci. Eng.* **2022**, *211*, 110216. [CrossRef]

14. Wang, K.; Li, X.; Wang, Y.; He, R. Numerical investigation of the erosion behavior in elbows of petroleum pipelines. *Powder Technol.* **2017**, *314*, 490–499. [CrossRef]

15. Khan, R.; Ya, H.H.; Shah, I.; Niazi, U.M.; Ahmed, B.A.; Irfan, M.; Glowacz, A.; Pilch, Z.; Brumercik, F.; Azeem, M.; et al. Influence of Elbow Angle on Erosion-Corrosion of 1018 Steel for Gas–Liquid–Solid Three Phase Flow. *Materials* **2022**, *15*, 3721. [CrossRef] [PubMed]

16. Duarte, C.A.R.; de Souza, F.J.; dos Santos, V.F. Numerical investigation of mass loading effects on elbow erosion. *Powder Technol.* **2015**, *283*, 593–606. [CrossRef]

17. Karimi, S.; Shirazi, S.A.; McLaury, B.S. Predicting fine particle erosion utilizing computational fluid dynamics. *Wear* **2017**, *376–377*, 1130–1137. [CrossRef]

18. Cui, B.; Chen, P.; Zhao, Y. Numerical simulation of particle erosion in the vertical-upward-horizontal elbow under multiphase bubble flow. *Powder Technol.* **2022**, *404*, 117437. [CrossRef]

19. Li, B.; Zeng, M.; Wang, Q. Numerical Simulation of Erosion Wear for Continuous Elbows in Different Directions. *Energies* **2022**, *15*, 1901. [CrossRef]

20. Zhu, H.; Qi, Y. Numerical investigation of flow erosion of sand-laden oil flow in a U-bend. *Process Saf. Environ. Prot.* **2019**, *131*, 16–27. [CrossRef]

21. Mazumder, Q.H. S-bend erosion in particulated multiphase flow with air and sand. *J. Comput. Multiph. Flows* **2016**, *8*, 157–166. [CrossRef]

22. Gnanavelu, A.; Kapur, N.; Neville, A.; Flores, J.F. An integrated methodology for predicting material wear rates due to erosion. *Wear* **2009**, *267*, 1935–1944. [CrossRef]

23. Mansouri, A.; Arabnejad, H.; Karimi, S.; Shirazi, S.A.; McLaury, B.S. Improved CFD modeling and validation of erosion damage due to fine sand particles. *Wear* **2015**, *338–339*, 339–350. [CrossRef]

24. Chochua, G.G.; Shirazi, S.A. A combined CFD-experimental study of erosion wear life prediction for non-Newtonian viscous slurries. *Wear* **2019**, *426–427*, 481–490. [CrossRef]

25. Pei, J.; Lui, A.; Zhang, Q.; Xiong, T.; Jiang, P.; Wei, W. Numerical investigation of the maximum erosion zone in elbows for liquid-particle flow. *Powder Technol.* **2018**, *333*, 47–59. [CrossRef]

26. Wee, S.K.; Yap, Y.J. CFD study of sand erosion in pipeline. *J. Pet. Sci. Eng.* **2019**, *176*, 269–278. [CrossRef]

27. Peng, W.; Cao, X. Numerical simulation of solid particle erosion in pipe bends for liquid–solid flow. *Powder Technol.* **2016**, *294*, 266–279. [CrossRef]

28. Grant, G.; Tabakoff, W. Erosion Prediction in Turbomachinery Resulting from Environmental Solid Particles. *J. Aircr.* **1975**, *12*, 471–478. [CrossRef]

Article

Influence of Elbow Angle on Erosion-Corrosion of 1018 Steel for Gas–Liquid–Solid Three Phase Flow

Rehan Khan [1,*], Hamdan H. Ya [2], Imran Shah [3], Usama Muhammad Niazi [4], Bilal Anjum Ahmed [1], Muhammad Irfan [5], Adam Glowacz [6], Zbigniew Pilch [7], Frantisek Brumercik [8], Mohammad Azeem [2], Mohammad Azad Alam [2] and Tauseef Ahmed [2]

[1] Department of Mechanical Engineering, College of Electrical and Mechanical Engineering, National University of Sciences and Technology, Islamabad 44000, Pakistan; bilal.anjum@ceme.nust.edu.pk
[2] Mechanical Engineering Department, Universiti Teknologi PETRONAS, Seri Iskandar 31750, Malaysia; hamdan.ya@utp.edu.my (H.H.Y.); mohammad_18000380@utp.edu.my (M.A.); mohammad_18000664@utp.edu.my (M.A.A.); tauseef_17007229@utp.edu.my (T.A.)
[3] Department of Aerospace Engineering, College of Aeronautical Engineering, National University of Sciences and Technology, Risalpur 24080, Pakistan; ishah@cae.nust.edu.pk
[4] Department of Mechanical Engineering Technology, National Skills University, Islamabad 44000, Pakistan; ukniaxi@gmail.com
[5] Electrical Engineering Department, College of Engineering, Najran University, Najran 61441, Saudi Arabia; miditta@nu.edu.sa
[6] Department of Automatic Control and Robotics, Faculty of Electrical Engineering, Automatics, Computer Science and Biomedical Engineering, AGH University of Science and Technology, Al. A. Mickiewicza 30, 30-059 Krakow, Poland; adglow@agh.edu.pl
[7] Department of Electrical Engineering, Cracow University of Technology, Warszawska 24 Str., 31-155 Cracow, Poland; zbigniew.pilch@pk.edu.pl
[8] Department of Design and Machine Elements, Faculty of Mechanical Engineering, University of Zilina, Univerzitna 1, 010 26 Zilina, Slovakia; frantisek.brumercik@fstroj.uniza.sk
* Correspondence: mrehan.khan@ceme.nust.edu.pk

check for **updates**

Citation: Khan, R.; Ya, H.H.; Shah, I.; Niazi, U.M.; Ahmed, B.A.; Irfan, M.; Glowacz, A.; Pilch, Z.; Brumercik, F.; Azeem, M.; et al. Influence of Elbow Angle on Erosion-Corrosion of 1018 Steel for Gas–Liquid–Solid Three Phase Flow. *Materials* **2022**, *15*, 3721. https://doi.org/10.3390/ma15103721

Academic Editors: Artur Czupryński and Claudio Mele

Received: 22 April 2022
Accepted: 20 May 2022
Published: 23 May 2022

Publisher's Note: MDPI stays neutral with regard to jurisdictional claims in published maps and institutional affiliations.

Abstract: Erosive wear due to the fact of sand severely affects hydrocarbon production industries and, consequently, various sectors of the mineral processing industry. In this study, the effect of the elbow geometrical configuration on the erosive wear of carbon steel for silt–water–air flow conditions were investigated using material loss analysis, surface roughness analysis, and microscopic imaging technique. Experiments were performed under the plug flow conditions in a closed flow loop at standard atmospheric pressure. Water and air plug flow and the disperse phase was silt (silica sand) with a 2.5 wt % concentration, and a silt grain size of 70 μm was used for performing the tests. The experimental analysis showed that silt impact increases material disintegration up to 1.8 times with a change in the elbow configuration from 60° to 90° in plug flow conditions. The primary erosive wear mechanisms of the internal elbow surface were sliding, cutting, and pit propagation. The maximum silt particle impaction was located at the outer curvature in the 50° position in 60° elbows and the 80° position in 90° elbows in plug flow. The erosion rate decreased from 10.23 to 5.67 mm/year with a change in the elbow angle from 90° to 60°. Moreover, the microhardness on the Vickers scale increased from 168 to 199 in the 90° elbow and from 168 to 184 in the 60° elbow.

Keywords: erosion; wear; corrosion; sand; plug flow; elbow

1. Introduction

Erosion of pipeline is the cumulative removal of material due to the target surface and impinging dispersed phase interaction. It is a critical and convolute issue in the hydrocarbon and mineral processing industry. Erosion can substantially reduce the service life of pipelines and increase the production cost [1]. The erosion induced in multiphase flow is very complicated and is essentially sustained by flow regimes, impact conditions, disperse phase properties, and flow characteristics [2,3]. Complex disperse phase and carrier fluid interactions take place inside pipelines that affect the erosion-induced damage [4]. Aside from the flow conditions, the parameters that considerably affect pipe erosive wear also include the geometrical configuration of pipelines.

Erosion–corrosion in ductile metals has been investigated by many researchers, who found that it was due to the cutting action of the abrasive particles [5,6]. Erosion-induced damage is very much influenced by dispersed phase properties, particle impact conditions, and properties of the target material [7,8].

Sedrez et al. [9] investigated erosion by liquid–solid flow for elbow configurations. The wear pattern identified that the maximum impaction was found at the outer wall near the end of the elbows. Moreover, their computational fluid dynamics (CFD) study and experimental results showed good agreement.

Recently, Owen et al. [10] designed a test methodology for erosion–corrosion analysis in 3D printed 90° elbows in a representative of field flow conditions; it appeared that high flow disturbance would be generated by the protruded samples, significantly influencing hydrodynamics in the flow through the elbow. This could significantly affect local turbulence inside the elbow pipe.

Vieira et al. [11] reported that for gas–sand flow, the highest erosion was identified at 45°. In addition, the sand size had no significant influence on the wear rate in gas–sand flows. It was found that the 300 micron sand degrades material between 1.9 and 2.5 times higher compared to the 150 micron sand. Wang et al. [12] performed a numerical analysis and found that the maximum wear hot spot was also influenced by the particle size due to the fact that the sedimentation will be enhanced with the increase in erodent size; the peak erosion location will be located adjacent to the elbow exit. Vieira et al. [13] observed that a sand size of 300 μm disintegrated 3.7 times more material compared to 20 μm; similarly, the 300 μm sand created 3.1 times more degradation compared to the 150 μm sand particles. In annular flow conditions, the highest erosion was identified at the axial angle of 45° in the outer curvature of the elbow. The influence of erodent size and flow viscosity on material degradation was investigated in [7,14].

X. Cao et al. [15] studied the effect of superficial carrier phase velocities on the erosive behavior of steel pipe bends in water–sand slug flow. They concluded that with the escalation in superficial velocity, the degradation of the maximum eroded specimen reduces. Zahedi et al. [16] observed that for annular flow conditions, erosive wear was incurred with the highest particle wall impaction at 40–50° at the outer radius.

Surprisingly, there is a dearth of research on the 90° elbow and 60° elbow configuration related to the study of the erosion mechanism of pipes and, more specifically, for silt particles under plug flow. In the plug flow pattern, the bubbles are smaller in size and drive more slowly in comparison with slug flow. This paper aimed to investigate the erosion mechanism for a 1018 carbon steel 90° elbow pipe and a 60° elbow pipe in water–sand plug flow conditions. In this work, a novel erosion test methodology was designed by using representative curve elbow specimens of 90° and 60° elbows; it appeared that in literature, the tests performed on flat specimens mounted inside the elbow influenced the hydrodynamics and increased turbulence inside the pipe. Because existing methods for evaluating the erosion–corrosion of elbows are inadequate, in this research, new experimental procedures are developed to quantify elbow erosion-induced damage under three phase flow conditions.

In this paper, the paint modeling method, microscopic imaging, mass loss quantifications, and hardness testing were used to evaluate the erosive wear of carbon steel 90° and 60° elbow pipelines in plug flow. Furthermore, the erosive wear mechanism of the impact of water–silt–air plug flow was elucidated. In this study, sand of 70 μm size was primarily used to simulate the field operating conditions, with water and air as carrier phases, including the internal erosion of hydrocarbon production and mineral processing industries in multiphase flow, where erosion is due to the sand production such as conditions encountered in oil and gas fields.

2. Experiment Procedure and Test Methods

The elbow specimen used in this experimental study was 1018 carbon steel (CS) used with the following composition (in weight %): 0.2% C, 0.26% Si, 0.52% Mn, 0.21% Cr, and 98.12% Fe. The specimens were obtained from the supplier in the form of 90° elbow pipes and 60° elbow pipes. The elbow pipe, specimens in the shape of an axially cut section, as shown in Figure 1, were machined using wire electric discharge machining (WEDM). The finely polished specimens were obtained by grinding and polishing procedures resulting in a low-level surface roughness. The 10×10 mm^2 sizes of the specimens after the test were cut from the different locations of the elbow for microscopic imaging, as shown in Figure 2. A total of 36 specimens were cut from the upper and bottom walls of the 90° elbows and 60° elbows at various locations. The Vickers hardness of the specimens was evaluated under 5 N load using a diamond indenter for a 15 s indentation time using a Leco LM 247AT microhardness tester. The worn surface of the 90° and 60° elbow specimens was studied with a backscattered electron microscope (Phenom ProX, Eindhoven, The Netherlands). Each 10×10 mm specimen mass loss was measured using a precision weighing scale to quantify the wear rate after the test. The device used to acquire surface roughness parameters was Mitutoyo SURFTEST SJ-210. Details about the testing procedure were published in our previous work [17,18].

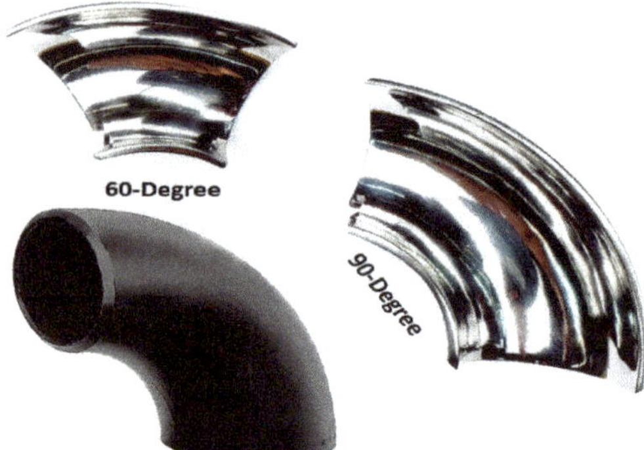

Figure 1. Elbow test specimen.

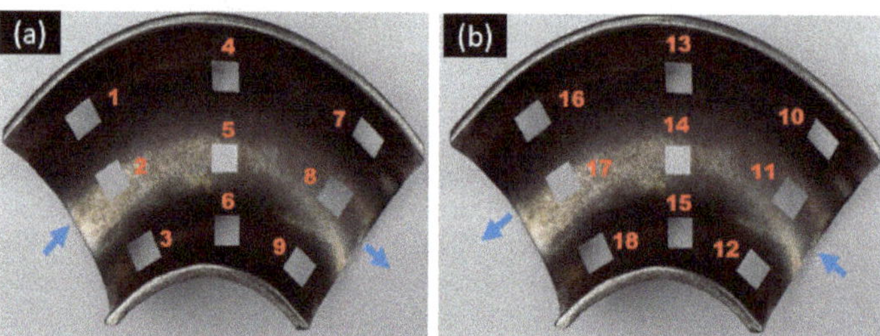

Figure 2. (a,b) Specimen location after test.

The locally fabricated erosion test flow loop was fabricated in University Technology PETRONAS, Seri Iskandar, Malaysia, using an abrasive pump that used a rubber liner to avoid wear of the pump. The silt particle (silica sand) was used as an erodent for all the experimental evaluations, as shown in Figure 3. The designed flow loop was semi-automated to simulate multiphase flow test conditions. The silt and water carrier phases were mixed in the slurry tank using a stirrer. The dispersed phase and water were then circulated using a variable speed pump in the flow pipelines, as shown in Figure 4. The liquid flow rate was measured by a magnetic flow meter through a 50.8 mm diameter PVC pipe. The test section was designed to mount 90° and 60° elbows for both multilayer paint modeling and erosion investigation for multiphase flow conditions.

Figure 3. SEM image of silt grain particles.

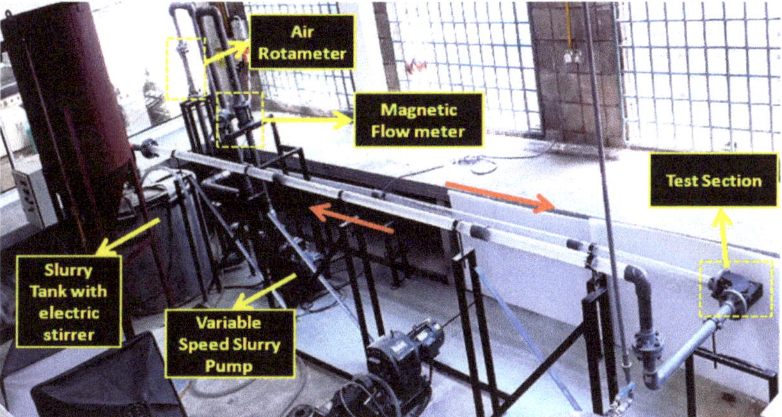

Figure 4. Layout of the experimental setup.

The literature review showed that various erosion test methods have been implemented to study elbow erosion under different flow conditions [19]. Some of the tests use the square sample in which the flat plate is placed at a different axial angle along the elbow pipe. This method leads to huge mass transfer shifts and inaccuracy in measurements and reduces the accuracy of measured data. To resolve this issue, the finely polished elbow sample cut axially into two sections was integrated with the specimen holder in this study, as shown in Figure 5. To quantify the localized erosion rate, the 36 specimens (10×10 mm^2) were cut from different locations on the 90° and 60° elbows, and a standard mass loss test was adopted for erosion rate measurement after the test. The initial mass of all the samples was measured before the test using separate specimens. The location numbers of the specimens are shown in Figure 2a,b. The test section designed in this study used the representative elbow configuration, which provided a better understanding of the erosion mechanism.

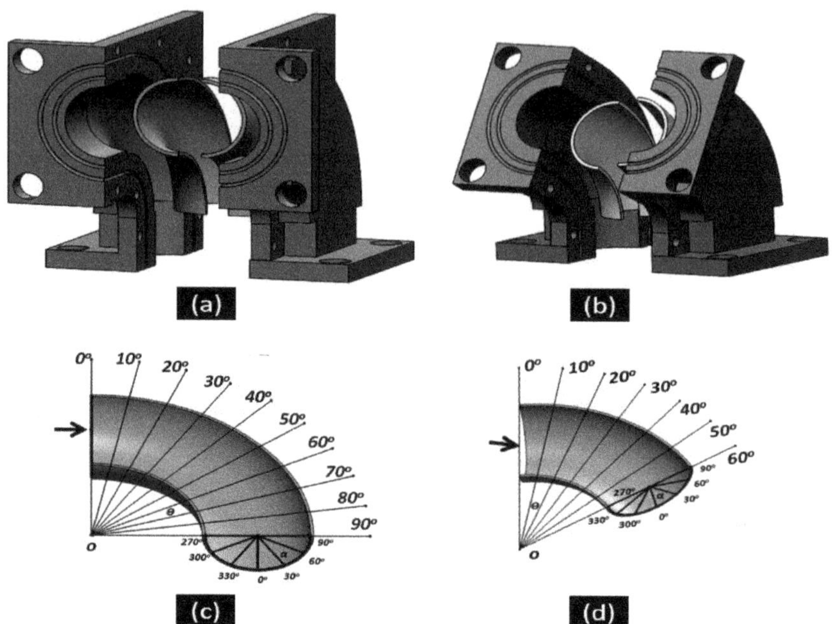

Figure 5. The 90° (**a**) and 60° (**b**) elbow test sections used for the erosion–corrosion studies; (**c**,**d**) definition of the axial angles.

3. Results and Discussion

3.1. Qualitative Paint Erosion Test

Paint erosion studies were executed on 90° and 60° elbows with a 70 μm particle size for a 90 min flow time. In the paint removal experiment, the inside area of the axially cut elbow specimen was coated with red colored enamel paint and silver colored acrylic paint applied by a spray gun. A digital paint thickness gauge was used to ensure the uniformity of each paint layer. Prior to the paint erosion test, it was necessary to make sure that the paint removal was not caused by flow conditions, and it must be ensured that it was exclusively due to the particle impaction. For tests under nonerosion conditions, it has been concluded that nonerosion flow conditions do not contribute to painting removal in the elbow specimens. Therefore, it can be deduced that the considered paint removal method qualitatively measures the particle impaction regions. Each paint removal test was performed three times to ensure accuracy; it was noticed that the paint erosion pattern tended to be similar for all tests.

The location of particle impaction in the pipe wall was evaluated after visualization of the paint-eroded regions. Figure 6 shows the paint removal patch in the upper and bottom 90° elbow sections with a 1.5 m/s (liquid velocity) and 0.7 m/s (air velocity) using silt of 70 μm. The high impaction region, on the 90° elbows, tended to be at a location approximately 45° and 90° in the bottom wall and the middle of 0° and 90° in the upper wall of the elbow, respectively. The paint removal marks were nonsymmetric in the top and bottom of the 60° and 90° elbows and the reason air with abrasive particles moving in the upper part and liquid was moving in the bottom section; thus, less erosion occurred in the bottom compared to the top. Figure 7 illustrates the paint erosion pattern in the 60° elbows with silt impact in the bottom and upper half sections for plug flow. For the 60° elbow, the paint was removed between 30° and 60° in the bottom wall and middle of 0° to 60° in the upper section, with greater paint removal pattern clearly seen towards the downstream section in all evaluated cases. In the plug flow regime, the plug body was the key source of erosive wear, because the maximum sand particles were transported by the continuous phase, i.e., water and the plug body had the highest water phase holdup. The highly turbulent plug front with abrasive particles at the elbow curvature can accelerate erosion-induced damage in the top part of pipelines which was evident in the experimental data collected in this study.

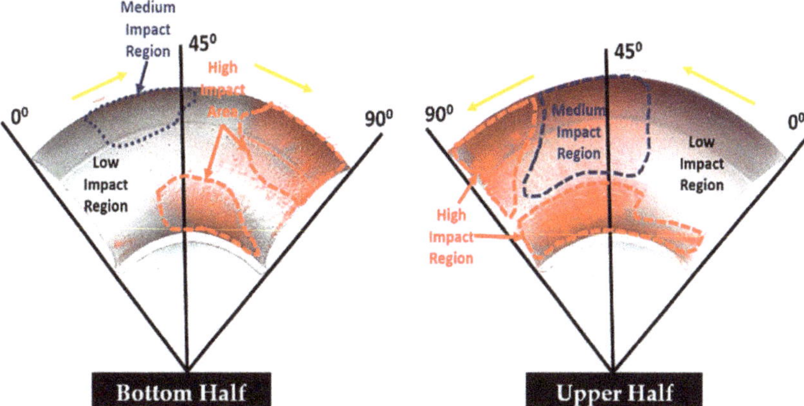

Figure 6. Erosion pattern on the 90° elbow coated with a two-layer paint.

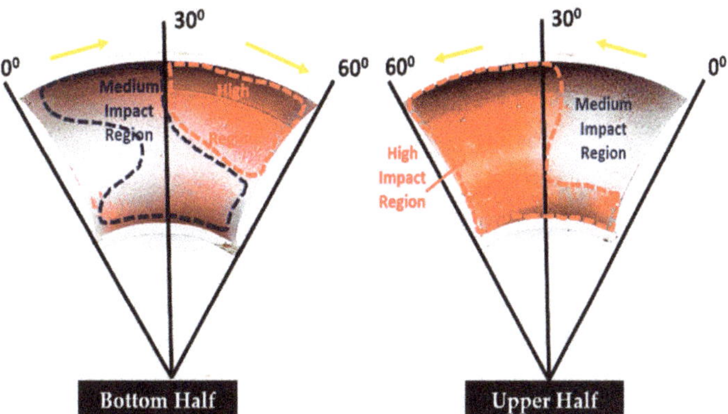

Figure 7. Erosion pattern on the 60° elbow coated with a two-layer paint.

3.2. Roughness Measurements and SEM Microscopic Imaging

In erosion, the degradation mechanism usually varied from upstream to downstream on the elbow's internal surface. Identifying the erosion mechanism is important, because it provides a pattern of the degree of wear at different locations. Therefore, after the test, samples were subjected to surface roughness evaluation and microscopic imaging to identify the wear mechanism due to the multiphase flow.

Surface roughness (Ra) values were measured in the flow direction for all 36 specimens cut from 90° and 60° elbows. Three measurements were conducted along the length of the surface of 10 mm of the cut specimen. In Figure 8a, the arithmetic surface roughness values (Ra) of the 1018 carbon steel depending on the location of the bottom elbow section are given. In Figure 8a,b, the arithmetic surface roughness values of the carbon steel varied dramatically as the flow approached from upstream to downstream. The maximum Ra value was observed between 45° and 90° axial angles at the upper and bottom half of the 90° elbows. Hence, it was concluded that carbon steel 90° elbow showed maximum erosion behavior near the outlet, as mentioned in the literature [17]. On the other hand, the Ra of the samples was maximum in the outer wall in both the top and bottom of the elbow pipe. In Figure 9a,b, arithmetic surface roughness values of the 60° elbow samples at the different locations inside the elbow are given. It can be seen that the surface roughness was significantly changed depending on the location. It was clearly observed that surface roughness increased in downstream locations. On the other hand, the maximum surface roughness was between the axial angle of 30° and 60°; however, it decreased in the upstream location. As a result, it can be said that the arithmetic roughness value of the carbon steel generally increased with increases in particle impingement. It was reported in the literature that erosion increases the surface roughness of metallic materials [20].

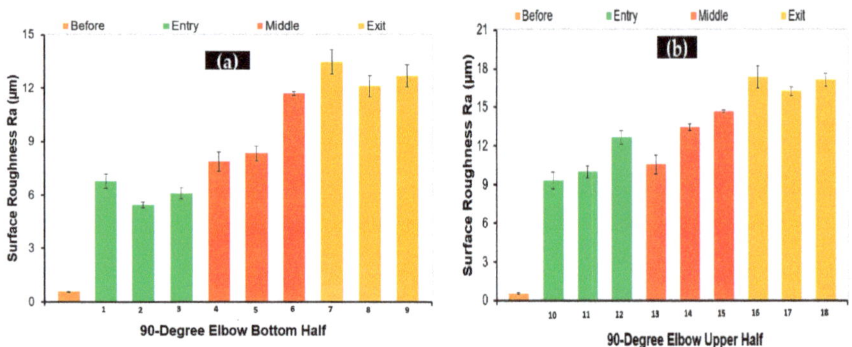

Figure 8. Arithmetic surface roughness values (Ra) before and after the test in 90° elbows: (**a**) bottom; (**b**) top.

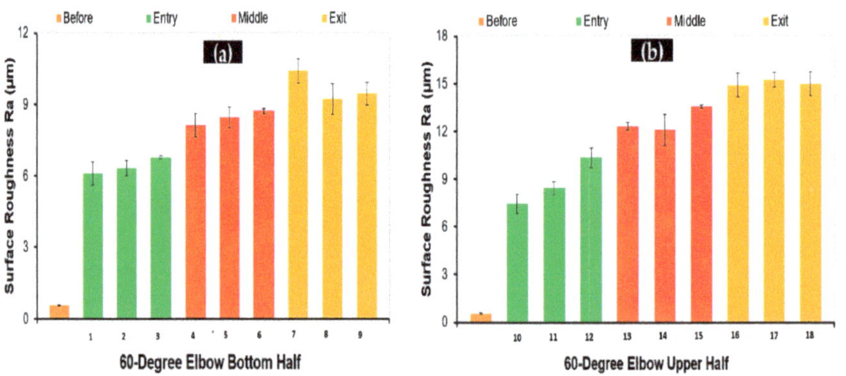

Figure 9. Arithmetic surface roughness values (Ra) before and after tests in the 60° elbows: (**a**) bottom; (**b**) top.

SEM micrographs of the carbon steel elbow specimen from the upper half and bottom half having maximum arithmetic surface roughness values (Ra) after erosion are shown in Figures 10 and 11. The SEM images showed the surface morphology after the carbon steel elbow was exposed to silt particles under plug flow conditions (a total of 10 h of erosion). Magnified images of sections 6, 7, 16, and 18 with maximum Ra values is shown in Figure 10. Upon impact with silt particles, it locally damaged the surface by the cutting and ploughing action at downstream sections during the erosion and formed pits with corrosion attack at outlet due to the high particle wall impactions.

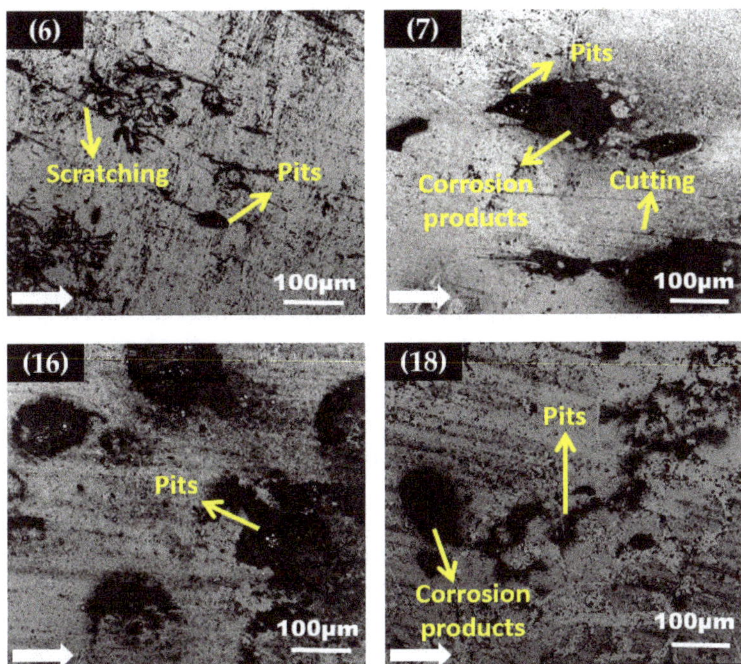

Figure 10. The backscattered electron (BSE) images of a carbon steel 90° elbow after the test.

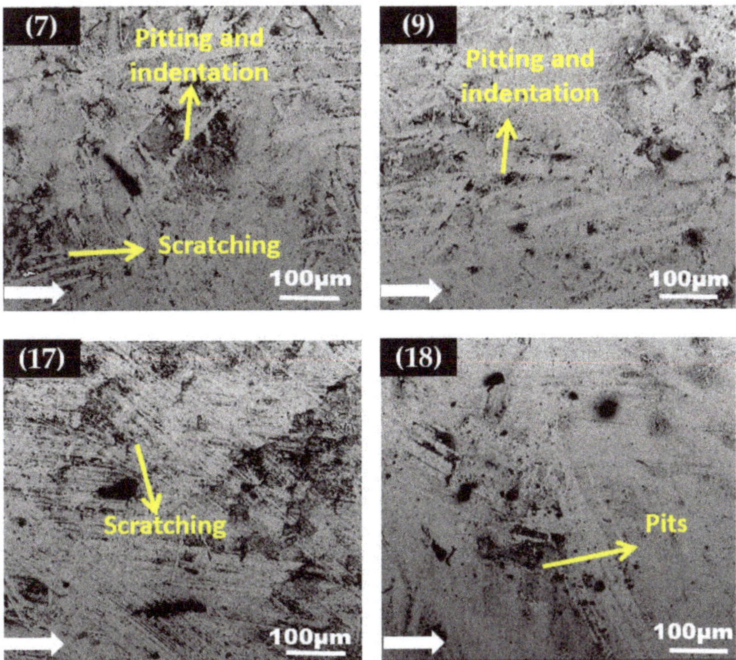

Figure 11. The backscattered electron (BSE) images of a carbon steel 60° elbow after the test.

At the top of the 90° elbow, more areas of pitting and cutting were observed, and this was due to the silt particles impacting with higher kinetic energy which disintegrated materials. Additionally, indentation was also seen at the outlet due to the plastic deformation. Such plastic deformation is usually due to the silt particles redirecting at a curvature and impacting the outer curvature of the elbow pipe. Scratching, pitting, and ploughing are the predominant erosion mechanisms in the downstream section of the elbow. In the outlet, more perforation sites with corrosion attacks were observed which suggests that erosion-corrosion pitting is the dominant erosion mechanism here. Figure 11 presents the erosion mechanism on a carbon steel sample observed after 10 h of testing with silt particles for 60° elbow. After the test, the downstream sections 9 and 18 have smooth areas on the target surface, but in sections 7 and 17 multiple particle impact was visible. In 90° elbow, more pits were detected downstream, pitting corrosion mainly most often at sensitive sites, such as at high particle-wall impaction zones. If these zones are predominant, pitting corrosion is confined to new locations on the surface. Therefore, the number of perforation pits sites will gradually coalesce with the stable pits. In comparison the 60° elbow SEM micrographs showed minimal particle impaction, resulting in the sliding and indentation with relatively small pits.

Significant disparities in the surface morphologies and development of the pattern of corrosion products at the elbow exit surface after 10 h of tests with silt particles were visible from the SEM microscopy analysis. It can be seen from micrographs that after exposure to the silt particles in the plug flow, the pitting corrosion profile was attributed to a large corrosion zone around the pits, which was covered by a corrosion products layer vicinity of elliptical pits and wide–narrow pits that were evident in the micrographs of the elbow exit section. Moreover, the pits in the 60° elbow grew individually, and the 90° elbows showed sensitive sites that grew up to elongated pits as identified in Figure 10.

The corrosion product concentrated around the pits could be seen in the SEM images. An intriguing observation was that corrosion was detected at the pits because the corrosion product washed out the onset of pits with the increase in particle impact. Notwithstanding, the extent of erosion-induced damage to the 90° elbow was more than those of the 60° elbow surface in the same flow conditions. In the 60° elbow, the erosive wear was less, because the change in the flow direction of the 60° (small angle) elbow was not as abrupt as for the 90° (wide angle) elbow. Apparently, fewer particles are prone to impact the 60° elbow's outer curvature compared to a 90° elbow outer wall. At the 60° elbow, the flow was redirected more smoothly, which causes the abrasive particles to follow the flow and impact the bottom part with less frequency as compared to the upper part of the elbow.

The EDS method is used for identification and quantifying elemental compositions after the test for sample #16 in the 90° elbow. The analytical identification of the elements (elemental composition) after the erosion imparted the presence of iron (Fe) and oxygen (O) atoms on sample #16 at the 90° elbows, as shown in Figure 12. In addition, the identification of Si on the eroded surface confirmed that the silt was embedded in the surface after the erosion.

3.3. Mass Loss

Figure 13a,b show the erosion rate for 90° and 60° elbow pipes under plug flow with silt particles. For both elbows, the maximum erosion rate occurred downstream near the outlet. In the 90° elbow, the corresponding maximum erosion rates were 10.23 mm/year compared to the 60° elbow which was 5.67 mm/year. Regardless of the elbow angle, Figure 13b shows that specimens at locations 16 and 18 of the 90° elbow's upper half provided the highest erosion, and specimens at locations 16 and 17 of the 60° elbow's upper half reflected the maximum wear rates compared with other positions. This identified that the location adjacent to an outlet for both the 90° and 60° elbows was likely to be eroded during the silt particle's impact under plug flow conditions. Moreover, the wear rate of the upper half elbow section was more than that of the bottom half. It can be concluded that in plug flow, the top of the elbow downstream is more prone to erosion due to the multiple particle impactions than in other positions. There was an approximately 1.8 times increase in the erosion rate of the maximum impaction region in the 90° elbow compared to the 60° elbow observed for carbon steel for identical flow conditions. The severe silt particle impaction was located at the outer curvature in the 50° position in 60° elbows and the 80° position in 90° elbows in plug flow.

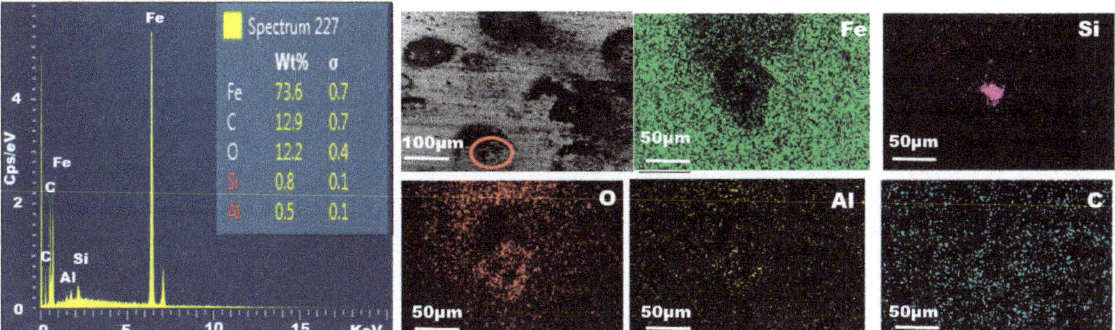

Figure 12. EDS spectra and elemental mapping after erosion in 90° elbow.

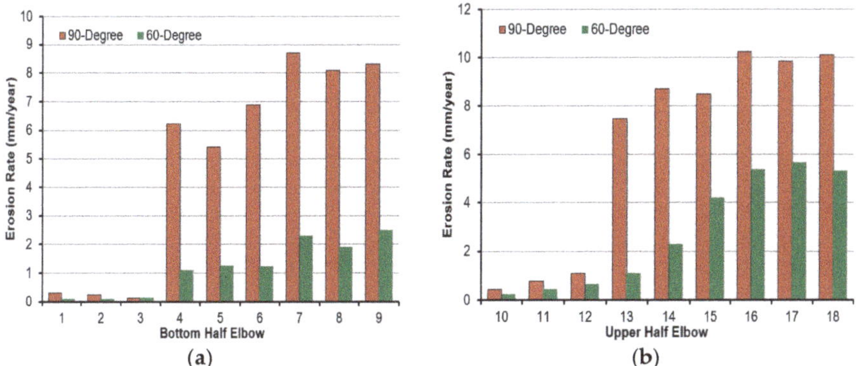

Figure 13. Mass loss in carbon steel elbow section after test: (**a**) bottom; (**b**) top.

3.4. Hardness Measurements

Figure 14 shows the results of microhardness evaluation at the different locations of the 90° and 60° elbows. As indicated in Figure 14, the hardness of the polished specimen after the test of carbon steel 1018 samples increased due to the impact of silt particles as flow approached downstream. Figure 14 shows a similar trend in the results for both the 90° elbow and the 60° elbow. It was clear from the results that the maximum hardness in both 90° and 60° elbows was, however, observed adjacent to the outlet, which was due to maximum particle impaction. The erosive wear leads to strain that hardened the target surface and the hardness of the carbon steel improved from 168 to 199 in the 90° elbow exit section and to 181 in the 60° elbow exit section on the Vickers hardness scale. The escalation in the hardness value after erosive wear in Figure 14 was accordant with the findings in a previous study [20].

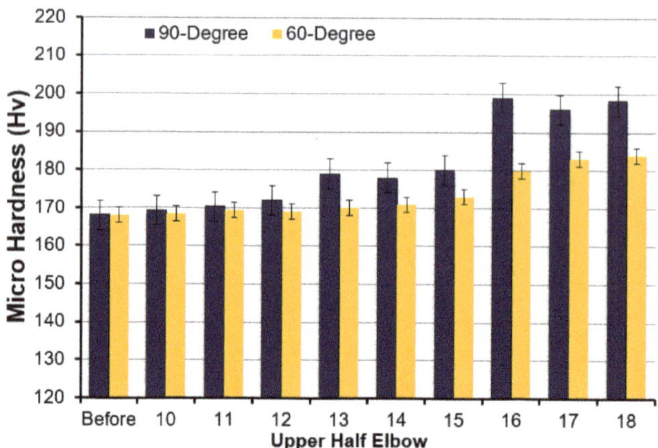

Figure 14. Microhardness of carbon steel elbows' upper half sections before and after the test.

4. Conclusions

This paper investigated the influence of elbow angle on the erosive behavior of carbon steel due to the impaction of silt particles in plug flow conditions. A total of 36 specimens were cut from the upper and bottom halves of the 90° elbow and the 60° elbow at various locations. Moreover, a paint removal method, mass loss analysis, microscopic imaging, surface roughness evaluation, and microhardness analysis were employed to study the relationship between the erosion distribution and the elbow angle. The following conclusions were drawn:

1. In plug flow, the erosive wear increased significantly with a change in elbow angle from 60° to 90°. Compared with the 60° elbow, there was an approximately 1.8 times increase in maximum erosion rate in 90° elbows for identical flow conditions;

2. At the top of the 90° and 60° elbows adjacent to the outlet, the erosion maximized due to the redirected flow, and the maximum silt particle impaction was identified at the outer curvature in the 50° position in the 60° elbow and the 80° position in the 90° elbow in plug flow. In the 60° elbow, the erosive wear was less because the change in the flow direction of the 60° (small angle) elbow was not as abrupt as for the 90° (wide angle) elbow;

3. The arithmetic mean surfaces roughness of the samples was dramatically influenced by elbow angle. The surface roughness values and microhardness obtained showed that the surface roughness and hardness of the samples were increased on the top of the elbow compared to the bottom part at the elbow exit. The silt particle impact on the surface of the 60° and 90° elbows in the top part and the subsequent surface damage through scratching, pitting, and material removal resulted in subsequent strain hardening of the surface, which resulted in increased surface roughness and hardness;

4. The microscopic study of the eroded sample showed that the primary causes of wear in the 90° elbow included pitting, ploughing, and cuttings. The microscopic images of the test specimens manifested that pitting, scratching, and indentation eventuated, which is an indication of plastic deformation due to the impact of the silt particles. The progressive effect of pitting, scratching, and indentation increased erosion in the exit of the 90° elbow pipe.

Author Contributions: R.K. designed the experimental setup; R.K. and H.H.Y. performed the analysis and wrote the manuscript; I.S. and U.M.N. guided the data analysis and validation; B.A.A., M.A., M.A.A. and T.A. guided in investigation and project management; the conceptualization, visualization, funding management, editing, and proofreading were performed by M.I., A.G., Z.P. and F.B. All authors have read and agreed to the published version of the manuscript.

Materials **2022**, 15, 3721

Funding: This research, which was carried out under the theme Department of Electrical Engineering E-2, was founded by the subsidies on science granted by the Polish Ministry of Science and Higher Education for journal fee payment.

Institutional Review Board Statement: Not applicable.

Informed Consent Statement: Not applicable.

Data Availability Statement: The data presented in this study are available upon request from the corresponding author.

Acknowledgments: The authors acknowledge the support from the Deanship of Scientific Research, Najran University, Kingdom of Saudi Arabia, for funding this work under the Research Collaboration funding program (grant no: NU/RC/SERC/11/2). The authors also acknowledge the YUTP-FRG (grant no. 0153AA-H19) from the Universiti Teknologi PETRONAS internal research fund for purchasing the research equipment and providing research facilities.

Conflicts of Interest: The authors declare no conflict of interest.

References

1. Li, Y.; Cao, J.; Xie, C. Research on the Wear Characteristics of a Bend Pipe with a Bump Based on the Coupled CFD-DEM. *J. Mar. Sci. Eng.* **2021**, *9*, 672. [CrossRef]
2. Zhao, X.; Cao, X.; Xie, Z.; Cao, H.; Wu, C.; Bian, J. Numerical study on the particle erosion of elbows mounted in series in the gas-solid flow. *J. Nat. Gas Sci. Eng.* **2022**, *99*, 104423. [CrossRef]
3. Wang, Q.; Huang, Q.; Wang, N.; Wen, Y.; Ba, X.; Sun, X.; Zhang, J.; Karimi, S.; Shirazi, S.A. An experimental and numerical study of slurry erosion behavior in a horizontal elbow and elbows in series. *Eng. Fail. Anal.* **2021**, *130*, 105779. [CrossRef]
4. Wang, Q.; Ba, X.; Huang, Q.; Wang, N.; Wen, Y.; Zhang, Z.; Sun, X.; Yang, L.; Zhang, J. Modeling erosion process in elbows of petroleum pipelines using large eddy simulation. *J. Pet. Sci. Eng.* **2022**, *211*, 110216. [CrossRef]
5. Khan, R.; Ya, H.H.; Pao, W.; bin Abdullah, M.Z.; Dzubir, F.A. Influence of Sand Fines Transport Velocity on Erosion-Corrosion Phenomena of Carbon Steel 90-Degree Elbow. *Metals* **2020**, *10*, 626. [CrossRef]
6. Parsi, M.; Najmi, K.; Najafifard, F.; Hassani, S.; McLaury, B.S.; Shirazi, S.A. A comprehensive review of solid particle erosion modeling for oil and gas wells and pipelines applications. *J. Nat. Gas Sci. Eng.* **2014**, *21*, 850–873. [CrossRef]
7. Mansouri, A.; Shirazi, S.A.; McLaury, B.S. Experimental and Numerical Investigation of the Effect of Viscosity and Particle Size on the Erosion Damage Caused by Solid Particles. In Proceedings of the ASME 2014 4th Joint US-European Fluids Engineering Division Summer Meeting collocated with the ASME 2014 12th International Conference on Nanochannels, Microchannels, and Minichannels, Chicago, IL, USA, 3 August 2014.
8. Javaheri, V.; Porter, D.; Kuokkala, V.-T. Slurry erosion of steel—Review of tests, mechanisms and materials. *Wear* **2018**, *408–409*, 248–273. [CrossRef]
9. Sedrez, T.A.; Rajkumar, Y.R.; Shirazi, S.A.; Khanouki, H.A.; McLaury, B.S. CFD predictions and experiments of erosion of elbows in series in liquid dominated flows. In Proceedings of the Fluids Engineering Division Summer Meeting, Montreal, QC, Canada, 15–20 July 2018; p. V003T017A001.
10. Owen, J.; Ducker, E.; Huggan, M.; Ramsey, C.; Neville, A.; Barker, R. Design of an elbow for integrated gravimetric, electrochemical and acoustic emission measurements in erosion-corrosion pipe flow environments. *Wear* **2019**, *428–429*, 76–84. [CrossRef]
11. Vieira, R.E.; Mansouri, A.; McLaury, B.S.; Shirazi, S.A. Experimental and computationcheckal study of erosion in elbows due to sand particles in air flow. *Powder Technol.* **2016**, *288*, 339–353. [CrossRef]
12. Wang, K.; Li, X.; Wang, Y.; He, R. Numerical investigation of the erosion behavior in elbows of petroleum pipelines. *Powder Technol.* **2017**, *314*, 490–499. [CrossRef]
13. Vieira, R.E.; Parsi, M.; Zahedi, P.; McLaury, B.S.; Shirazi, S.A. Sand erosion measurements under multiphase annular flow conditions in a horizontal-horizontal elbow. *Powder Technol.* **2017**, *320*, 625–636. [CrossRef]
14. Naz, M.Y.; Shukrullah, S.; Noman, M.; Yaseen, M.; Naeem, M.; Sulaiman, S.A. Effect of Water-Mixed Polyvinyl Alcohol Viscosity on Wear Response of Carbon Steel Exposed to an Eroding Medium. *J. Mater. Eng. Perform.* **2021**, *30*, 2066–2073. [CrossRef]
15. Cao, X.; Li, X.; Fan, Y.; Peng, W.; Xie, Z. Effect of superficial air and water velocities on the erosion of horizontal elbow in slug flow. *Powder Technol.* **2020**, *364*, 785–794. [CrossRef]
16. Zahedi, P.; Parsi, M.; Asgharpour, A.; McLaury, B.S.; Shirazi, S.A. Experimental investigation of sand particle erosion in a 90° elbow in annular two-phase flows. *Wear* **2019**, *438–439*, 203048. [CrossRef]
17. Khan, R.; Ya, H.H.; Pao, W.; Majid, M.A.A.; Ahmed, T.; Ahmad, A.; Alam, M.A.; Azeem, M.; Iftikhar, H. Effect of Sand Fines Concentration on the Erosion-Corrosion Mechanism of Carbon Steel 90° Elbow Pipe in Slug Flow. *Materials* **2020**, *13*, 4601. [CrossRef] [PubMed]
18. Khan, R.; Ya, H.H.; Pao, W.; Khan, A. Erosion–Corrosion of 30°, 60°, and 90° Carbon Steel Elbows in a Multiphase Flow Containing Sand Particles. *Materials* **2019**, *12*, 3898. [CrossRef] [PubMed]

19. Xu, Y.; Liu, L.; Zhou, Q.; Wang, X.; Tan, M.Y.; Huang, Y. An Overview of Major Experimental Methods and Apparatus for Measuring and Investigating Erosion-Corrosion of Ferrous-Based Steels. *Metals* **2020**, *10*, 180. [CrossRef]
20. Elemuren, R.; Tamsaki, A.; Evitts, R.; Oguocha, I.N.A.; Kennell, G.; Gerspacher, R.; Odeshi, A. Erosion-corrosion of 90° AISI 1018 steel elbows in potash slurry: Effect of particle concentration on surface roughness. *Wear* **2019**, *430–431*, 37–49. [CrossRef]

Article

Effect of Microstructure and Hardness on Cavitation Erosion and Dry Sliding Wear of HVOF Deposited CoNiCrAlY, NiCoCrAlY and NiCrMoNbTa Coatings

Mirosław Szala *, Mariusz Walczak and Aleksander Świetlicki *

Department of Materials Engineering, Faculty of Mechanical Engineering, Lublin University of Technology, 20-618 Lublin, Poland; m.walczak@pollub.pl
* Correspondence: m.szala@pollub.pl (M.S.); aleksander.swietlicki@pollub.edu.pl (A.Ś.)

Abstract: Metallic coatings based on cobalt and nickel are promising for elongating the life span of machine components operated in harsh environments. However, reports regarding the ambient temperature tribological performance and cavitation erosion resistance of popular MCrAlY (where M = Co, Ni or Co/Ni) and NiCrMoNbTa coatings are scant. This study comparatively investigates the effects of microstructure and hardness of HVOF deposited CoNiCrAlY, NiCoCrAlY and NiCrMoNbTa coatings on tribological and cavitation erosion performance. The cavitation erosion test was conducted using the vibratory method following the ASTM G32 standard. The tribological examination was done using a ball-on-disc tribometer. Analysis of the chemical composition, microstructure, phase composition and hardness reveal the dry sliding wear and cavitation erosion mechanisms. Coatings present increasing resistance to both sliding wear and cavitation erosion in the following order: NiCoCrAlY < CoNiCrAlY < NiCrMoNbTa. The tribological behaviour of coatings relies on abrasive grooving and oxidation of the wear products. In the case of NiCrMoNbTa coatings, abrasion is followed by the severe adhesive smearing of oxidised wear products which end in the lowest coefficient of friction and wear rate. Cavitation erosion is initiated at microstructure discontinuities and ends with severe surface pitting. CoNiCrAlY and NiCoCrAlY coatings present semi brittle behavior, whereas NiCrMoNbTa presents ductile mode and lesser surface pitting, which improves its anti-cavitation performance. The differences in microstructure of investigated coatings affect the wear and cavitation erosion performance more than the hardness itself.

Keywords: cavitation corrosion; wear; surface engineering; roughness; nickel; cobalt; tribology; hardness; erosion rate; failure analysis; MCrAlY

Citation: Szala, M.; Walczak, M.; Świetlicki, A. Effect of Microstructure and Hardness on Cavitation Erosion and Dry Sliding Wear of HVOF Deposited CoNiCrAlY, NiCoCrAlY and NiCrMoNbTa Coatings. *Materials* **2022**, *15*, 93. https://doi.org/10.3390/ma15010093

Academic Editors: Artur Czupryński and Claudio Mele

Received: 3 December 2021
Accepted: 21 December 2021
Published: 23 December 2021

Publisher's Note: MDPI stays neutral with regard to jurisdictional claims in published maps and institutional affiliations.

1. Introduction

Thermal spraying processes are commonly used methods for metallic materials restoration and modification of surface layer properties that are applied in many industries. The process is used in the aerospace industry, automotive engineering, and maritime sectors [1–3]. One of the most popular thermal spraying methods is the HVOF (high velocity oxygen-fuel). The HVOF method is most often used because the process is characterized by a high particle velocity, relatively low temperature and a short time of particle exposure in the stream, which results in a low content of oxides and minimal porosity [4,5]. In comparison to other thermal spraying techniques such as arc, flame spraying, or APS coatings deposited using the HVOF, they are usually characterized by high adhesion to the substrate and low porosity [6–9]. This technology allows for the extension of the service life of materials, the improving of the mechanical properties, and the increase in the operational performance of the engineering devices. These features determine the economic benefits of HVOF-deposited coatings. Systematic broadening of the coatings industrial applications followed by the diversity of the feedstock materials makes the thermally sprayed coatings a crucial subject of recent scientific papers.

A broad range of materials is deposited via the HVOF method, such as metals, ceramics and composites. Overall, the main application of thermally sprayed coatings is protection against different types of wear [10–13], and more sophisticated implementation involves the prevention from cavitation erosion [14–16]. Besides, industrial applications combine these two deterioration processes, such as pressure values, diesel liners or specialised chemical equipment. As an example, MCrAlY coatings (where M = Co, Ni or Co/Ni) are widely used in the aviation industry, as MCrAlY is used as the bond coat of TBCs or as a standalone protective coating to increase resistance to high temperatures and high pressure occurring in gas turbines [17,18]. Therefore, the wear resistance of MCrAlY and NiCoMo coatings is most often analysed in publications in terms of high-temperature environments [19–22]. Few studies describe the behaviour of these coatings at room temperature, and to the best of the authors' knowledge, there is no comparative analysis for the MCrAlY and NiCrMo coatings in terms of dry sliding wear and cavitation erosion behavior.

Similarly to tribological deterioration, cavitation erosion (CE) can harm the service life of machinery and equipment [23–26]. Therefore, HVOF coatings constituted of nickel and/or cobalt-based feedstock powders are often used as a protective layer [16,27–29]. However, in terms of cavitation erosion, it is believed that cobalt-based materials present one of the highest cavitation erosion resistances [28,30,31]. On the other hand, the nickel-based powders also give interesting anti-cavitation erosion results [32–34]. Still, the effect of the metallic matrix type, influencing microstructure and other properties, is not clearly stated, especially in the case of HVOF Ni/Co-based coatings, including MCrAlY and NiCrMo. Besides, nickel-based coatings present desirable properties in anti-wear and cavitation, preventing applications due to the availability of various deposition methods, high adhesion of the coating to the substrate, and resistance to corrosion required in the wet environment. Furthermore, plastic deformation of nickel-based coatings may also reduce the rate of CE damage of the material by absorbing the energy generated by cavitation pressure waves and micro-jets [35,36]. Therefore, it seems interesting to investigate the CE resistance of HVOF coatings containing nickel and clarify the effect of coatings properties on the CE behavior.

Furthermore, depositing a protective MCrAlY and NiCrMo coating on a base metal substrate is usually studied concerning their typical high-temperature operation conditions [36,37], while their sliding behaviour at ambient temperature and anti-cavitation performance are not thoroughly investigated by the literature of the object [38]. Nevertheless, a well-known fact is that cobalt-based HVOF metallic [28,39,40] and Co-WC, CoCr-WC [41–43] cermet coatings present superior tribological performance and resistance to cavitation erosion. The anti-cavitation properties of Ni/Co-based HVOF deposits are not completely discussed. Mainly, limited studies describe the research on CE resistance for MCrAlY, NiCoMo deposits. Therefore, this study comparatively analysed the coatings microstructure and properties with a resistance to cavitation erosion and dry sliding wear at ambient temperature. The results give useful remarks for broadening the knowledge regarding dry sliding wear and cavitation erosion failure mechanisms of MCrAlY and NiCrMo coatings.

This work aimed to investigate the effect of microstructure and hardness of HVOF sprayed metallic coatings on their tribological and cavitation erosion performance. Moreover, sliding wear and cavitation erosion mechanisms were studied. The research analysed commercial HVOF coatings deposited from popular nickel-based feedstock powders: CoNiCrAlY, NiCoCrAlY and NiCrMoNbTa.

2. Materials and Methods

2.1. Materials

In this article, three coatings thermally sprayed via the HVOF (High Velocity Oxygen Fuel) method using popular commercial powders were examined to state the effect of nickel content on tribological and cavitation erosion (CE) behaviour. (Ni,Co)CrAlY and NiCrAlNbTa feedstock powders were sprayed on 2 mm thick Inconel 617 sandblasted substrates and thickness of coatings yielding 100 ± 25 µm. The nominal chemical composition of substrate and feedstock powders are given in Tables 1 and 2, respectively.

Table 1. Nominal chemical composition of Inconel 617.

	Ni	Cr	Mo	Cu	Co	C	Mn	Si	S	Fe	Ti	Al	B
Wt.%	44.5	20–24	8–10	0–0.5	10–15	0.05–0.15	0–1	0–1	0–0.015	0–3	0–0.6	0.8–1.5	0–0.006

Table 2. Nominal chemical composition of feedstock powders used for HVOF spraying.

CoNiCrAlY		NiCoCrAlY		NiCrMoNbTa	
Element	(wt.%)	Element	(wt.%)	Element	(wt.%)
Ni	29.00–35.00	Ni	Bal.	Ni	Bal.
Co	Bal.	Co	22.00	Co	-
Cr	18.00–24.00	Cr	17.00	Cr	21.50
Al	5.00–11.00	Al	12.50	Al	-
Y	0.10–0.80	Y	0.55	Y	-
Mo	-	Mo	-	Mo	9.00
Nb + Ta	-	Nb + Ta	-	Nb + Ta	3.70
Fe	-	Fe	-	Fe	2.50

The CoNiCrAlY coating was fabricated using cobalt-based powder Amdry 9954 manufactured by Oerlikon Metco (Pfäffikon, Switzerland), which is intended to protect surfaces against oxidation and corrosion at high temperatures above 850 °C. The second coating of NiCoCrAlY nickel-based powder is made of a powder called Ni-191-4 (PRAXAIR), which is also designed to produce high-temperature layers, protecting against oxidation and high-temperature corrosion. The last tested NiCrMoNbTa coating was sprayed using spheroidal powder with the trade name Diamalloy 1005 (Oerlikon Metco), which is dedicated to surface regeneration, protection of the surface against oxidation and corrosion at high temperatures above 982 °C. MCrAlY and NiCrMoNbTa coatings are usually studied concerning their typical high-temperature operating conditions. The other practical applications are not identified. Thus, this paper comparatively analyses their tribological and anti-cavitation performance.

2.2. Research Methodology

The coatings were investigated using scanning electron microscopy (SEM), X-ray phase analysis (XRD) and profilometer measurements. The surface morphology of the as-sprayed coatings and cross-section microstructures of the coatings were examined using the SEM-EDS method. The phase composition of the coatings was investigated by the X-ray diffraction method. The phase composition studies were carried out according to the procedure described in a previous paper [38]. The surface roughness Ra parameter was determined using the surface profiler (Surtronic S-128, Taylor-Hobson, Leicester, UK) according to the ISO 4287 standard. Finally, the hardness of the tested coatings was measured following the PN-EN ISO 6507-1 standard. The tests were carried out on polished transverse specimens. Investigations were done with the use of the load of 0.9807 N and the dwelling time of 10 s. At the latest, ten indentations were made to obtain statistical accuracy.

In order to compare the wear resistance of as-sprayed HVOF coatings with other engineering materials tested by our group [44–46] the ball-on-disc method was employed. Wear tests were carried out on a CS-Instruments ball tribotester. The idea of the wear test is presented in Figure 1. As a counter-sample (ball), six mm diameter balls made of tungsten carbide-cobalt (WC–Co) cemented carbide (supplied by CSM Instruments, Needham Heights, MA, USA) were used. The tests were carried out under an applied load of F = 10 N. The tests were carried out with a linear speed of 0.05 m/s for a radius of 3 mm and a sliding distance of 200 m. Test conditions prevent local abrasion of coatings to the substrate. The friction coefficient was also determined. Additionally, the wear factor K was determined using the procedure and the Equation (1) described in a previous paper [44].

$$K = \frac{volume\ loss\ \left[\text{mm}^3\right]}{load\ [N]\cdot sliding\ distance\ [\text{m}]} \tag{1}$$

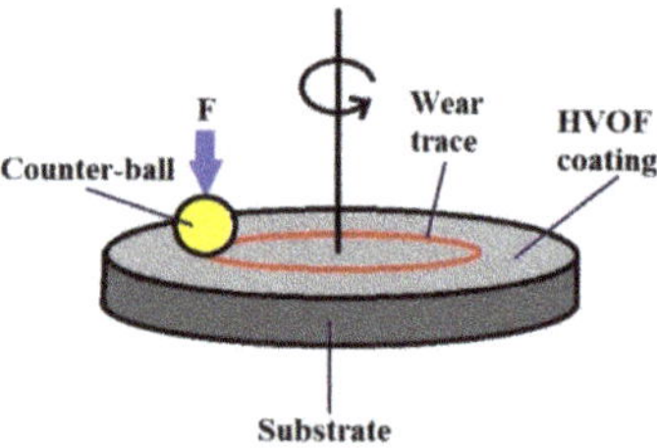

Figure 1. Scheme of ball-on-disc wear testing.

Additionally, this research aimed to test the wear resistance of the as-sprayed coatings, i.e., investigated coatings differ in surface roughness. Therefore sliding wear resistance (estimated by wear factor) was validated using the weighting method. HVOF coatings mass loss was measured with an accuracy of 0.1 mg. Finally, the sliding wear mechanism was comparatively investigated using the SEM-EDS method.

The cavitation erosion resistance tests were carried out in accordance with the ASTM G32 standard [47]. The sonotrode tip distance from the sample was 1 mm $\pm$ 0.05 mm, the medium in which cavitation was induced was distilled water, see Figure 2. The tip working area equals 1.92 cm^2. The analysis of the resistance to cavitation erosion consisted of systematic measurements of the weight loss of the tested samples with an accuracy of 0.01 mg. The cavitation testing conditions conform to the previous research conditions done for APS [45], HVOF [48], cold spray [49] and electrostatic [50] deposits investigated by our group. In the current study, the total exposure time lasted 3 h. The study cavitation erosion mechanisms, surfaces of test samples were finished by grinding with emery papers # 240, # 320 and # 600 to achieve the surface roughness of Ra < 5.28 μm, Rt < 38.3 μm and Rz < 28.5 μm. At stated time intervals of the cavitation erosion testing, the eroded surfaces roughness measurements and SEM microscopic analysis were carried out.

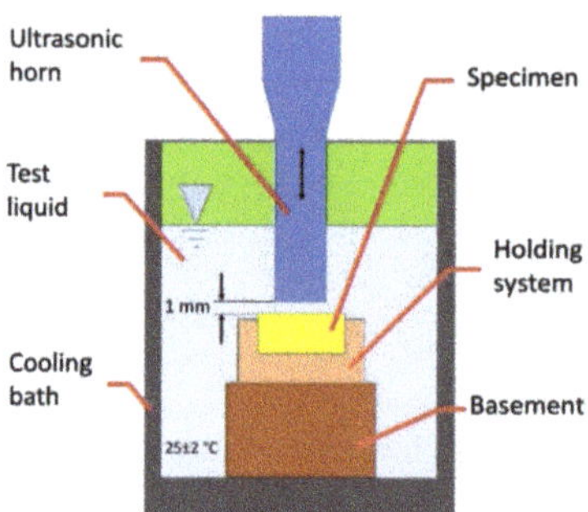

Figure 2. Schematic representation of the ultrasonic vibratory system used for cavitation.

3. Results and Discussion

3.1. Microstructure and Coatings Properties

The tests carried out showed the presence of a microstructure characteristic for thermally sprayed coatings. The as-sprayed surfaces are characterised by the presence of splats, unmelted particles and oxides (Figure 3). Much more uniform surface roughness was identified for NiCrMoNbTa than for MCrAlY's (Figure 4). Besides, the mean value of the Ra roughness parameter was the highest for the CoNiCrAlY coating (7.94 µm) and presented the widest spread of results. The obtained value compares to the mean roughness of the same type of HVOF coating (Ra = 7.41 µm) reported by Rajasekaran et al. [51]. As-sprayed NiCoCrAlY presents surface roughness even lower than those reported by the literature for the same type of MCrAlY [52]. The smoothest surface roughness of the Ra (4.56 µm) was obtained for NiCrMoNbTa. Thus, the Ra roughness ranges between values reported by Oladijo et al. [53] and Al-Fadhli et al. [54] reported for HVOF sprayed Inconel 625 coatings (similar chemical composition to NiCrMoNbTa). According to the literature, the roughness of MCrAlY bond coat affects adhesion between the metallic bond coat and the ceramic topcoat and is an important factor for the extended TBC life [52,55]. Whereas one of the typical applications of NiCrMoNbTa is surface regeneration. Thus, the low coating roughness of NiCrMoNbTa is optimal for the restoration and repair technology of superalloy components.

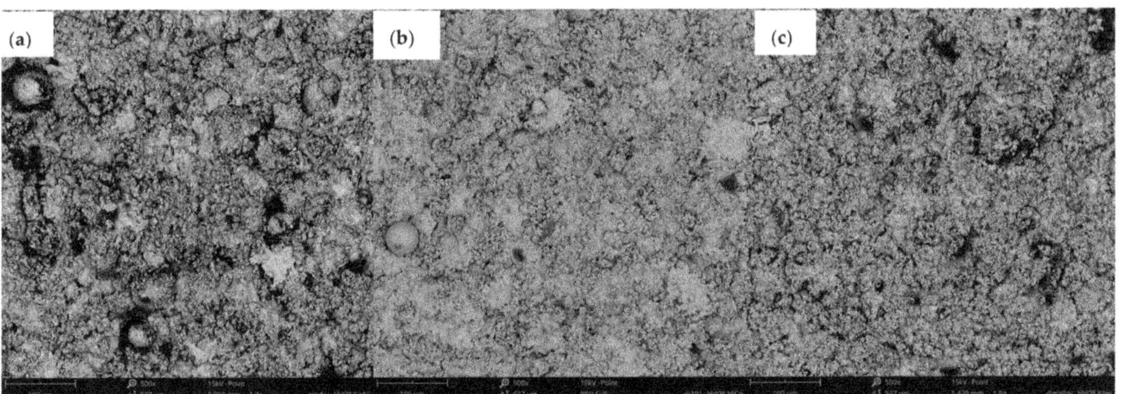

Figure 3. Surface morphology of as-sprayed coatings: (**a**) CoNiCrAlY; (**b**) NiCoCrAlY and (**c**) NiCrMoNbTa.

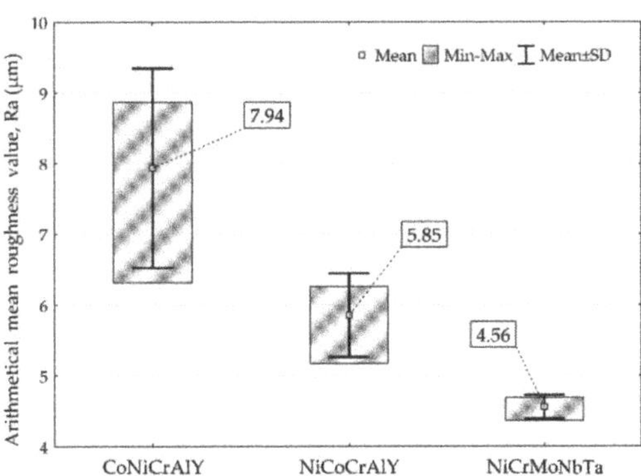

Figure 4. Comparison of the roughness of as-sprayed coatings.

The cross-section microstructures (Figure 5) show a sufficient interlocking between the applied coating and the substrate. The coatings have a lamellar structure, porosity, and oxides at the clear boundaries between the lamellas, and incompletely melted powder particles are visible. In the case of the NiCoCrAlY coating (Figure 5a), a much more dense structure was observed than in the case of MCrAlY coatings. The SEM-EDS investigations (Figure 5 and Table 3) reveal the percentage content of the main elements in the range of the nominal feedstock powders (Table 2). Due to the low nominal content and EDS method limitations, the yttrium was not detected, and the percentage content of other chemical elements varied. Both nickel and cobalt constitute the coatings metallic matrix and are crucial for phase composition and overall properties and anti-wear performance. Thus, the X-ray phase composition was investigated, see Figure 6. In the case of coating CoNiCrAlY the γ-Co,Ni phase with fcc structure was identified. On the other hand, the literature of the subject [56–59] suggests the presence of an intermetallic phase β-(Co,Ni)Al with bcc structure, although this phase wasn't confirmed in CoNiCrAlY sample. The second HVOF coating NiCoCrAlY has a two-phase composition consisting of γ-Ni(Co,Cr) and β-NiAl, which follows findings obtained for MCrAlY coating deposited by HVOF [60,61]. The third sample, NiCrMoNbTa presents a single-phase structure based on γ-Ni(NiCr) which corresponds to the results obtained for HVOF fabricated NiCrMoNb-matrix composites [12,62]. The coatings' microstructure and phase composition differs and, of course, influences coatings hardness, see Figure 7. Coatings hardness exceeds Inconel 617 substrate, namely 248.1 HV. The highest average hardness was noted for the NiCoCrAlY coating, about 393 HV. The lowest hardness was found for NiMoCrNbTa and it was 342 HV. In sum, the hardness of CoNiCrAlY exceeds those reported for NiCoCrAlY, which agrees with the trends given in the literature [21]. On the other hand, even though the mean hardness differs, it should be noted that analysis of hardness variability confirms overlapping of standard deviations, as shown in Figure 7. On the whole, the coatings nominal chemical composition affects morphology, phase composition, and hardness which are essential factors to clarify tribological and cavitation erosion performance.

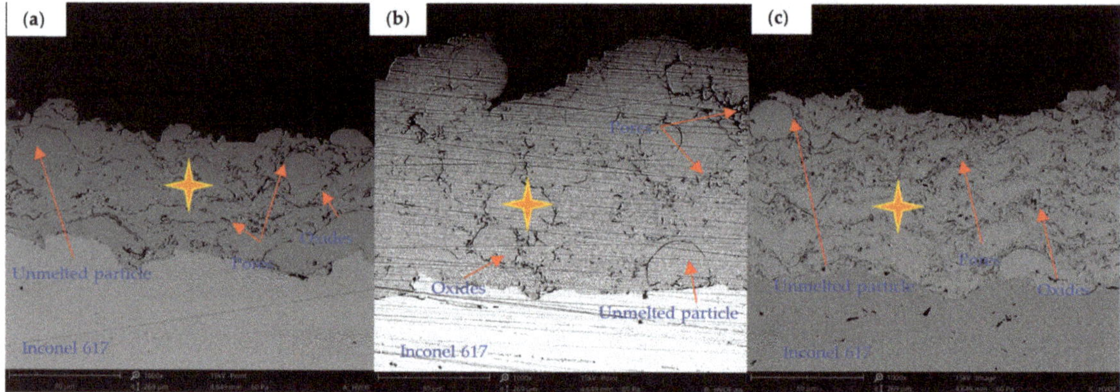

Figure 5. Coatings cross sections: (**a**) CoNiCrAlY, (**b**) NiCoCrAlY, (**c**) NiCrMoNbTa, SEM-EDS (spot analysis are given in Table 3).

Table 3. Chemical composition of coatings estimated in spots marked in Figure 5, SEM-EDS.

CoNiCrAlY		NiCoCrAlY		NiCrMoNbTa	
Element	(wt.%)	Element	(wt.%)	Element	(wt.%)
Ni	35.15	Ni	47.49	Ni	64.56
Co	37.33	Co	22.14	Co	-
Cr	19.83	Cr	20.03	Cr	18.71
Al	7.68	Al	10.34	Mo	9.34
Y	-	Y	-	Nb	4.32
				Ta	2.81
				Fe	0.25

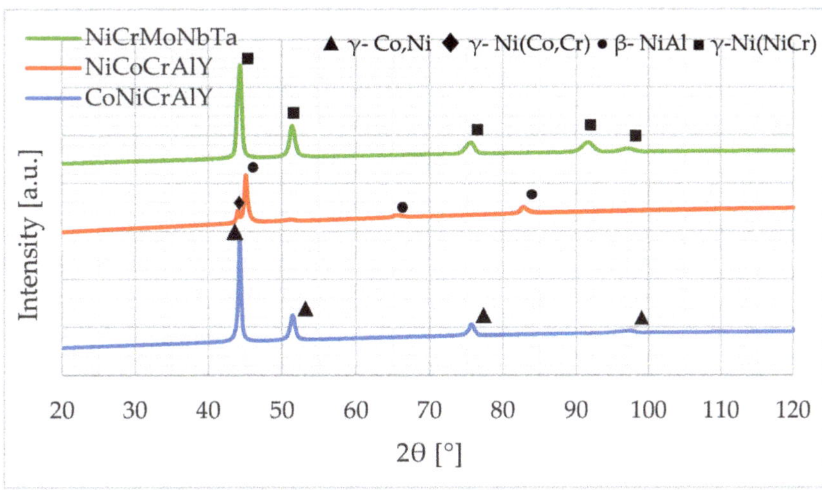

Figure 6. Comparison of phase composition of coatings, XRD.

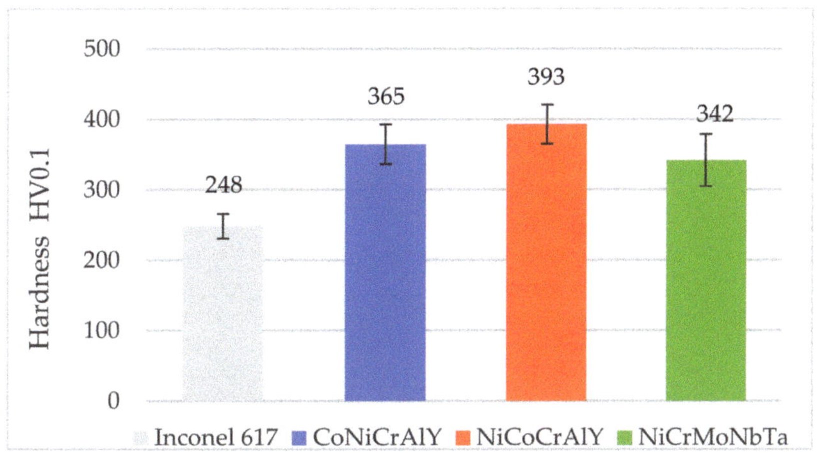

Figure 7. Hardness variability of coatings and the substrate.

3.2. Tribological Testing

The coefficient of friction (COF) was determined for the tested coatings. The results are presented in Figure 8, while mass loss and wear factor K are shown in Figures 9 and 10, respectively. The tribological tests were conducted for as-sprayed surfaces. Therefore, the accuracy of tribological results was positively validated by comparing mass losses (evaluated using the weighting method) with the volume losses (wear factors), see Figures 9 and 10. In other words, utilised test conditions allow obtaining reliable tribological results for the as-sprayed coatings. Nonetheless, the investigated HVOF coatings present inferior wear behaviour compared to other MCrAlY, and NiCrMo mostly tested at elevated temperatures [13,22,63], as well as to ceramic [10,45], metallic [44,46] materials tested at the same load at ambient temperature.

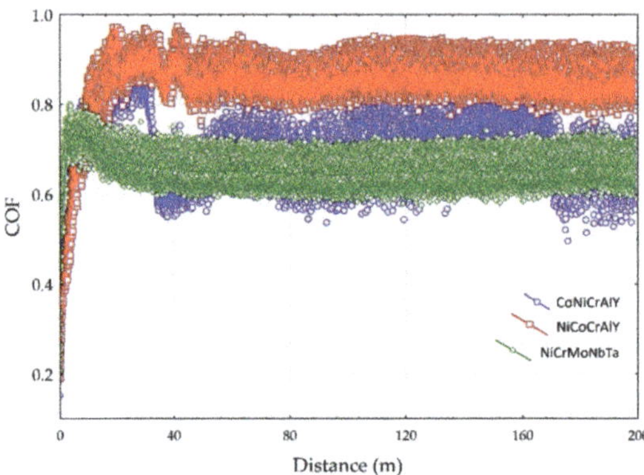

Figure 8. Coefficient of friction (COF) of the tested coatings.

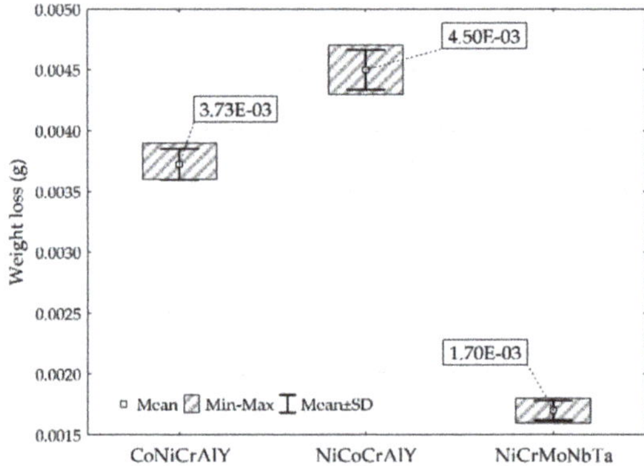

Figure 9. Mass losses of tested coatings.

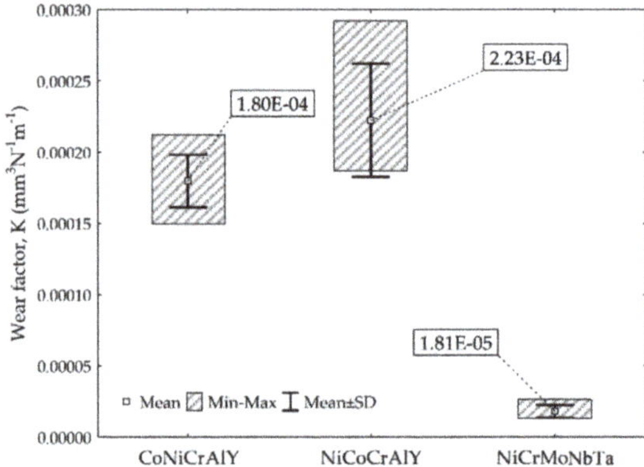

Figure 10. Sliding wear result of investigated coatings.

The influence of mean hardness on the wear resistance of the coatings was stated. NiCoCrAlY coating with the highest mean hardness 393 HV0.1 showed the worst resistance to wear. Moreover, the coating with the lowest mean hardness 342 HV0.1 showed the highest resistance to wear. It contradicts the literature, which reports that higher hardness increases material wear resistance [64]. It should be noted that differences were identified for the mean hardness and the spread of hardness results overlaps. Therefore, it can be concluded that the microstructure of investigated coatings affects the wear performance more than the hardness itself. Even though the NiCoCrAlY coating has the highest mean hardness, the presence of the second phase could be responsible for its poor tribological performance. Besides, the NiCrMoNbTa coating has the lowest COF and material loss. At the same time, the NiCoCrAlY coating had the highest COF and K factor, which can be explained by the presence of the second phase β-NiAl. Other coatings present a single-phase structure. The wear factor of NiCrMoNbTa was almost 13 times lower than the one obtained for coating NiCoCrAlY. This can be explained by the microstructure of the coating, which affects the wear mechanisms. The NiCrMoNbTa coating is much more affected by the adhesive smearing of wear debris than the MCrAlY coatings. In the case of all samples, the abrasive wear grooving and the tribochemical wear, namely oxidation of wear products, were confirmed by the SEM-EDS investigations, as shown in Figure 11. In the case of sliding testing of as-sprayed HVOF coatings, the hard WC counter-ball smashes the rough surface, and the produced wear debris acts as a third body, which is confirmed by other researchers [65]. Then the sliding wear mechanism changes to abrasion affected by grooving. Moreover, the adhesive transfer of smeared wear products is visible in CoNiCrAlY and NiCrMoNbTa coatings, shown in Figure 11a,b, which affects the decrease of wear factor and COF in comparison to NiCoCrAlY. Finally, their wear mechanism is supported by fatigue which is observed especially for wear debris. Fatigue took place as a result of repeated movement of the counter-sample on the surface of the coating and initiated microcracks, which proceeds and finally causes the material detachment and transfer through the wear tack. In the case of the NiCrMoNbTa coating shown in Figure 11c, the abrasive grooving followed by adhesive smearing of oxidised wear products. In sum, the NiCrMoNbTa coating present superior ambient dry sliding performance than the MCrAlY one.

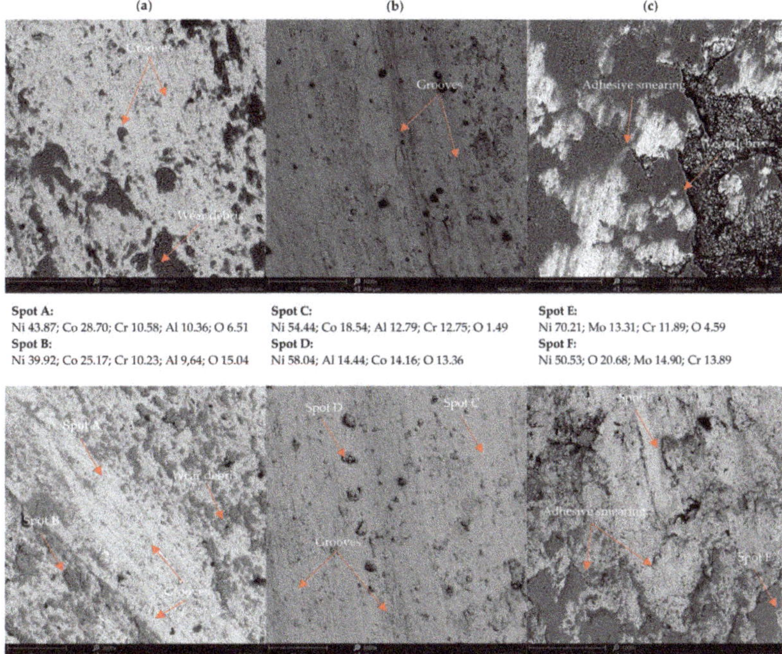

Spot A:
Ni 43.87; Co 28.70; Cr 10.58; Al 10.36; O 6.51
Spot B:
Ni 39.92; Co 25.17; Cr 10.23; Al 9,64; O 15.04

Spot C:
Ni 54.44; Co 18.54; Al 12.79; Cr 12.75; O 1.49
Spot D:
Ni 58.04; Al 14.44; Co 14.16; O 13.36

Spot E:
Ni 70.21; Mo 13.31; Cr 11.89; O 4.59
Spot F:
Ni 50.53; O 20.68; Mo 14.90; Cr 13.89

Figure 11. Wear traces and spot chemical analysis: (**a**) CoNiCrAlY, (**b**) NiCoCrAlY, (**c**) NiCrMoNbTa wear trace, SEM-EDS in wt.%.

3.3. Cavitation Erosion (CE) Resistance

The cavitation erosion (CE) results, presented in Figure 12a, indicate an almost twice lower material loss of NiCrMoNbTa than for the MCrAlY coatings. All investigated materials present negligible incubation time (Figure 12b), following the results obtained for plastic and ceramics coatings [45,50]. The cavitation damage varies in the erosion rates. The highest erosion rates were obtained for NiCoCrAlY and CoNiCrAlY, and the lowest for NiCrMoNbTa coating. The literature reports that in the case of similar types of materials, higher hardness facilitates cavitation erosion resistance. Additionally, research papers [30,66,67] report that in the case of different materials systems, hardness alone is not a direct indicator of cavitation erosion performance. The microstructure of investigated coatings' is not similar and affects the hardness. Figure 13 confirms that the mass loss correlates with the decreasing mean hardness and the increase of the surface roughness of the eroded area. The highest material loss was obtained for the hardest coating (mean hardness of 393 HV0.1). The softest NiCrMoNbTa (mean hardness of 342 HV0.1) presents the lowest mass loss. As discussed in previous sections, the analysis of hardness variability indicates overlapping of standard deviations (see Figure 7), even though it seems that overall hardness can be employed as an indicator of material deformability. Moreover, the previous study regarding NiCrSiB alloys [32] confirms it. Softer nickel-based phases were prone to plastic deformation, which effectively mitigated cavitation damage. Overall, in the case of investigated HVOF metallic coatings containing nickel, hardness and deformability are essential factors for CE resistance. However, these factors derive from the microstructure of coatings, and microstructure features are responsible for cavitation erosion behavior.

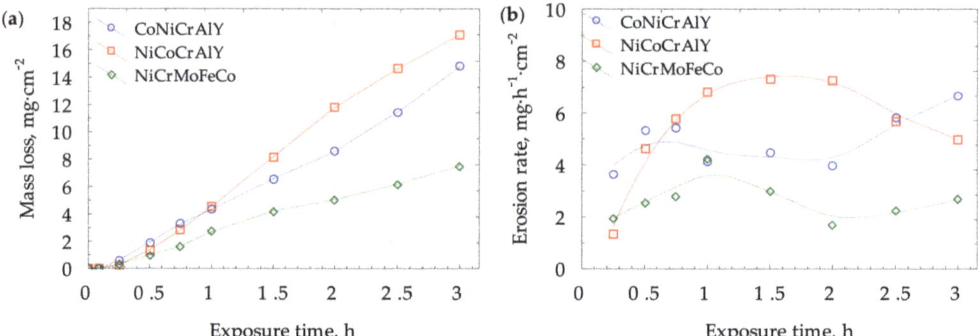

Figure 12. Cavitation erosion curves: (**a**) mass loss; (**b**) erosion rate plots.

The erosion mechanism was analysed by combining SEM microscopic observations (Figure 14) with coating properties and profilometric measurements (Figure 13). Therefore, the presence of microstructure discontinuities, such as unmelted particles and pores results in the creation of deep pits and facilitates material removal, especially at the initial stages of erosion, see Figures 12b and 14. At greater exposure times, severe pitting is observed. On the contrary, the well-melted splats undergo plastic deformation, which slows down the erosion rate (see Figure 14c). Finally, brittle cracking is visible in the NiCoCrAlY coating, which leads to material detachment. Profilometric measurements of eroded surfaces (Figure 13) confirmed that the NiCrMoNbTa coating was characterised by more uniform surface roughness and lesser pitting. Its plastic deformation slowed down the large craters' growth and material loss. Consequently, a relatively soft NiCrMoNbTa coating shows an ability to consume the cavitation loads for plastic deformation (visible in Figure 14), while other coatings present much more brittle behaviour resulting in accelerating material detachment.

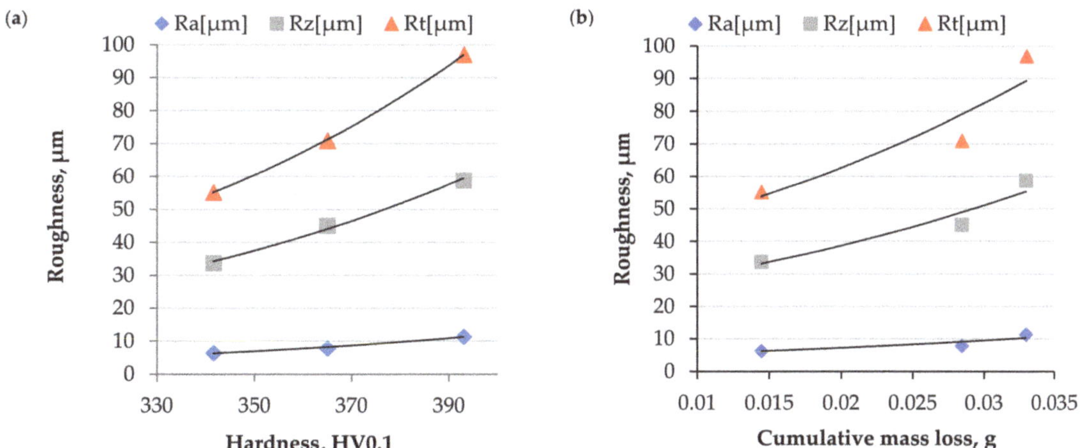

Figure 13. Effect of hardness (**a**) and material loss (**b**) on eroded surface roughness, 3 h of exposure.

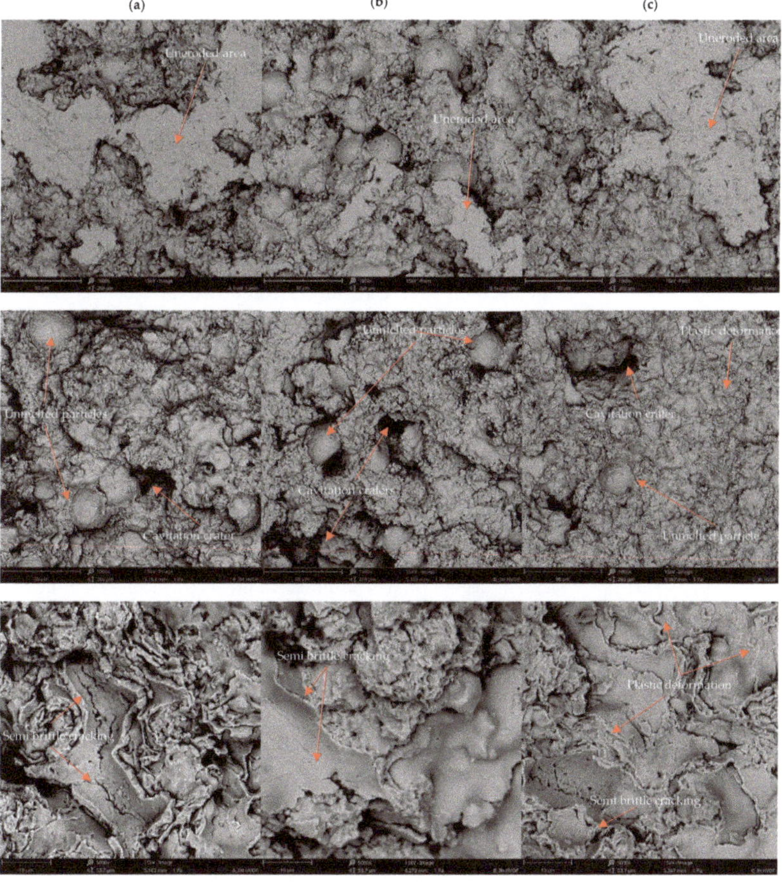

Figure 14. Comparison of eroded surfaces: (**a**) CoNiCrAlY, (**b**) NiCoCrAlY, (**c**) NiCrMoNbTa observed at 15 min and 180 min of cavitation erosion testing.

The severe erosion of the coatings relies on the formation of deep pits, and the coatings subject to more significant erosive damage were characterised by higher roughness parameters (see Figure 13). High nickel content facilitates the deformability of NiCrMoNbTa coating, resulting in ductile erosive behaviour. Contrary to this, MCrAlYs are characterised by lower deformability (higher mean hardness). Therefore, their failure mechanisms have a semi brittle mode which speeds erosion damage. It seems that microstructure features affect mechanical properties and determine the CE behaviour. Uniform microstructure and high deformability are crucial factors for increasing the CE resistance of thermally sprayed metallic coatings.

4. Conclusions

The thermally sprayed coatings, namely CoNiCrAlY, NiCoCrAlY, and NiCrMoNbTa, were comparatively examined in terms of the effect of microstructure and hardness on sliding wear and cavitation erosion resistance. As a result, the following conclusions can be drawn:

- The coatings present increasing resistance to both wear and cavitation erosion in the following order: NiCoCrAlY < CoNiCrAlY < NiCrMoNbTa.
- The NiCrMoNbTa and CoNiCrAlY coatings' microstructure is dominated by a single-phase matrix and presents mean hardness of 342 HV0.1 and 365 HV0.1, respectively. Even though the NiCoCrAlY coating presents the highest mean hardness (393 HV0.1), it shows the worst tribological and cavitation erosion performance. The differences in microstructure of investigated coatings affect the wear and cavitation erosion performance more than the hardness itself.
- Superior sliding wear behaviour, i.e., lowest coefficient of friction and wear factor K, was noted for NiCrMoNbTa coatings. This coating presents abrasive grooving wear, adhesive smearing, and fatigue of oxidised wear products, while the NiCoCrAlY wear mechanism relies mainly on the abrasive grooving and oxidation of wear products which results in the highest wear rate and COF.
- The coatings' microstructure affects mechanical properties and determines the CE behaviour. Cavitation erosion is initiated at microstructure discontinuities and ends up with severe surface pitting. MCrAlY coatings present semi brittle behavior, while NiCrMoNbTa coating shows ductile mode and lesser surface pitting.
- In the case of cavitation erosion, the hardness correlates well with the erosion results, i.e., softer coatings present higher deformability and display better performance under cavitation load by absorbing it for plastic deformation. Thus, NiCrMoNbTa coatings present the lowest cavitation erosion rate.

Author Contributions: Conceptualization, M.S.; methodology, M.S. and M.W.; software, M.S., M.W. and A.Ś.; validation, M.S and M.W.; formal analysis, M.S. and M.W.; investigation, M.S., M.W. and A.Ś.; resources, M.S. and M.W.; data curation, M.S. and M.W.; writing—original draft preparation, M.S. and A.Ś.; writing—review and editing, M.S.; visualization, M.S., M.W. and A.Ś.; supervision, M.S. and M.W.; project administration, M.S.; funding acquisition, M.S. All authors have read and agreed to the published version of the manuscript.

Funding: The project/research was financed in the framework of the project Lublin University of Technology-Regional Excellence Initiative, funded by the Polish Ministry of Science and Higher Education (contract no. 030/RID/2018/19). The APC publication fee was fully covered by the authors' vouchers.

Institutional Review Board Statement: Not applicable.

Informed Consent Statement: Not applicable.

Data Availability Statement: Data is contained within the article.

Acknowledgments: The authors would like to thank Leszek Łatka for his helpful advice on thermally sprayed coatings properties and inspiring discussions.

Conflicts of Interest: The authors declare no conflict of interest.

References

1. Łatka, L. Thermal barrier coatings manufactured by suspension plasma spraying—A review. *Adv. Mat. Sci.* **2018**, *18*, 95–117. [CrossRef]
2. Winnicki, M. Advanced Functional Metal-Ceramic and Ceramic Coatings Deposited by Low-Pressure Cold Spraying: A Review. *Coatings* **2021**, *11*, 1044. [CrossRef]
3. Łatka, L.; Szala, M.; Macek, W.; Branco, R. Mechanical Properties and Sliding Wear Resistance of Suspension Plasma Sprayed YSZ Coatings. *Adv. Sci. Technol. Res. J.* **2020**, *14*, 307–314. [CrossRef]
4. Han, Y.; Zhu, Z.; Zhang, B.; Chu, Y.; Zhang, Y.; Fan, J. Effects of process parameters of vacuum pre-oxidation on the microstructural evolution of CoCrAlY coating deposited by HVOF. *J. Alloys Compd.* **2018**, *735*, 547–559. [CrossRef]
5. Davis, J.R. *Handbook of Thermal Spray Technology*; ASM International: Russell Township, OH, USA, 2004.
6. Łatka, L.; Pawłowski, L.; Winnicki, M.; Sokołowski, P.; Małachowska, A.; Kozerski, S. Review of functionally graded thermal sprayed coatings. *Appl. Sci.* **2020**, *10*, 5153. [CrossRef]
7. Czupryński, A. Flame spraying of aluminum coatings reinforced with particles of carbonaceous materials as an alternative for laser cladding technologies. *Materials* **2019**, *12*, 3467. [CrossRef]
8. Jonda, E.; Łatka, L. Comparative analysis of mechanical properties of WC-based cermet Coatings sprayed by HVOF onto AZ31 magnesium alloy substrates. *Adv. Sci. Technol. Res. J.* **2021**, *15*, 57–64. [CrossRef]
9. Górka, J.; Czupryński, A. The properties and structure of arc sprayed coatings alloy of Fe-Cr-Ti-Si-Mn. *Int. J. Mod. Manuf. Technol.* **2016**, *8*, 35–40.
10. Michalak, M.; Sokołowski, P.; Szala, M.; Walczak, M.; Łatka, L.; Toma, F.-L.; Björklund, S. Wear behavior analysis of Al_2O_3 coatings manufactured by APS and HVOF spraying processes using powder and suspension feedstocks. *Coatings* **2021**, *11*, 879. [CrossRef]
11. Mahade, S.; Mulone, A.; Björklund, S.; Klement, U.; Joshi, S. Investigating load-dependent wear behavior and degradation mechanisms in Cr_3C_2–NiCr coatings deposited by HVAF and HVOF. *J. Mater. Res. Technol.* **2021**, *15*, 4595–4609. [CrossRef]
12. Daniel, J.; Grossman, J.; Houdková, Š.; Bystrianský, M. Impact wear of the protective Cr_3C_2-based HVOF-sprayed coatings. *Materials* **2020**, *13*, 2132. [CrossRef] [PubMed]
13. Bolelli, G.; Vorkötter, C.; Lusvarghi, L.; Morelli, S.; Testa, V.; Vaßen, R. Performance of wear resistant MCrAlY coatings with oxide dispersion strengthening. *Wear* **2020**, *444–445*, 203116. [CrossRef]
14. Dorfman, M.R. Thermal spray coatings. In *Handbook of Environmental Degradation of Materials*, 1st ed.; Kutz, M., Ed.; Elsevier: Amsterdam, The Netherlands, 2005; pp. 469–488.
15. Mayer, A.R.; Bertuol, K.; Siqueira, I.B.A.F.; Chicoski, A.; Váz, R.F.; de Sousa, M.J.; Pukasiewicz, A.G.M. Evaluation of cavitation/corrosion synergy of the Cr_3C_2-25NiCr coating deposited by HVOF process. *Ultrason. Sono-Chem.* **2020**, *69*, 105271. [CrossRef] [PubMed]
16. Ronzani, A.G.; Pukasiewicz, A.G.M.; da Silva Custodio, R.M.; de Vasconcelos, G.; de Oliveira, A.C.C. Cavitation resistance of tungsten carbide applied on AISI 1020 steel by HVOF and remelted with CO_2 laser. *J. Braz. Soc. Mech. Sci. Eng.* **2020**, *42*, 316. [CrossRef]
17. Taylor, T.A.; Overs, M.P.; Gill, B.J.; Tucker, R.C. Experience with MCrAl and thermal barrier coatings produced via inert gas shrouded plasma deposition. *J. Vac. Sci. Technol. A Vac. Surf. Film.* **1985**, *3*, 2526–2531. [CrossRef]
18. Soltani, R.; Samadi, H.; Garcia, E.; Coyle, T.W. *Development of Alternative Thermal Barrier Coatings for Diesel Engines*; SAE Technical Paper 2005-01-0650; SAE International: Warrendale, PA, USA, 2005. [CrossRef]
19. Chun, G.; Jianmin, C.; Rungang, Y.; Jiansong, Z. Microstructure and high temperature wear resistance of laser cladding $NiCoCrAlY/ZrB_2$ coating. *Rare Met. Mater. Eng.* **2013**, *42*, 1547–1551. [CrossRef]
20. Wang, W.; Li, J.; Ge, Y.; Kong, D. Structural characteristics and high-temperature tribological behaviors of laser cladded $NiCoCrAlY–B_4C$ composite coatings on Ti6Al4V alloy. *Trans. Nonferrous Met. Soc. China* **2021**, *31*, 2729–2739. [CrossRef]
21. Pereira, J.; Zambrano, J.; Licausi, M.; Tobar, M.; Amigó, V. Tribology and high temperature friction wear behavior of MCrAlY laser cladding coatings on stainless steel. *Wear* **2015**, *330–331*, 280–287. [CrossRef]
22. Cao, S.; Ren, S.; Zhou, J.; Yu, Y.; Wang, L.; Guo, C.; Xin, B. Influence of composition and microstructure on the tribological property of SPS sintered MCrAlY alloys at elevated temperatures. *J. Alloy. Compd.* **2018**, *740*, 790–800. [CrossRef]
23. Zakrzewska, D.E.; Krella, A.K. Cavitation erosion resistance influence of material properties. *Adv. in Mater. Sci.* **2019**, *19*, 18–34. [CrossRef]
24. Vuoristo, P. Thermal Spray Coating Processes. In *Comprehensive Materials Processing*, 1st ed.; Hashmi, S., Ed.; Elsevier: Amsterdam, The Netherlands, 2014; Volume 4.
25. Oksa, M.; Turunen, E.; Suhonen, T.; Varis, T.; Hannula, S.P. Optimization and characterization of high velocity oxy-fuel sprayed coatings: Techniques, materials, and applications. *Coatings* **2011**, *1*, 17–52. [CrossRef]
26. Tzanakis, I.; Bolzoni, L.; Eskin, D.G.; Hadfield, M. Evaluation of cavitation erosion behavior of commercial steel grades used in the design of fluid machinery. *Metall. Mater. Trans. A* **2017**, *48*, 2193–2206. [CrossRef]
27. Hattori, S.; Mikami, N. Cavitation erosion resistance of Stellite alloy weld overlays. *Wear* **2009**, *267*, 1954–1960. [CrossRef]

28. Szala, M.; Chocyk, D.; Skic, A.; Kamiński, M.; Macek, W.; Turek, M. Effect of nitrogen ion implantation on the cavitation erosion resistance and cobalt-based solid solution phase transformations of HIPed Stellite 6. *Materials* **2021**, *14*, 2324. [CrossRef] [PubMed]
29. Jonda, E.; Łatka, L.; Pakieła, W. Comparison of different cermet coatings sprayed on magnesium alloy by HVOF. *Materials* **2021**, *14*, 1594. [CrossRef] [PubMed]
30. Caccese, V.; Light, K.H.; Berube, K.A. Cavitation erosion resistance of various material systems. *Ships Offshore Struct.* **2006**, *1*, 309–322. [CrossRef]
31. Grist, E. *Cavitation and the Centrifugal Pump: A Guide for Pump Users*; CRC Press: Boca Raton, FL, USA, 1998.
32. Szala, M.; Walczak, M.; Hejwowski, T. Factors influencing cavitation erosion of NiCrSiB hardfacings deposited by oxy-acetylene powder welding on grey cast iron. *Adv. Sci. Technol. Res. J.* **2021**, *15*, 376–386. [CrossRef]
33. Lin, J.; Wang, Z.; Lin, P.; Cheng, J.; Zhang, X.; Hong, S. Microstructure and cavitation erosion behavior of FeNiCrBSiNbW coating prepared by twin wires arc spraying process. *Surf. Coat. Technol.* **2014**, *240*, 432–436. [CrossRef]
34. Kekes, D.; Psyllaki, P.; Vardavoulias, M.; Vekinis, G. Wear micro-mechanisms of composite WC-Co/Cr-NiCrFeBSiC coatings. Part II: Cavitation erosion. *Tribol. Ind.* **2014**, *36*, 375–383.
35. Zheng, J.; Yang, D.; Gao, Y. Effect of vacuum heat treatment on the high-temperature oxidation resistance of NiCrAlY coating. *Coatings* **2020**, *10*, 1–12. [CrossRef]
36. Sadeghimeresht, E.; Reddy, L.; Hussain, T.; Huhtakangas, M.; Markocsan, N.; Joshi, S. Influence of KCl and HCl on high temperature corrosion of HVAF-sprayed NiCrAlY and NiCrMo coatings. *Mater. Des.* **2018**, *148*, 17–29. [CrossRef]
37. Singh, G.; Bala, N.; Chawla, V. High Temperature oxidation behaviour and characterization of NiCrAlY-B4C coatings deposited by HVOF. *Mater. Res. Express* **2019**, *6*, 086436. [CrossRef]
38. Szala, M.; Walczak, M.; Łatka, L.; Gancarczyk, K.; Özkan, D. Cavitation erosion and sliding wear of MCrAlY and NiCrMo coatings deposited by HVOF thermal spraying. *Adv. Mater. Sci.* **2020**, *20*, 26–38. [CrossRef]
39. Ciubotariu, C.-R.; Secosan, E.; Marginean, G.; Frunzaverde, D.; Campian, V.C. Experimental study regarding the cavitation and corrosion resistance of Stellite 6 and self-fluxing remelted coatings. *Strojniški Vestnik—J. Mech. Eng.* **2016**, *62*, 154–162. [CrossRef]
40. Budzyński, P.; Kamiński, M.; Turek, M.; Wiertel, M. Impact of nitrogen and manganese ion implantation on the tribological properties of Stellite 6 alloy. *Wear* **2020**, *456–457*, 203360. [CrossRef]
41. Varis, T.; Suhonen, T.; Laakso, J.; Jokipii, M.; Vuoristo, P. Evaluation of residual stresses and their influence on cavitation erosion resistance of high kinetic HVOF and HVAF-sprayed WC-CoCr coatings. *J. Therm. Spray Technol.* **2020**, *29*, 1365–1381. [CrossRef]
42. Ding, X.; Ke, D.; Yuan, C.; Ding, Z.; Cheng, X. Microstructure and cavitation erosion resistance of HVOF deposited WC-Co coatings with different sized WC. *Coatings* **2018**, *8*, 307. [CrossRef]
43. Huang, Q.; Qin, E.; Li, W.; Wang, B.; Pan, C.; Wu, S. The cavitation resistance of WC-CoCr cermet coating deposited by HVOF for hydraulic application. *J. Therm. Spray Eng.* **2020**, *3*, 68–73. [CrossRef]
44. Walczak, M.; Szala, M. Effect of shot peening on the surface properties, corrosion and wear performance of 17-4PH steel produced by DMLS additive manufacturing. *Archiv. Civ. Mech. Eng.* **2021**, *21*, 157. [CrossRef]
45. Łatka, L.; Michalak, M.; Szala, M.; Walczak, M.; Sokołowski, P.; Ambroziak, A. Influence of 13 wt.% TiO$_2$ content in alumina-titania powders on microstructure, sliding wear and cavitation erosion resistance of APS sprayed coatings. *Surf. Coat. Technol.* **2021**, *410*, 126979. [CrossRef]
46. Drozd, K.; Walczak, M.; Szala, M.; Gancarczyk, K. Tribological behavior of AlCrSiN-coated tool steel K340 versus popular tool steel grades. *Materials* **2020**, *13*, 4895. [CrossRef]
47. ASTM International. *ASTM G32-16 Standard Test Method for Cavitation Erosion Using Vibratory pparatus*; ASTM International: West Conshohocken, PA, USA, 2016.
48. Szala, M.; Walczak, M. Cavitation erosion and sliding wear resistance of HVOF coatings. *Weld. Technol. Rev.* **2018**, *90*. [CrossRef]
49. Szala, M.; Łatka, L.; Walczak, M.; Winnicki, M. Comparative study on the cavitation erosion and sliding wear of cold-sprayed Al/Al$_2$O$_3$ and Cu/Al$_2$O$_3$ coatings, and stainless steel, aluminium alloy, copper and brass. *Metals* **2020**, *10*, 856. [CrossRef]
50. Szala, M.; Świetlicki, A.; Sofińska-Chmiel, W. Cavitation erosion of electrostatic spray polyester coatings with different surface finish. *Bull. Pol. Acad. Sci. Tech. Sci.* **2021**, *69*, e137519. [CrossRef]
51. Rajasekaran, B.; Mauer, G.; Vaßen, R. Enhanced characteristics of HVOF-sprayed MCrAlY bond coats for TBC applications. *J. Therm. Spray Tech.* **2011**, *20*, 1209–1216. [CrossRef]
52. Yuan, K.; Yu, Y.; Wen, J.-F. A study on the thermal cyclic behavior of thermal barrier coatings with different MCrAlY roughness. *Vacuum* **2017**, *137*, 72–80. [CrossRef]
53. Oladijo, O.P.; Collieus, L.L.; Obadele, B.A.; Akinlabi, E.T. Correlation between residual stresses and the tribological behaviour of Inconel 625 coatings. *Surf. Coat. Technol.* **2021**, *419*, 127288. [CrossRef]
54. Al-Fadhli, H.Y.; Stokes, J.; Hashmi, M.S.J.; Yilbas, B.S. The erosion-corrosion behaviour of high velocity oxy-fuel (HVOF) thermally sprayed Inconel-625 coatings on different metallic surfaces. *Surf. Coat. Technol.* **2006**, *200*, 5782–5788. [CrossRef]
55. Nowak, W.; Naumenko, D.; Mor, G.; Mor, F.; Mack, D.E.; Vassen, R.; Singheiser, L.; Quadakkers, W.J. Effect of processing parameters on MCrAlY bondcoat roughness and lifetime of APS-TBC systems. *Surf. Coat. Technol.* **2014**, *260*, 82–89. [CrossRef]
56. Saeidi, S.; Voisey, K.T.; McCartney, D.G. Mechanical properties and microstructure of VPS and HVOF CoNiCrAlY coatings. *J. Therm. Spray Tech.* **2011**, *20*, 1231–1243. [CrossRef]
57. Mohammadi, M.; Javadpour, S.; Kobayashi, A.; Jenabali Jahromi, S.A.; Shirvani, K. Thermal shock properties and microstructure investigation of LVPS and HVOF-CoNiCrAlYSi coatings on the IN738LC superalloy. *Vacuum* **2013**, *88*, 124–129. [CrossRef]

58. Fossati, A.; Ferdinando, M.D.; Lavacchi, A.; Scrivani, A.; Giolli, C.; Bardi, U. Improvement of the oxidation resistance of CoNiCrAlY bond coats sprayed by high velocity oxygen-fuel onto nickel superalloy substrate. *Coatings* **2010**, *1*, 3–16. [CrossRef]
59. Abu-warda, N.; Tomás, L.M.; López, A.J.; Utrilla, M.V. High temperature corrosion behavior of Ni and Co base HVOF coatings exposed to NaCl-KCl salt mixture. *Surf. Coat. Technol.* **2021**, *418*, 127277. [CrossRef]
60. Elshalakany, A.B.; Osman, T.A.; Hoziefa, W.; Escuder, A.V.; Amigó, V. Comparative study between high-velocity oxygen fuel and flame spraying using MCrAlY coats on a 304 stainless steel substrate. *J. Mater. Res. Technol.* **2019**, *8*, 4253–4263. [CrossRef]
61. Bai, M.; Song, B.; Reddy, L.; Hussain, T. Preparation of MCrAlY–Al2O3 Composite coatings with enhanced oxidation resistance through a novel powder manufacturing process. *J. Therm. Spray Tech.* **2019**, *28*, 433–443. [CrossRef]
62. Fantozzi, D.; Matikainen, V.; Uusitalo, M.; Koivuluoto, H.; Vuoristo, P. Effect of carbide dissolution on chlorine induced high temperature corrosion of HVOF and HVAF sprayed Cr_3C_2-NiCrMoNb coatings. *J. Therm. Spray Tech.* **2018**, *27*, 220–231. [CrossRef]
63. Hao, E.; An, Y.; Zhao, X.; Zhou, H.; Chen, J. NiCoCrAlYTa Coatings on nickel-base superalloy substrate: Deposition by high velocity oxy-fuel spraying as well as investigation of mechanical properties and wear resistance in relation to heat-treatment duration. *Appl. Surf. Sci.* **2018**, *462*, 194–206. [CrossRef]
64. Gahr, K.-H.Z. *Microstructure and Wear of Materials*, 1st ed.; Elsevier: Amsterdam, The Netherlands, 1987.
65. Brunetti, C.; Belotti, L.P.; Miyoshi, M.H.; Pintaúde, G.; D'Oliveira, A.S.C.M. Influence of Fe on the room and high-temperature sliding wear of NiAl coatings. *Surf. and Coat. Technol.* **2014**, *258*, 160–167. [CrossRef]
66. Jiang, X.; Overman, N.; Smith, C.; Ross, K. Microstructure, hardness and cavitation erosion resistance of different cold spray coatings on stainless steel 316 for hydropower applications. *Mater. Today Commun.* **2020**, *25*, 101305. [CrossRef]
67. Roa, C.V.; Valdes, J.A.; Larrahondo, F.; Rodríguez, S.A.; Coronado, J.J. Comparison of the resistance to cavitation erosion and slurry erosion of four kinds of surface modification on 13-4 Ca6NM hydro-machinery steel. *J. Materi Eng. Perform.* **2021**, *30*, 7195–7212. [CrossRef]

Article

Effect of Parylene C on the Corrosion Resistance of Bioresorbable Cardiovascular Stents Made of Magnesium Alloy 'Original ZM10'

Makoto Sasaki [1,*], Wei Xu [1,*], Yuki Koga [2], Yuki Okazawa [3], Akira Wada [2], Ichiro Shimizu [4] and Takuro Niidome [1]

[1] Faculty of Advanced Science and Technology, Kumamoto University, Kumamoto 860-8555, Japan; niidome@kumamoto-u.ac.jp
[2] Japan Medical Device Technology, Co., Ltd., Kumamoto 861-2202, Japan; y.koga@jmdt.co.jp (Y.K.); a.wada@jmdt.co.jp (A.W.)
[3] Fuji Light Metal, Co., Ltd., Kumamoto 869-0192, Japan; yuuki-okazawa@fujisash.net
[4] Department of Mechanical Engineering, Okayama University of Science, Okayama 700-0005, Japan; shimizu@mech.ous.ac.jp
* Correspondence: sasakimakoto@kumamoto-u.ac.jp (M.S.); xuwei@kumamoto-u.ac.jp (W.X.); Tel.: +81-96-342-3288 (M.S.); +81-96-342-3668 (W.X.)

Abstract: Magnesium (Mg) alloy has attracted significant attention as a bioresorbable scaffold for use as a next-generation stent because of its mechanical properties and biocompatibility. However, Mg alloy quickly degrades in the physiological environment. In this study, we investigated whether applying a parylene C coating can improve the corrosion resistance of a Mg alloy stent, which is made of 'Original ZM10', free of aluminum and rare earth elements. The coating exhibited a smooth surface with no large cracks, even after balloon expansion of the stent, and improved the corrosion resistance of the stent in cell culture medium. In particular, the parylene C coating of a hydrofluoric acid-treated Mg alloy stent led to excellent corrosion resistance. In addition, the parylene C coating did not affect a polymer layer consisting of poly(ε-caprolactone) and poly(D,L-lactic acid) applied as an additional coating for the drug release to suppress restenosis. Parylene C is a promising surface coating for bioresorbable Mg alloy stents for clinical applications.

Keywords: corrosion resistance; magnesium alloy; bioresorbable stent; parylene C; surface coating

Citation: Sasaki, M.; Xu, W.; Koga, Y.; Okazawa, Y.; Wada, A.; Shimizu, I.; Niidome, T. Effect of Parylene C on the Corrosion Resistance of Bioresorbable Cardiovascular Stents Made of Magnesium Alloy 'Original ZM10'. *Materials* **2022**, *15*, 3132. https://doi.org/10.3390/ma15093132

Academic Editors: Claudio Mele and Artur Czupryński

Received: 20 March 2022
Accepted: 25 April 2022
Published: 26 April 2022

Publisher's Note: MDPI stays neutral with regard to jurisdictional claims in published maps and institutional affiliations.

1. Introduction

Coronary vascular disease is a major cause of death in the developed world [1]. Percutaneous coronary intervention combining balloon angioplasty and stent implantation is a primary treatment for coronary heart disease [2]. Drug-eluting stents are now widely used because they can reduce in-stent restenosis and target lesion revascularization compared with bare-metal stents [3]. Sirolimus (SRL) is an antiproliferative drug widely used in drug-eluting stent applications [4]. However, drug-eluting stents have been reported to impair endothelial regeneration, leading to an increased risk of very late stent thrombosis in patients after stent implantation [5]. Thus, bioresorbable stents have been developed to address the problems associated with conventional metallic stents [6]. The bioresorbable stents made of biodegradable polymers or bioresorbable metals have become major trends. Compared with biodegradable polymer stents, bioresorbable metals such as magnesium, zinc, and iron have better mechanical properties and are expected as new-generation stent materials [7].

Magnesium (Mg) alloy is expected as the new generation metal for use in the development of bioresorbable stents because it exhibits good mechanical properties and its corrosion products are non-toxic [8]. However, Mg alloy exhibits rapid corrosion in the physiological environment, limiting its successful clinical application [9]. A vascular stent is required to maintain its mechanical strength for at least 6 months and to be resorbed over 6 months to 2 years [10,11]. Therefore, enhancing the anti-corrosion properties of Mg alloy is key to its clinical application as a bioresorbable stent.

To improve the corrosion resistance of Mg alloy, the addition of rare earth element to Mg alloy [12,13], surface coating [14,15], and chemical conversion treatment [16,17] have been widely studied. The rare earth elements refer to the metals yttrium (Y), dysprosium (Dy), neodymium (Nd), gadolinium (Gd), etc. However, the suitability and biocompatibility of rare earth elements are still considered carefully [18]. Conventional Mg alloy, i.e., WE43, contains rare earth elements and is suspected to show cytotoxicity [19]. Moreover, AZ31B contains aluminum, which is not a rare earth element but is suspected to be related to Alzheimer's disease [20]. In this study, we used 'Original ZM10', free of rare earth elements and aluminum, and it is appropriate for medical applications with higher biocompatibility.

In our previous studies, both surface coating with biodegradable polymers and chemical treatment with hydrofluoric acid (HF) have been reported to be effective strategies for improving the corrosion resistance of Mg alloy stents [21,22]. Surface coating with biodegradable polymers is used not only for corrosion resistance, but also for loading drugs to inhibit in-stent restenosis [23]. HF treatment is an effective approach for improving the corrosion resistance of Mg alloys. An HF-treated Mg alloy stent can support the vessel wall in a coronary artery for several days, but it is not possible for a gradually degrading stent to maintain its radial strength for several months [22]. Therefore, another approach is required to improve the durability of Mg alloy stents.

It has been reported that poly(chloro-para-xylynene) (parylene C) coating has the potential to reduce the surface wettability of Mg alloy owing to its dense structure and hydrophobic surface [24]. Parylene C is widely used for various medical devices, such as conventional stents and catheters [25,26], because of its durability and biocompatibility [27]. It can also be formed on the surface of various biomaterials by chemical vapor deposition (CVD) [28].

However, surface coating of a bioresorbable Mg alloy stent with parylene C has not yet been reported. As stated above, HF treatment was effective for improving the corrosion resistance of a Mg alloy stent. The objective of this study was to examine the effect of the parylene C coating on the corrosion resistance of Mg alloy stent, particularly to evaluate if it could be resistant to shape changes caused by crimping on the balloon catheter and subsequent expansion.

2. Materials and Methods

2.1. Chemicals

Hydrofluoric acid (HF, 46% *w*/*w* aqueous solution), tetrahydrofuran (THF), and Eagle's minimal essential medium (E-MEM) were purchased from Wako Pure Chemical Industries Ltd. (Osaka, Japan). Fetal bovine serum (FBS) was purchased from Cosmo Bio Co., LTD. (Tokyo, Japan). Poly(D,L-lactide) (PDLLA) and poly(ε-caprolactone) (PCL) were purchased from LACTEL Absorbable Polymers (Birmingham, AL, USA). Mg alloy 'Original ZM10' was developed and manufactured by Fuji Light Metal Co., Ltd. The component was analyzed with an inductively coupled plasma optical emission spectrometry (ICP-OES Agilent 720, Agilent Technologies, Santa Clara, CA, USA) as shown in Table S1.

2.2. Stents

The Mg alloy stent made of 'Original ZM10' was designed, fabricated, and mirror-polished by Fuji Light Metal Co., Ltd. (Kumamoto, Japan) and Japan Medical Device Technology Co., Ltd. (Kumamoto, Japan). The specifications of the stent used in this study are shown in Table S2. The structure of our original stent is shown in Figure S1.

The bare Mg alloy stents were immersed in 46% (*w*/*w*) HF aqueous solution for 24 h at room temperature. Then, they were rinsed with deionized water and acetone in turn, and we finally dried the samples at 55 °C under vacuum [22].

2.3. Parylene C Coating

The parylene C coating of the HF-treated Mg alloy stent was performed by Specialty Coating Systems coating service (Specialty Coating Systems Inc., Indianapolis, IN, USA) [24]. The thickness of the parylene C coating layer was measured by reflection spectroscopy using an optical thickness meter (OPTM-F2, Otsuka Electronics Co., Ltd., Osaka, Japan).

2.4. Polymer Coating

Additional polymer coatings were applied after the parylene C coating. First, the stent was coated with a soft polymer, poly(ε-caprolactone) (PCL), as a base layer, and poly(D,L-lactic acid) (PDLLA) with sirolimus (SRL; a conventional antiproliferation drug) as a top layer, as described in our previous report [21]. Briefly, for the base layer, PCL was dissolved in THF to obtain a 0.5 wt% solution. Then, 400 ± 30 μg of polymer was applied to each stent at 0.02 mL/min using an ultrasonic spray coater, Exacta Coat Ultrasonic Spraying System (Milton, NY, USA). For the top layer, PDLLA and SRL were dissolved in THF to obtain a 0.5 wt% and 5 mg/mL solution, respectively. Then, 250 ± 30 μg and 1.0 μg/mm^2 of polymer and SRL, respectively, were applied to each stent at 0.02 mL/min using the ultrasonic spray coater.

2.5. Contraction and Expansion of the Stent

We used a manual crimp tool (13JSK, Blockwise Engineering, Chicago, IL, USA) to crimp the coating stent on a balloon-tipped catheter (Vega SDD, Arthesys, Saclay, France) to 1.2 mm in outer diameter, as described in our previous studies [21,22]. Then, before the expansion, the crimped stent was immersed in 37 °C cell culture medium (E-MEM) containing 10% fetal bovine serum (FBS) for 2 min to simulate the clinical stent implantation process. For the expansion of the stent, we inflated the catheter with an Encore™ 26 Inflator (Boston Scientific, Marlborough, MA, USA) and expanded the stents to 3 mm in outer diameter.

2.6. Surface Observation

The surface morphology of the Mg alloy stents before and after balloon expansion were observed by field emission scanning electron microscopy (FE-SEM, JSM-7100F, JEOL, Tokyo, Japan) with a hot electron gun. The observe condition was at an accelerating voltage of 15 kV and a working distance of 15 mm by a secondary electron (SE) detector.

2.7. Evaluation of Corrosion Behavior

After expansion of the polymer-coated Mg alloy stents, the stents were immersed in 10 mL cell culture medium (E-MEM + 10% FBS) in an incubator with a condition of 5% CO_2, 37 °C under 100 rpm agitation for 12 months. The supernatant medium of 100 μL was collected after immersion for 1 month and 6 months, using a Magnesium B-test Wako test kit (Wako Pure Chemical Industries Ltd., Osaka, Japan) to measure the amount of Mg ions released from the Mg alloy stent, following the manufacturer's protocol [29]. Mean values were calculated from measurements for five different stents. After the incubation, the stents were removed from the medium and dried in a vacuum overnight.

Elemental analysis of the cross-sections of the stents after incubation for 1, 6, and 12 months was carried out using an X-ray energy dispersive spectroscopy (EDS) detector installed into the FE-SEM (JSM-7100F, JEOL, Tokyo, Japan). The EDS mappings were recorded with a JEOL resolution silicon drift detector and indicated the location of elements analyzed.

2.8. SRL Elution from the Polymer Layer

After the expansion of the stent coated with SRL-loaded polymer using the same method as described in Section 2.5, the stent was then immersed in 10 mL PBS in an incubator at 37 °C in the dark under 100 rpm agitation for 64 days. To evaluate the SRL release from the polymer, 1 mL supernatant solution after immersion for 1, 3, 7, 14, 21,

28, and 64 days was measured with an ultraviolet/visible spectrometer at 278 nm. The amount of SRL eluting was calculated from three different samples. The SRL elution rate was calculated based on the SRL loading density of each stent.

3. Results and Discussion

3.1. Parylene C Coating of Mg Alloy Stents

The thickness of the parylene C coating layer was approximately 0.5 ± 0.15 μm (data not shown). Normally, a coronary stent is made of flexible tube-like meshes that is crimped onto a balloon-tipped catheter, and then the balloon is deflated and the stent is expanded at a narrowed coronary artery [30]. Considering the changes in the surface structure of the balloon-expanded stent is therefore essential. The surfaces of the Mg alloy stents were observed using a scanning electron microscope (SEM). The bare Mg alloy stent had a smooth surface both before and after the balloon expansion (Figure 1a). The HF-treated Mg alloy stent had a smooth surface before the balloon expansion but showed small wrinkles after the expansion, with no large cracks (Figure 1b). Following parylene C coating, the stent surface was smooth before the expansion, while after the expansion small wrinkles, similar to those shown by the HF-treated Mg alloy stent, were observed, with no large cracks. Because the parylene C coating was very thin (~0.5 μm) and flexible, it was able to expand and follow the structural changes of the HF-treated stent without peeling off (Figure 1c).

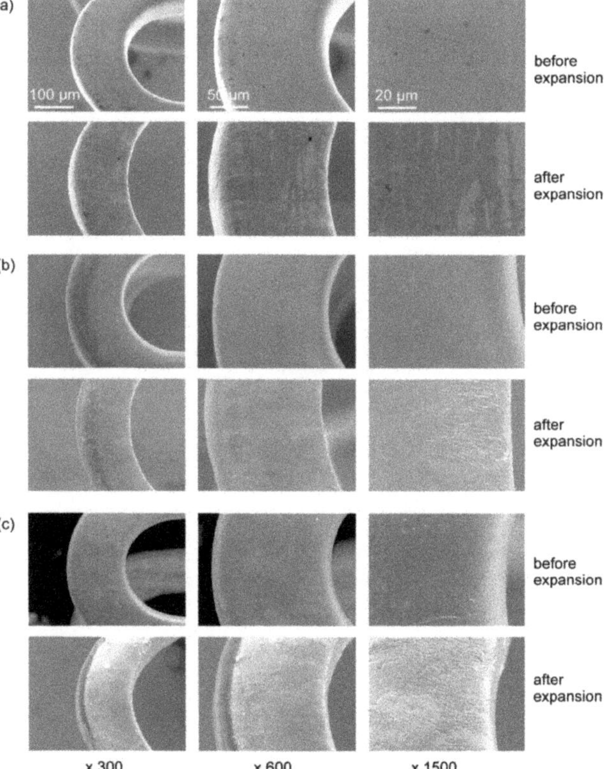

Figure 1. Surface morphology of original Mg alloy stents observed by SEM (**a**), HF-treated stents (**b**), and HF-treated Mg alloy stents coated with parylene C (**c**) before and after balloon expansion at different magnifications.

3.2. Corrosion Behavior of the Parylene-Coated Mg Alloy Stents after Expansion

The corrosion behavior of the stents after the balloon expansion was examined over a 1-month and 6-month immersion in cell culture medium at 37 °C (Figure 2 and Table S3). Mg ion release (%) represents the relative corrosion rate of Mg alloy stent calculated from the total amount of Mg alloy stent [29]. After 1-month immersion, the bare Mg alloy stent indicated a rapid corrosion rate calculated as 38%, while the HF-treated Mg alloy stent exhibited a corrosion rate of 13%. The HF treatment of Mg alloy stent enhanced the corrosion resistance due to a protective layer of MgF_2 and $Mg(OH)_2$ formed on the Mg alloy stent surface, which prevented rapid corrosion [22,23]. Additional coating of the HF-treated Mg alloy stents with parylene C resulted in complete resistance (almost 0% corrosion). To further evaluate the corrosion resistance of the parylene C coating on HF-treated Mg alloy stent, it was immersed in cell culture medium for 6 months (Figure 2B). Compared with HF-treated Mg alloy stents, parylene coating still exhibited excellent corrosion resistance over a 6-month period. The HF-treated Mg alloy stents showed the corrosion rate of 78% was due to the MgF_2 content with high mechanical strength inducing cracks on the surface after balloon expansion [22,31]. Additional coating with parylene C perfectly improved this disadvantage. Kandala et al. also studied the corrosion behavior of Mg alloy (AZ31) stents with parylene C coating (without HF treatment) in the cell culture medium for 3 days [32]. Compared with bare Mg alloy stent, parylene C-coated Mg alloy stent improved the corrosion rate from 0.45 mm/year to 0.12 mm/year. However, the balloon expansion of parylene C-coated stents accelerated the corrosion of Mg alloy stents from 0.05 mm/year to 0.15 mm/year due to defects and openings in the parylene C coating [33]. Therefore, tight adhesion between MgF_2 and the parylene C layer was thought to form a strong and stable water barrier that was not affected by the balloon expansion of the stent struts.

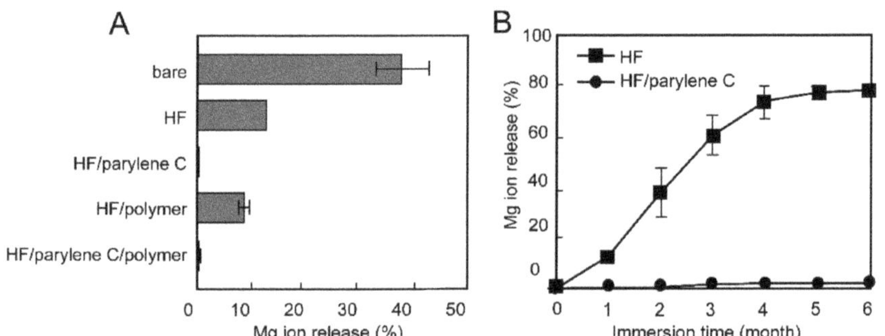

Figure 2. Mg ion release rate from Mg alloy stents with balloon expansion after 1-month (**A**) and 6-month (**B**) incubation in cell culture medium. The data represent the mean value for n = 5 and the error bars show the standard deviations of the means.

Drugs to suppress restenosis are generally introduced in additional polymer layers; therefore, we examined the effect of subsequent PCL and PDLLA polymer coatings on the performance of parylene C in protecting the HF-treated Mg alloy stent from corrosion. In our previous study, coating with PCL and PDLLA was found to be the best approach for enhancing the corrosion resistance of the Mg alloy stent and achieving long-lasting drug release from the polymer layer [21]. The polymer-coated HF-treated Mg alloy stents showed similar corrosion resistance to that of the HF-treated Mg alloy stents without polymer coating. The polymer-coated, HF-treated, and parylene C-coated stent showed almost complete resistance to corrosion. This indicates that the additional polymer coating did not affect the complete corrosion resistance provided by HF treatment and parylene C-coating over a 1-month period.

We examined further corrosion of the polymer and parylene C-coated HF-treated Mg alloy stent after 1-, 6-, and 12-month incubations by SEM images and EDS mapping of their cross-sections (Figure 3). At 1 month, the SEM images and Mg signal from the EDS mapping showed no significant degradation of the Mg alloy. At this stage, F and O signals were observed on the surface of the strut and corresponded to the HF treated layer and the polymer layer, respectively. After 6 and 12 months, the area of the Mg signal had decreased, and a small amount of the O signal was observed on the Mg signal. The O signal that emerged is attributed to corrosion products of the Mg alloy. P and Ca signals also indicated corrosion products, corresponding to magnesium phosphate and calcium phosphate [34]. The source of phosphate and calcium was the incubation medium (E-MEM + 10% FBS) for the corrosion resistance experiment. According to the EDS mapping, the shallow corrosion on the Mg stent surface under the coating layer was observed, even after immersion for 12 months. In addition, it was confirmed that the coating layer remained on the Mg alloy stents because of a slow corrosion rate after the 12-month incubation.

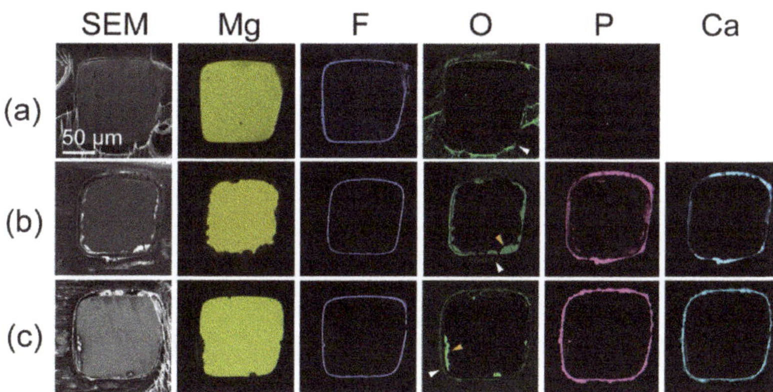

Figure 3. EDS mapping for cross-sections of parylene C and polymer-coated HF-treated stents immersed for 1 month (**a**), 6 months (**b**), and 12 months (**c**) in the cell culture medium. The white and orange arrow heads indicate O signal from polymer layer and corrosion products of the Mg alloy, respectively.

3.3. SRL Release from the Polymer Layer

For clinical application of the cardiovascular stent, drug release from the polymer outer layer is key for suppressing restenosis after implantation. We therefore examined the release of the drug SRL from the polymer layer (Figure 4). The polymer layer had eluted approximately 60% of the SRL at 14 days, 70% at 28 days, and 90% at 64 days. Cypher—a drug-eluting stent made of stainless steel coated with parylene C, polyethylene-co-vinyl acetate, and poly-*n*-butyl methacrylate—eluted 80% of the loaded SRL over 1 month [35]. Ultimaster—a drug-eluting stent made of Co-Cr alloy coated with PDLLA and poly(L-lactide–co-ε-caprolactone)—showed 90% elution at 3 months [36]. The elution profile of the reported stent is therefore considered to be similar to those of Cypher and Ultimaster.

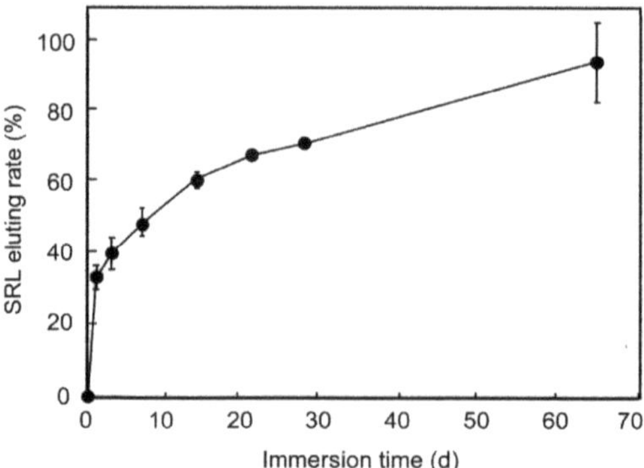

Figure 4. SRL-release profiles from the parylene C- and polymer-coated HF-treated stents after balloon expansion, in 37 °C PBS over 64 days. The data represent the mean value for n = 3 and error bars show the standard deviations of the means.

4. Conclusions

This study demonstrated that parylene C coating is an efficient method to improve the corrosion resistance of Mg alloy stent. According to a long-term evaluation of corrosion behavior, the parylene C coating is expected to be a promising surface coating of Mg alloy stents for clinical applications. The following conclusions are obtained:

(1) Parylene C coating on a HF-treated Mg alloy stent exhibited a smooth surface without cracks after balloon expansion. Because the parylene C coating was very thin (~0.5 μm) and flexible, it was able to expand and follow the structural changes of the HF-treated Mg stent without peeling off.

(2) Combining the parylene C coating with the HF treatment of the Mg alloy stent showed complete corrosion resistance for at least 1 month in physiological conditions. The tight adhesion between the MgF_2 and parylene C layers was thought to form a strong and stable water barrier even when the structure of the struts was deformed.

(3) The parylene C coating did not affect subsequent layers of biodegradable polymer coating applied to a long-lasting release of antiproliferative drug for 64 days to suppress restenosis.

Supplementary Materials: The following supporting information can be downloaded at: https://www.mdpi.com/article/10.3390/ma15093132/s1, Table S1: Elemental composition (wt.%) of 'Original ZM10'; Table S2: Specifications of the stent; Table S3: Mg ion release (%) from Mg alloy stent after balloon expansion; Figure S1: Structure of Mg alloy stents, 3.0 × 20 mm with six-crown two-link design and square-shaped struts. (A) Developed structure of the stent. (B) Structure of one unit of the stent. (C) Microscopy images of the stent before (upper) and after (lower) balloon expansion.

Author Contributions: Conceptualization, M.S., W.X. and T.N.; data curation, Y.K., Y.O. and I.S.; writing—original draft preparation, M.S, W.X. and T.N.; writing—review and editing, M.S., W.X., A.W., I.S. and T.N. All authors have read and agreed to the published version of the manuscript.

Funding: This research was funded by a research grant from the New Energy and Industrial Technology Development Organization (NEDO) (P14033-28J1011).

Institutional Review Board Statement: Not applicable.

Informed Consent Statement: Not applicable.

Data Availability Statement: Not applicable.

Conflicts of Interest: Y.K. and A.W. are full-time employees of Japan Medical Device Technology Co., Ltd., Y.O. is a full-time employee of Fuji Light Metal Co., Ltd. All other authors have no conflict of interest and did not receive grant or financial support from industry or from any other source to prepare this manuscript.

References

1. Gersh, B.J.; Sliwa, K.; Mayosi, B.M.; Yusuf, S. Novel therapeutic concepts: The epidemic of cardiovascular disease in the developing world: Global implications. *Eur. Heart J.* **2010**, *31*, 642–648. [CrossRef]
2. Kappetein, A.P.; Dawkins, K.D.; Mohr, F.W.; Morice, M.C.; Mack, M.J.; Russell, M.E.; Pomar, J.; Serruys, P.W.J.C. Current percutaneous coronary intervention and coronary artery bypass grafting practices for three-vessel and left main coronary artery disease. Insights from the SYNTAX run-in phase. *Eur. J. Cardiothorac. Surg.* **2006**, *29*, 486–491. [CrossRef]
3. Valgimigli, M.; Tebaldi, M.; Borghesi, M.; Vranckx, P.; Campo, G.; Tumscitz, C.; Cangiano, E.; Minarelli, M.; Scalone, A.; Cavazza, C.; et al. Two-year outcomes after first- or second-generation drug-eluting or bare-metal stent implantation in all-comer patients undergoing percutaneous coronary intervention. *JACC Cardiovasc. Interv.* **2014**, *7*, 20–28. [CrossRef]
4. Abraham, R.T.; Wiederrecht, G.J. Immunopharmacology of rapamycin. *Annu. Rev. Immunol.* **1996**, *14*, 483–510. [CrossRef]
5. Gomez-Lara, J.; Brugaletta, S.; Jacobi, F.; Ortega-Paz, L.; Nato, M.; Roura, G.; Romaguera, R.; Ferreiro, J.L.; Teruel, L.; Gracida, M.; et al. Five-year optical coherence tomography in patients with ST-segment-elevation myocardial infarction treated with bare-metal versus everolimus-eluting stents. *Circ. Cardiovasc. Interv.* **2016**, *9*, e003670. [CrossRef]
6. Xu, W.; Sasaki, M.; Niidome, T. Sirolimus release from biodegradable polymers for coronary stent application: A review. *Pharmaceutics* **2022**, *14*, 492. [CrossRef]
7. Omar, W.A.; Kumbhani, D.J. The current literature on bioabsorbable stents: A review. *Curr. Atheroscler. Rep.* **2019**, *21*, 54. [CrossRef]
8. Aljihmani, L.; Alic, L.; Boudjemline, Y.; Hijazi, Z.M.; Mansoor, B.; Serpedin, E.; Qaraqe, K. Magnesium-based bioresorbable stent materials: Review of reviews. *J. Bio-Tribo-Corros.* **2019**, *5*, 26. [CrossRef]
9. Xin, Y.; Hu, T.; Chu, P.K. In vitro studies of biomedical magnesium alloys in a simulated physiological environment: A review. *Acta Biomater.* **2011**, *7*, 1452–1459. [CrossRef]
10. Gastaldi, D.; Sassi, V.; Petrini, L.; Vedani, M.; Trasatti, S.; Migliavacca, F. Continuum damage model for bioresorbable magnesium alloy devices—Application to coronary stents. *J. Mech. Behav. Biomed. Mater.* **2011**, *4*, 352–365. [CrossRef] [PubMed]
11. Etave, F.; Finet, G.; Boivin, M.; Boyer, J.C.; Rioufol, G.; Thollet, G. Mechanical properties of coronary stents determined by using finite element analysis. *J. Biomech.* **2001**, *34*, 1065–1075. [CrossRef]
12. Xie, J.; Zhang, J.; You, Z.; Liu, S.; Guan, K.; Wu, R.; Wang, J.; Feng, J. Towards developing Mg alloys with simultaneously improved strength and corrosion resistance via RE alloying. *J. Magnes. Alloys* **2021**, *9*, 41–56. [CrossRef]
13. Li, H.; Wang, P.; Lin, G.; Huang, J. The role of rare earth elements in biodegradable metals: A review. *Acta Biomater.* **2021**, *129*, 33–42. [CrossRef]
14. Zhang, Z.Q.; Wang, L.; Zeng, M.Q.; Zeng, R.C.; Lin, C.G.; Wang, Z.L.; Chen, D.C.; Zhang, Q. Corrosion resistance and superhydrophobicity of one-step polypropylene coating on anodized AZ31 Mg alloy. *J. Magnes. Alloys* **2021**, *9*, 1443–1457. [CrossRef]
15. Saberi, A.; Bakhsheshi-Rad, H.R.; Abazari, S.; Ismail, A.F.; Sharif, S.; Ramakrishna, S.; Daroonparvar, M.; Berto, F. A comprehensive review on surface modifications of biodegradable magnesium-based implant alloy: Polymer coatings opportunities and challenges. *Coatings* **2021**, *11*, 747. [CrossRef]
16. Guo, Y.; Li, G.; Xu, Z.; Xu, Y.; Yin, L.; Yu, Z.; Zhang, Z.; Lian, J.; Ren, L. Corrosion resistance and biocompatibility of calcium phosphate coatings with a micro-nanofibrous porous structure on biodegradable magnesium alloys. *ACS Appl. Bio Mater.* **2022**, *5*, 1528–1537. [CrossRef]
17. da Conceicao, T.F.; Scharnagl, N.; Blawert, C.; Dietzel, W.; Kainer, K.U. Surface modification of magnesium alloy AZ31 by hydrofluoric acid treatment and its effect on the corrosion behavior. *Thin Solid Films* **2010**, *518*, 5209–5218. [CrossRef]
18. Weng, W.; Biesiekierski, A.; Li, Y.; Dargusch, M.; Wen, C. A review of the physiological impact of rare earth elements and their uses in biomedical Mg alloys. *Acta Biomater.* **2021**, *130*, 80–97. [CrossRef]
19. Dargusch, M.S.; Balasubramani, N.; Yang, N.; Johnston, S.; Ali, Y.; Venezuela, J.; Carluccio, J.; Lau, C.; Allavena, R.; Liang, D.; et al. In vivo performance of a rare earth free Mg-Zn-Ca alloy manufactured using twin roll casting for potential applications in the cranial and maxillofacial fixation devices. *Bioact. Mater.* **2022**, *12*, 85–96. [CrossRef]
20. Rondeau, V.; Commenges, D.; Jacqmin-Gadda, H.; Dartigues, J.F. Relation between aluminum concentrations in drinking water and Alzheimer's disease: An 8-year follow-up study. *Am. J. Epidemiol.* **2000**, *152*, 59–66. [CrossRef]
21. Xu, W.; Yagoshi, K.; Koga, Y.; Sasaki, M.; Niidome, T. Optimized polymer coating for magnesium alloy-based bioresorbable scaffolds for long-lasting drug release and corrosion resistance. *Colloids Surf. B* **2018**, *163*, 100–106. [CrossRef] [PubMed]
22. Xu, W.; Sato, K.; Koga, Y.; Sasaki, M.; Niidome, T. Corrosion resistance of HF-treated Mg alloy stents following balloon-expansion and its improvement through biodegradable polymer coating. *J. Coat. Technol.* **2020**, *17*, 1023–1032. [CrossRef]
23. Liu, X.L.; Zhen, Z.; Liu, J.; Xi, T.F.; Zheng, Y.D.; Guan, S.K.; Zheng, Y.F.; Cheng, Y. Multifunctional MgF$_2$/polydopamine coating on Mg alloy for vascular stent application. *J. Mater. Sci. Technol.* **2015**, *31*, 733–743. [CrossRef]

24. Surmeneva, M.A.; Vladescu, A.; Cotrut, C.M.; Tyurin, A.I.; Pirozhkova, T.S.; Shuvarin, I.A.; Elkin, B.; Surmenev, R.A. Effect of parylene C coating on the antibiocorrosive and mechanical properties of different magnesium alloys. *Appl. Surf. Sci.* **2018**, *427*, 617–627. [CrossRef]
25. Wolf, K.V.; Zong, Z.; Meng, J.; Orana, A.; Rahbar, N.; Balss, K.M.; Papandreou, G.; Maryanoff, C.A.; Soboyejo, W. An investigation of adhesion in drug-eluting stent layers. *J. Biomed. Mater. Res. A* **2008**, *87*, 272–281. [CrossRef]
26. Schurtz, G.; Delhaye, C.; Hurt, C.; Thieuleux, H.; Lemesle, G. Biodegradable polymer Biolimus-eluting stent (Nobori®) for the treatment of coronary artery lesions: Review of concept and clinical results. *Med. Devices* **2014**, *7*, 35–43. [CrossRef]
27. Chang, T.Y.; Yadav, V.G.; Leo, S.D.; Mohedas, A.; Rajalingam, B.; Chen, C.L.; Selvarasah, S.; Dokmeci, M.R.; Khademhosseini, A. Cell and protein compatibility of parylene-C surfaces. *Langmuir* **2007**, *23*, 11718–11725. [CrossRef]
28. Tan, C.P.; Craighead, H.G. Surface engineering and patterning using parylene for biological applications. *Materials* **2010**, *3*, 1803–1832. [CrossRef]
29. Xu, L.; Yamamoto, A. Characteristics and cytocompatibility of biodegradable polymer film on magnesium by spin coating. *Colloids Surf. B* **2012**, *93*, 67–74. [CrossRef]
30. Pan, C.; Han, Y.; Lu, J. Structural design of vascular stents: A review. *Micromachines* **2021**, *12*, 770. [CrossRef]
31. Zieliński, M.; Czajka, B.; Pietrowski, M.; Tomska-Foralewska, I.; Alwin, E.; Kot, M.; Wojciechowska, M. MgO-MgF$_2$ system obtained by sol–gel method as an immobilizing agent of the electrolyte applied in the high temperature cells. *J. Sol-Gel Sci. Technol.* **2017**, *84*, 368–374. [CrossRef]
32. Kandala, B.S.P.K.; Zhang, G.; An, X.; Pixley, S.; Shanov, V. Effect of surface-modification on in vitro corrosion of biodegradable magnesium-based helical stent fabricated by photo-chemical etching. *Med. Res. Arch.* **2020**, *8*, 3. [CrossRef]
33. Kandala, B.S.P.K.; Zhang, G.; Hopkins, T.M.; An, X.; Pixley, S.; Shanov, V. In vitro and in vivo testing of zinc as a biodegradadble material for stents fabricated by photo-chemical etching. *Appl. Sci.* **2019**, *9*, 4503. [CrossRef]
34. Gnedenkov, A.S.; Sinebryukhov, S.L.; Filonina, V.S.; Egorkin, V.S.; Ustinov, A.Y.; Sergienko, V.I.; Gnedenkov, S.V. The detailed corrosion performance of bioresorbable Mg-0.8Ca alloy in physiological solutions. *J. Magnes. Alloys* **2022**, *1*, 51. [CrossRef]
35. Rizas, K.D.; Mehilli, J. Stent polymers Do they make a difference? *Circ. Cardiovasc. Interv.* **2016**, *9*, e002943. [CrossRef]
36. Stefanini, G.G.; Taniwaki, M.; Windecker, S. Coronary stents: Novel developments. *Heart* **2012**, *100*, 989–990. [CrossRef]

Article

Influence of Co Content and Chemical Nature of the Co Binder on the Corrosion Resistance of Nanostructured WC-Co Hardmetals in Acidic Solution

Tamara Aleksandrov Fabijanić [1,*], Marin Kurtela [1], Matija Sakoman [1] and Mateja Šnajdar Musa [2]

[1] Faculty of Mechanical Engineering and Naval Architecture, University of Zagreb, Ivana Lučića 5, 10000 Zagreb, Croatia; marin.kurtela@fsb.hr (M.K.); matija.sakoman@fsb.hr (M.S.)

[2] Department of Polytechnics, University of Rijeka, Sveučilišna Avenija 4, 51000 Rijeka, Croatia; mateja.snajdar@uniri.hr

* Correspondence: tamara.aleksandrov@fsb.hr; Tel.: +385-9845-3916

Abstract: The electrochemical corrosion resistance of nanostructured hardmetals with grain sizes $d_{WC} < 200$ nm was researched concerning Co content and the chemical nature of the Co binder. Fully dense nanostructured hardmetals with the addition of grain growth inhibitors GGIs, VC and Cr_3C_2, and 5 wt.%Co, 10 wt.%Co, and 15 wt.%Co were developed by a one cycle sinter-HIP process. The samples were detailly characterized in terms of microstructural characteristics and researched in the solution of $H_2SO_4 + CO_2$ by direct and alternative current techniques, including electrochemical impedance spectroscopy. Performed analysis revealed a homogeneous microstructure of equal and uniform grain size for different Co contents. The importance of GGIs content adjustment was established as a key factor of obtaining a homogeneous microstructure with WC grain size retained at the same values as in starting mixtures of different Co binder content. From the conducted research, Co content has shown to be the dominant influential factor governing electrochemical corrosion resistance of nanostructured hardmetals compared to the chemical composition of the Co binder and WC grain size. Negative values of E_{corr} measured for 30 min in 96% $H_2SO_4 + CO_2$ were obtained for all samples indicating material dissolution and instability in acidic solution. Higher values of R_p and lower values of i_{corr} and v_{corr} were obtained for samples with lower Co content. In contrast, the anodic Tafel slope increases with increasing Co content which could be attributed to more pronounced oxidation of the higher Co content samples. Previously researched samples with the same composition but different chemical composition of the binder were introduced in the analysis. The chemical composition of the Co binder showed an influence; samples with lower relative magnetic saturation related to lower C content added to the starting mixtures and more W dissolved in the Co binder during the sintering process showed better corrosion resistance. WC-5Co sample with significantly lower magnetic saturation value showed approximately 30% lower corrosion rate. WC-10Co sample with slightly lower relative magnetic saturation value and showed approximately 10% lower corrosion rate. Higher content of Cr_3C_2 dissolved in the binder contributed to a lower corrosion rate. Slight VC increase did not contribute to corrosion resistance. Superior corrosion resistance is attributed to W and C dissolved in the Co binder, lower magnetic saturation, or WC grain size of the sintered sample.

Keywords: nanostructured hardmetals; Co content; GGIs; chemical nature of Co binder; grain size; electrochemical corrosion resistance; $H_2SO_4 + CO_2$

Citation: Aleksandrov Fabijanić, T.; Kurtela, M.; Sakoman, M.; Šnajdar Musa, M. Influence of Co Content and Chemical Nature of the Co Binder on the Corrosion Resistance of Nanostructured WC-Co Hardmetals in Acidic Solution. *Materials* **2021**, *14*, 3933. https://doi.org/10.3390/ma14143933

Academic Editor: Artur Czupryński

Received: 30 April 2021
Accepted: 30 June 2021
Published: 14 July 2021

1. Introduction

Hardmetals contain tungsten carbide WC particles joined by a binder, most commonly cobalt Co, by a liquid phase sintering process. The properties of the obtained composite derive directly from its constituents; hard and brittle carbides and softer and more ductile binder [1]. By connecting these two components, superior mechanical, physical, and chemical properties are achieved. In recent years, the development of hardmetals is based mainly on the application of ultrafine and nano WC particles (grain size less than 0.5 μm) which require the addition of grain growth inhibitors GGIs to retain the WC grain size in the sintered product. Consequently, a homogeneous microstructure and significantly improved mechanical properties (hardness, wear-resistance, and strength) are achieved. Furthermore, achieving a homogeneous microstructure with a WC grain size in the nano area (<0.2 μm) allows application at higher cutting speeds, lower tolerance, and longer tool life. Due to superior mechanical properties, hardmetals are used in various applications with certain limitations in chemically aggressive environments because of relatively poor corrosion resistance [1].

The corrosion mechanism of conventional WC-Co hardmetals in the neutral and acidic solution is governed by the reduction of the Co binder. At the same time, WC particles are not affected by the corrosion attack [2–4]. Zheng et al. found that the binder dissolution started from the center of binder pools in the acid media, independent of binder chemical nature, and spreads to the edges until the binder phase was consumed entirely [4]. Accordingly, it is expected that the corrosion rate will increase with increasing Co content in the starting mixture.

Besides Co binder content, the corrosion mechanisms in hardmetals depend on many other factors such as surface characteristics and integrity, corrosive environment, and hardmetal microstructure [5–7]. Hardmetal microstructure, including WC grain size, binder composition/binder chemical nature, grain growth inhibitors GGIs, and porosity, influence the corrosion behavior of hardmetals [3–10]. Researchers have reported different experimental variations concerning the relationship between microstructure and corrosion resistance.

The chemical nature of the Co binder is represented by magnetic saturation. It depends on the C content added to the starting hardmetal mixture and technological processes of consolidation, among which the most important are sintering parameters and atmosphere [11–13]. Co binder is advantageous because of the relatively large carbon contents that give the preferred two-phase WC-Co composition, commonly called the carbon window [14,15]. During sintering, the Co binder is alloyed with tungsten (W) and carbon (C); other constituents such as GGIs also add alloying elements to the binder [14]. If a higher amount of tungsten is dissolved in the Co binder, lower relative magnetic saturation values would be obtained [3], and the formation of the brittle η-phase carbides M_6C and $M_{12}C$ would occur in the microstructure of hardmetals. It was found from previous research that hardmetals with lower relative magnetic saturation values show lower values of corrosion current density (i_{corr}) and critical current density (i_{crit}) measured by potentiodynamic polarization [2,6,7,16].

Regarding the influence of the WC grain size, different conclusions can be found in the literature. Li Zhang et al. found pseudo-passivation behavior of conventional hardmetals with the WC grain sizes ranging from 1.2 μm to 8.2 μm in sulfuric acid H_2SO_4 and better corrosion resistance of the coarse WC grain sizes [3]. On the other hand, Imasato et al. found that the corrosion rate of the WC-Co alloy with a smaller WC grain size was lower than coarse WC grain size both in acid and neutral solution. WC-Co alloy with a smaller WC grain size showed a higher corrosion resistance in the polarization test because of the low current densities of active dissolution and passivated region in the polarization curve [17]. Also, they found that the amount of dissolved metals in neutral and acidic solutions decreased with decreasing WC grain size and carbon content. Most published papers refer to conventional hardmetals, while there is still a lack of results published on nanostructured hardmetals with a WC grain size less than 200 nm.

The paper summarizes long-term research on the corrosion resistance of nanostructured hardmetals. From previous research it was found that the chemical nature of the binder has a more substantial influence on the electrochemical corrosion resistance compared to Co content in the starting mixture in neutral and acidic solution, which was quite surprising and not in line with conventional WC-Co hardmetals [6,7]. The presented research was performed to bring more exact conclusions on the influence of Co content and other microstructural characteristics on the corrosion resistance of nanostructured hardmetals.

2. Materials and Methods

Different starting mixtures were prepared to investigate the influence of Co content and chemical nature of the Co binder on the corrosion resistance of nanostructured hardmetals with a grain size $d_{WC} < 200$ nm. WC powder produced by H.C. Starck Tungsten (Goslar, Germany) with an average grain size d_{BET} of 95 nm and a specific surface area (BET) of 3.92 m^2/g, classified as real nanopowder with a grain size less than 100 nm, was used as a carbide phase. Grain growth inhibitors GGIs, vanadium carbide VC and chromium carbide Cr_3C_2 were added to the starting mixtures. VC powder has an average grain size d_{BET} of 350 nm and a specific surface area (BET) of 3.0 m^2/g, while Cr_3C_2 has an average grain size d_{BET} of 450 nm and a specific surface area (BET) of 2.0 m^2/g. Besides controlling the WC grain growth, VC and Cr_3C_2 increase hardness and reduce the rate of corrosion. At the same time, Cr significantly lowers the initial melting point and broadens the melting range, particularly at low carbon levels [12–14,16]. The amount of VC and Cr_3C_2 differs for each mixture; it was optimized to withhold WC powder size in the sintered samples and increased Co binder content. Half micron cobalt HMP Co, produced by Umicore (Brussels, Belgium), was used as a binder. Three mixtures with different Co content; 5, 10, and 15 wt.%Co were prepared. The consolidation process consisted of powder mixture homogenization in a horizontal ball mill (Zoz GmbH, Wenden, Germany). Compacting was performed by uniaxial die press type CA-NCII 250 (Osterwalder AG, Lyss, Switzerland). Final consolidation to total density was achieved by one cycle sinter-HIP process by furnace FPW280/600-3-2200-100 PS (FCT Anlagenbau GmbH, Sonneberg, Germany) at 1350 °C for 30 min, followed by 100 bars Argon 4.8 pressure for 45 min. The characteristics of the starting mixtures are presented in Table 1.

Table 1. The characteristics of the starting mixtures.

Mixture	Starting WC Powder	Grain Size d_{BET}, nm	Specific Surface, m^2/g	Co, wt.%	GGI, wt.%
WC-5Co				5	0.3% VC 160 (H. C. Starck) 0.5% Cr_3C_2 160 (H. C. Starck)
WC-10Co	WC DN 4-0 (H.C. Starck)	95	3.92	10	0.5% VC 160 (H. C. Starck) 0.75% Cr_3C_2 160 (H. C. Starck)
WC-15Co				15	0.75% VC 160 (H. C. Starck) 1.13% Cr_3C_2 160 (H. C. Starck)

The goal was to develop fully dense samples with optimal microstructural characteristics with no irregularities such as η-phase in the structure. Previous research found that C and GGIs content and WC grain size have a more substantial influence on the corrosion resistance of nanostructured hardmetals than Co content [6,7]. For the mentioned reason, special care was taken to obtain the optimal and comparable microstructural characteristics of consolidated samples.

The samples were detailly characterized, especially in terms of microstructural characteristics. The characterization of samples consisted of density measurements (Metler Toledo) according to ISO 3369:2006, the specific saturation magnetization (Setaram Instrumentation, Sigmameter) according to D6025, and the coercive field strength measurement (Foerster, Koerzimat 1.096) according to ISO 3326. Diamond disc and pastes were used to grind and polish the samples' surface before microstructural characterization and electrochemical measurements. Microstructural characterization consisted of porosity, free carbon, and η-phase evaluation. It was performed by comparing the sample's surfaces with photomicrographs from the standard ISO 4499-4:2016. For that purpose, an optical microscope (Olympus, Shinjuku City, Tokyo, Japan) was used. A field emission scanning electron microscope FESEM (Zeiss, Oberkochen, Germany) was used for WC grain size measurement by the linear intercept method and detection of irregularities such as abnormal growth and WC grains grouping or Co lakes. X-ray diffraction XRD analysis was used to identify the phases present in consolidated samples and exclude the occurrence of η-phase.

After detailed characterization of the samples and determination of optimal microstructural characteristics, the corrosion measurements were performed. The surface of the samples was placed into the corrosion cell filled with $H_2SO_4 + CO_2$ (pH = 0.6). As reference electrode, saturated calomel electrode SCE (SCHOTT Instruments GmbH, Mainz, Germany) with a potential of + 0.242 V according to the standard hydrogen electrode was selected. Graphite wires were used as a counter electrode. The samples were first researched by direct current techniques DC, the open-circuit potential E_{corr}, the linear polarization resistance (LPR), and the Taffel extrapolation method. Corrosion potential E_{corr} versus SCE was recorded for 30 min. LPR was carried out in the potential range from −0.02 V vs. open circuit potential to 0.02 V vs. open circuit potential with a scan rate of 0.167 mV/s. Tafel extrapolation was conducted in the potential range from −0.25 V vs. open circuit potential to 0.25 V vs. open circuit potential, total points 1001 with a scan rate of 0.167 mV/s. Immediately after the DC techniques, the samples were researched by alternating current (AC) techniques, more precisely by electrochemical impedance spectroscopy (EIS).

The EIS start frequency was 100,000 Hz, and the end frequency was 0.001 Hz; the amplitude was in the range of 10 mV root-mean-square (RMS). The recorded measurements were analyzed by software SoftCorr III (AMETEK Scientific Instruments, Princeton applied research, Berwyn, PA, USA). A convenient electrical equivalent circuit (EEC) was selected by fitting the results of measurements and presented in Nyquist and Bode plots. At each excitation frequency, an imaginary impedance component Zim is drawn according to the actual impedance component Zre. The impedance and the phase shift curves were plotted against the excitation frequency. Both DC and AC techniques were performed on the potentiostat AMETEK, Princeton applied research, model VersaSTAT3, and the results were recorded and analyzed by software SoftCorr III (AMETEK Scientific Instruments, Princeton applied research, Berwyn, PA, USA).

3. Results

3.1. Microstructural Characteristics of Consolidated Samples

Characteristics of consolidated samples with different Co content are presented in Table 2. This section may be divided by subheadings. It should provide a concise and precise description of the experimental results, their interpretation, as well as the experimental conclusions that can be drawn.

Table 2. Characteristics of consolidated samples.

Sample	Density, g/cm^3	Relative Density, %	Magnetic Saturation, μTm^3/kg	Rel. Magnetic Saturation, %	Coercive Force, kA/m	ISO Porosity			d_{WC}, nm
						A	B	C	
WC-5Co	14.91	100.0	8.4	92	52.0	A00	B00	C00	187
WC-10Co	14.31	100.0	14.8	79	40.0	A00	B00	C00	198
WC-15Co	13.84	100.0	22.3	79	37.0	A00	B00	C00	192

Full densification was achieved for all samples. The samples are characterized by the lowest possible degree of porosity, A00, B00, and C00, meaning no uncombined/free carbon or η-phase were revealed on the sample's surface. A high density of samples is related to Co liquid phase, which is spreading onto the surrounding WC particles. Binder propagation is associated with Laplace forces acting along the wetting front between Co binder and WC grains while rearranging the WC particles and reducing the mean distance between neighboring particles, resulting in densification [16,18]. It may be concluded that the WC particles were rearranged, and Co binder filled the micropores between the neighboring WC grains resulting in a theoretical/full density of the samples.

The amount of W dissolved in the Co binder phase can be assessed by measuring the magnetic saturation. The saturation value of Co decreases linearly with the addition of tungsten W and is not affected by the carbon content in the solution [19]. Typical relative magnetic saturation/percentage saturation ranges from 80–100% [19], while percentage saturation values higher than 70% indicate two-phase WC-Co microstructure. The highest percentage of 91% was measured for the WC-5Co sample, while 79% was measured for WC-10Co and WC-15Co samples. The two-phase WC-Co microstructure of researched samples is confirmed by optical analysis. Based on coercive force measurement it was estimated the WC grain size. Measured values indicate that all samples fall in the nano range. Two-phase WC-Co microstructure was confirmed by optical analysis, XRD analysis where only WC with the hexagonal crystal structure and Co with FCC cubic crystal structure were identified. Investigation of microstructure revealed homogeneous, uniform distribution of WC grains, without abnormal grain growth due to optimal GGIs content added to the starting mixtures. It was necessary to adjust the content of GGIs for different Co content to achieve a homogeneous and comparable microstructure with retained WC grain size of the starting powders in the sintered samples. Co binder was uniformly distributed, and no Co lakes occurred. Microstructure images and XRD patterns of samples are presented in Figures 1–3.

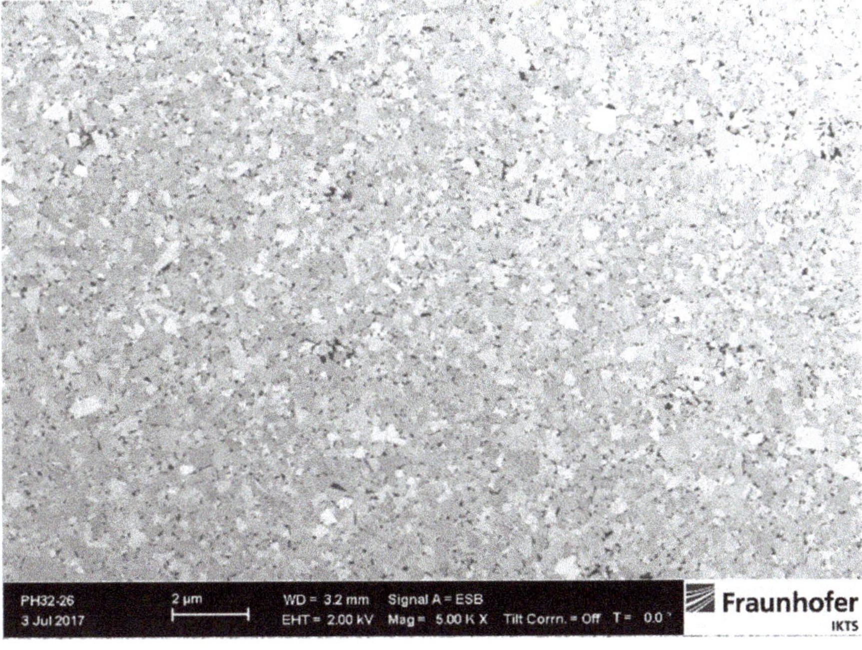

(a)

Figure 1. *Cont.*

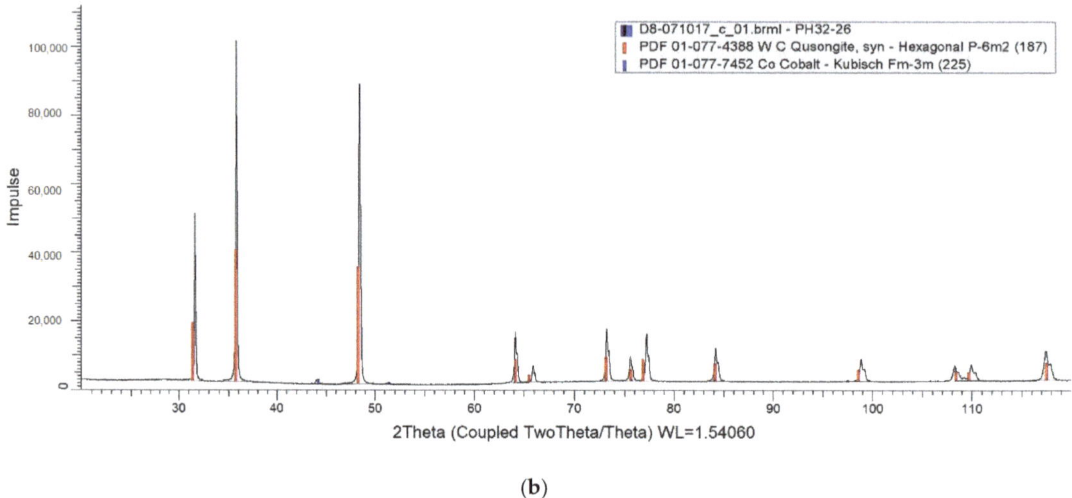

(**b**)

Figure 1. Microstructure and XRD pattern of WC-5Co sample [20]. (**a**) SEM image of microstructure; (**b**) XRD pattern.

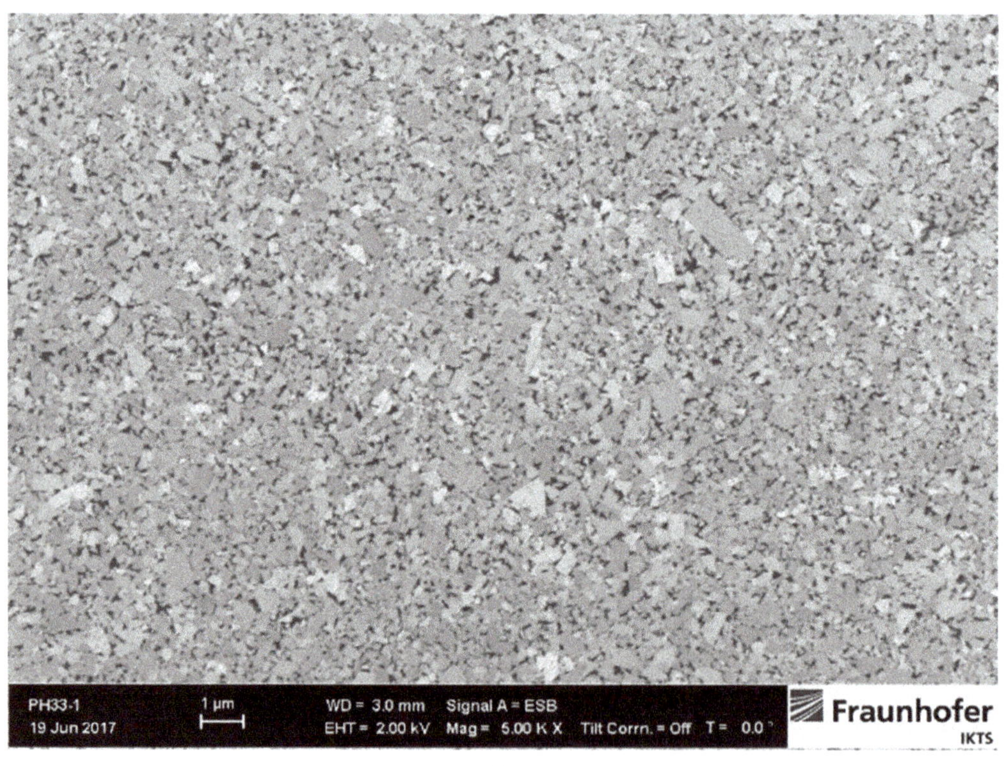

(**a**)

Figure 2. *Cont.*

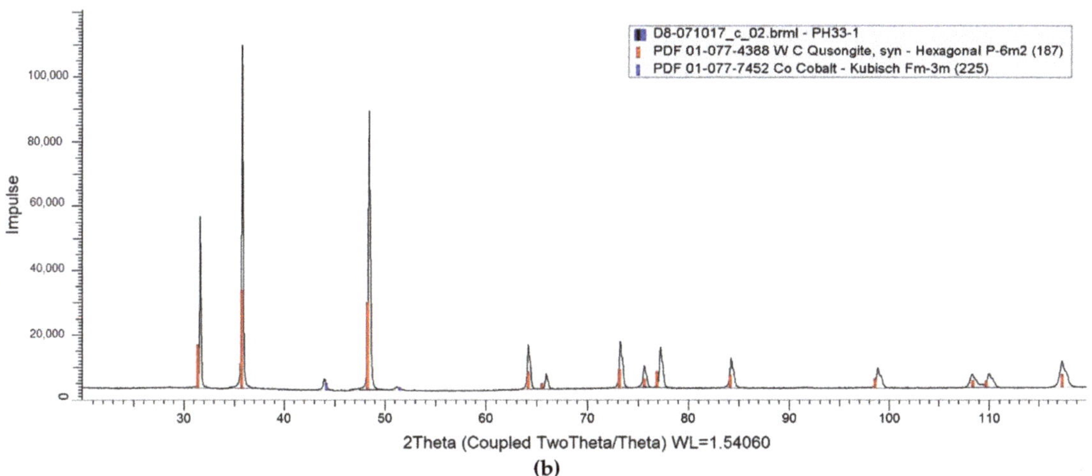

(b)

Figure 2. Microstructure and XRD pattern of WC-10Co sample [21]: (**a**) SEM image of microstructure; (**b**) XRD pattern.

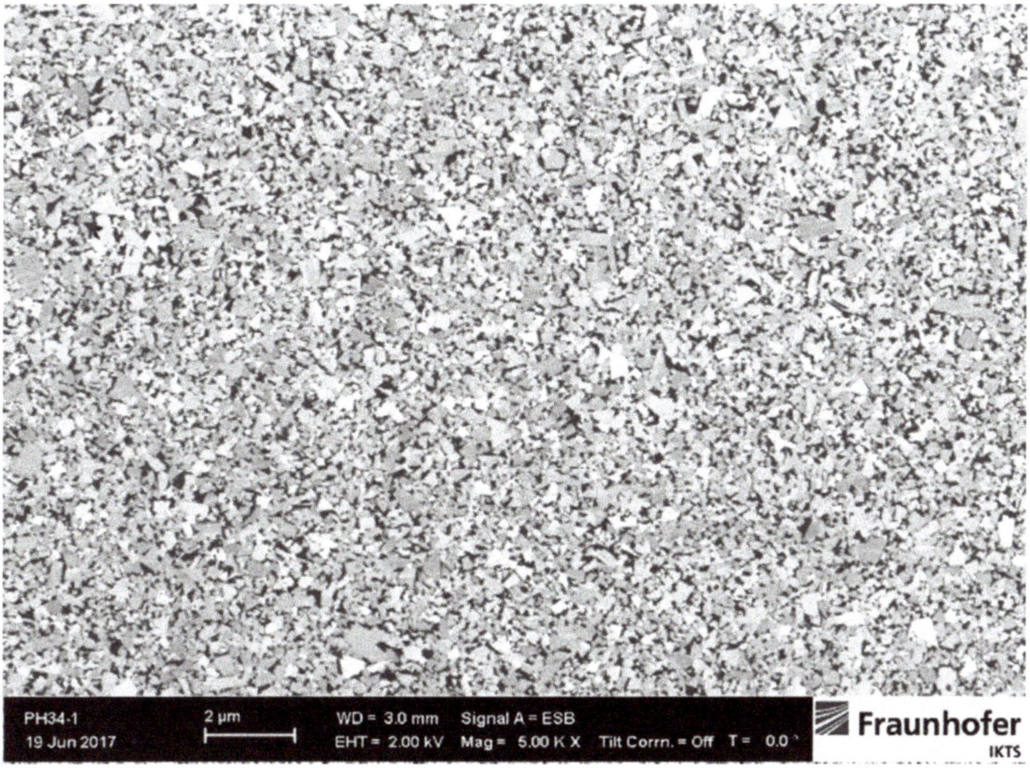

(a)

Figure 3. *Cont.*

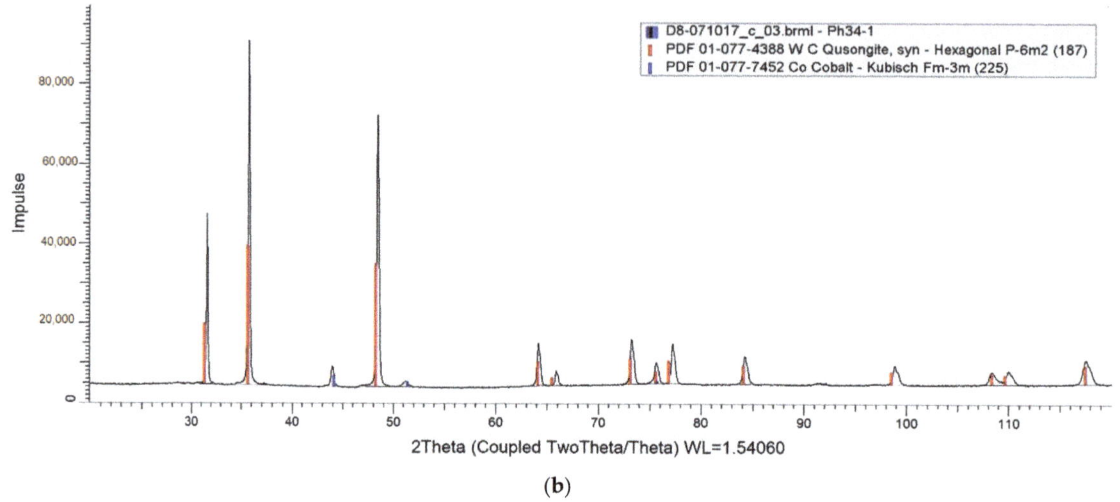

(**b**)

Figure 3. Microstructure and XRD pattern of WC-15Co sample [20]: (**a**) SEM image of microstructure; (**b**) XRD pattern.

3.2. Results of DC Techniques

The results of electrochemical DC techniques are presented in Table 3.

Table 3. Electrochemical DC techniques results.

Sample	Ts [°C]	E_{corr} vs. SCE [mV]	R_p [Ωcm^2]	β_a [mV/dec]	β_c [mV/dec]	i_{corr} [$\mu A/cm^2$]	v_{corr} [mm/y]
WC-5Co	20 ± 2	−249	654.5	75.31	90.37	20.7	0.1748
WC-10Co	20 ± 2	−308	452.8	98.34	97.67	36.6	0.3888
WC-15Co	20 ± 2	−291	349.9	120.19	86.95	50.8	0.4162

Ts—measured temperature; E_{corr}—corrosion potential; R_p—polarization resistance; βa—a slope of anodic Tafel curve; βc—a slope of cathodic Tafel curve; i_{corr}—corrosion current density; v_{corr}—corrosion rate.

3.3. Results of Electrochemical Impedance Spectroscopy EIS

EIS measurements aimed to investigate the corrosion behavior at the interface between the sample surface and electrolyte solution and determine the samples' corrosion rate. The results are presented in Table 4.

Table 4. Electrochemical impedance spectroscopy EIS technique results.

Sample	Ts [°C]	R_s [Ωcm^2]	Q	n_1	R_p/R_{ct} [Ωcm^2]
WC-5Co	20 ± 2	4.022	$1.761 \cdot 10^{-3}$	0.745	$1.101 \cdot 10^{-3}$
WC-10Co	20 ± 2	4.504	$2.213 \cdot 10^{-3}$	0.725	$8.068 \cdot 10^{-2}$
WC-15Co	20 ± 2	5.797	$2.552 \cdot 10^{-3}$	0.683	$4.657 \cdot 10^{-2}$

Ts—measured temperature; R_s—solution resistance between the working electrode and the reference electrode in a three-electrode cell; Q—Constant Phase Element (CPE); n_1—constant; R_{ct}—polarization resistance or resistance to charge transfer on the electrode/electrolyte interface.

As already mentioned in Section 2, a convenient and optimal electrical equivalent circuit (EEC) was selected using software SoftCorr III by fitting the measurements' results and presented in Nyquist and Bode plots. At each excitation frequency, an imaginary impedance component Zim is drawn according to the actual impedance component Zre. The impedance and the phase shift curves were plotted against the excitation frequency. The selected ECC model is shown in Figure 4.

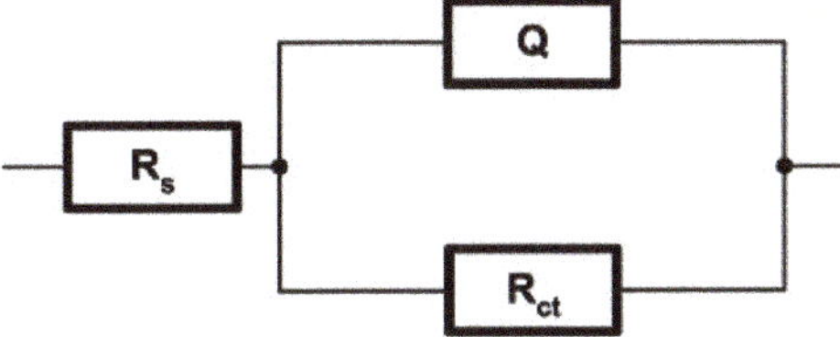

Figure 4. Selected EEC [22].

The same model R(QR), which best corresponds to the processes and reactions on the sample's surface, was selected for all samples. It is essential to mention that the same EEC model was established in previous research performed on near nanostructured WC-11Co and WC-11Ni samples [22]. The mentioned indicates the repeatability of the corrosion process between nanostructured hardmetal and acidic solutions. The impedance of a constant phase element is defined as:

$$Q = \left[Y(j\omega)^n \right]^{-1} \tag{1}$$

where Y and n $(-1 \leq n \leq 1)$ are constants independent of the angular frequency (ω) and temperature. For the value in the range $0.6 < n \leq 1$, CPE has the physical meaning of capacitance, an ideal inductor for $n = -1$, and an ideal resistor for $n = 0$.

4. Analysis and Discussion

4.1. Influence of Co Content on the Corrosion Resistance of Nanostructured WC-Co Hardmetals

From conducted research, it can be concluded that the corrosion potential E_{corr} of samples changes depending on the Co content. E_{corr} of WC-5Co and WC-10Co samples changed from more negative to more positive values indicating that the surfaces of the samples are getting passivated, and a reduction occurred. The corrosion potential curves are unstable and show random fluctuations. A drop from more positive to more negative values was detected for the WC-15Co sample, indicating sample oxidation in contact with the acidic electrolyte and reducing protons at the surface. The changes of E_{corr} are not significant, and the corrosion potential variations of each sample occurred in a narrow range. Negative values of E_{corr} measured for 30 min in 96% H_2SO_4 + CO_2 were obtained for all samples indicating material dissolution and instability in acidic solution. The E_{corr} vs. time curves are presented in Figure 5, and sample Tafel extrapolation curves in Figure 6.

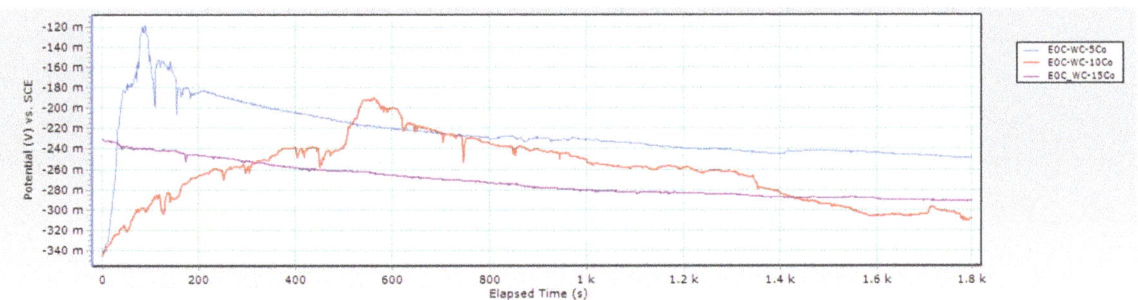

Figure 5. Corrosion potential E_{corr} of samples.

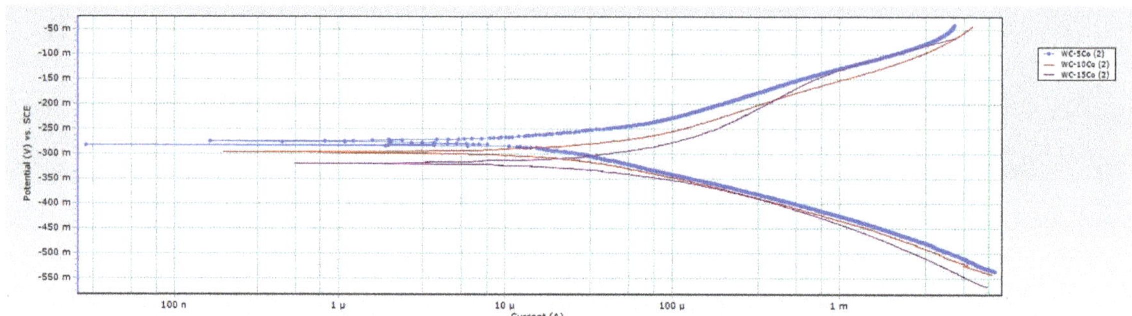

Figure 6. Tafel extrapolation curves.

Higher values of R_p and lower values of i_{corr} were obtained for samples with lower Co content. Accordingly, the corrosion rate in acidic solution increases with increasing Co content due to selective dissolution of the Co matrix. The cathodic Tafel slopes of samples show a similar trend. In contrast, the anodic Tafel slope increases with increasing Co content which could be attributed to more pronounced oxidation of the higher Co content samples. The dependence of polarization resistance R_p and corrosion rate v_{corr} for different Co contents is presented graphically in Figure 7.

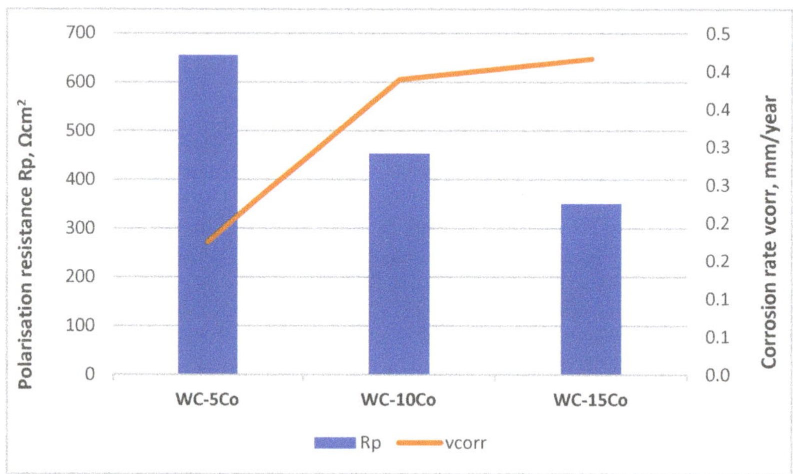

Figure 7. The dependence of R_p and v_{corr} concerning Co content.

Even though the addition of the refractory metal carbides, VC, Cr_3C_2 in the starting mixtures was increased with increasing Co content to maintain the WC powder size, Co content showed stronger influence on the electrochemical corrosion resistance of nanostructured hardmetal samples with optimal microstructural characteristics. The lowest v_{corr} of 0.1748 mm/y was obtained for the WC-5Co sample, while the highest v_{corr} of 0.4162 mm/y was measured for the WC-15Co sample.

Figures 8–10 present the Nyquist and Bode diagrams for the WC-5Co, WC-10Co, and WC-15Co samples, obtained using the corresponding EEC simulation model of EIS results. The highest Rp of $1.101 \cdot 103$ Ωcm^2 was measured for the WC-5Co sample, followed by Rp of $8.068 \cdot 102$ Ωcm^2 measured for the WC-10Co sample, and Rp of $4.657 \cdot 102$ Ωcm^2 measured for the WC-15Co sample. Higher Rp values were detected for the samples with lower Co content, indicating better corrosion resistance (Figure 11) which corresponds to the results obtained by DC linear polarization techniques. It can be seen from Figures 8–10 that the radius r of the capacitive semi-circles in the Nyquist plots differ for samples with different Co content. The diameter of the capacitive loop decreased with increasing Co content, indicating better charge transfer resistance on the electrode/electrolyte interface. Subsequently, the decrease in the diameter of the capacitance loop may be ascribed to the weaker protective ability of the sample surface.

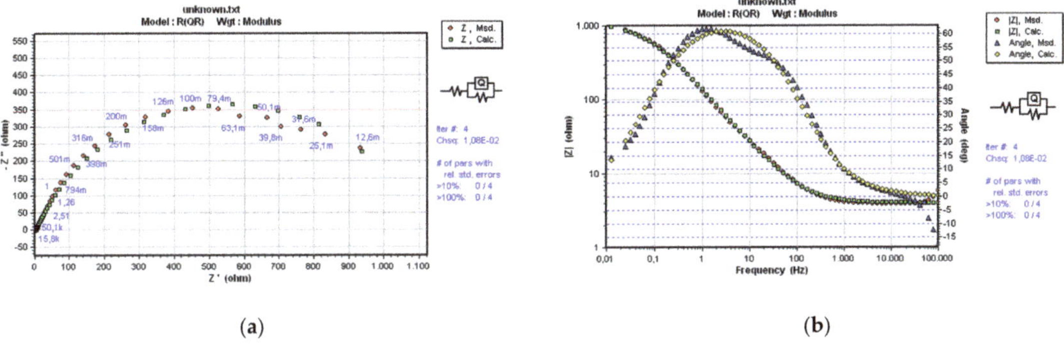

Figure 8. Nyquist (**a**) and Bode (**b**) plots of WC-5Co.

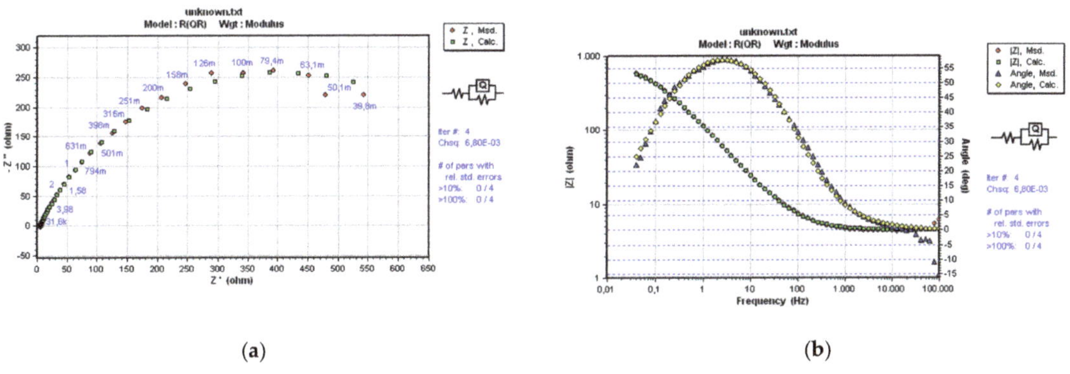

Figure 9. Nyquist (**a**) and Bode (**b**) plots of WC-10Co.

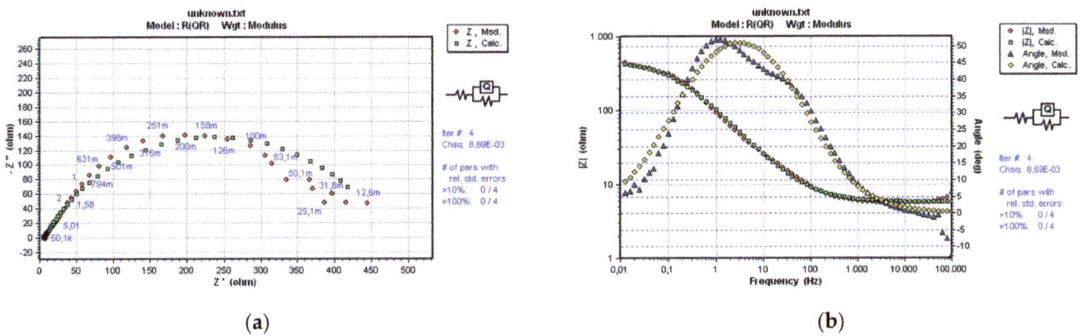

Figure 10. Nyquist (**a**) and Bode (**b**) plots of WC-15Co.

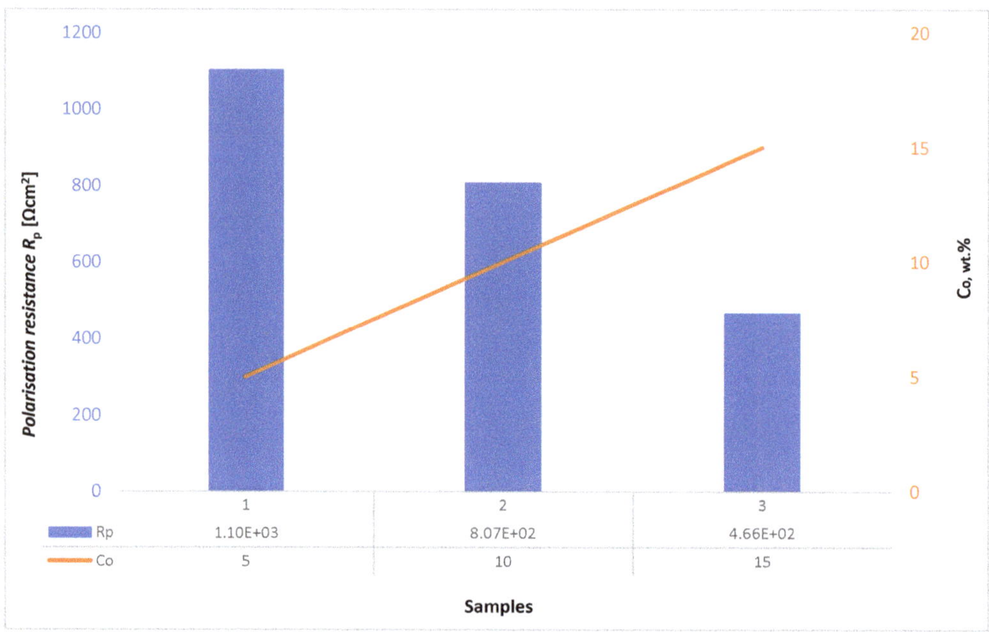

Figure 11. The dependence of R_p concerning Co content.

Nanostructured hardmetals with optimal microstructural characteristics exhibited behavior similar to previously researched conventional hardmetals with coarser WC grain size. Electrochemical corrosion resistance decreases with increasing Co content in a corrosive, acidic environment due to predominant active Co binder reduction, also known as Co leaching.

4.2. Influence of Co Binder Chemical Nature on the Corrosion Resistance of Nanostructured WC-Co Hardmetals

The chemical nature of the Co binder can be characterized by magnetic saturation. It depends on the C content added to the starting hardmetal mixture and consolidation procedure, where sintering parameters and atmosphere have a crucial influence. As referred in the Introduction, it was found from previous research that C content added to the starting mixture can significantly influence the electrochemical corrosion resistance in both neutral (3.5% NaCl with pH = 6.6) and acidic (96% H_2SO_4 + CO_2 with pH = 0.6) environments [6,7]. Comparing WC-5Co samples of the same composition, better corrosion resistance was observed for samples with lower C-added content, lower magnetic saturation, and coarser WC grain size. The opposite behavior governed by the C content and magnetic saturation was noted for WC-15Co samples [7]. Compared to C content, GGIs content, and grain size, the Co content showed less impact on the electrochemical corrosion resistance in both acid and neutral solutions. There was no clear trend of increasing corrosion current densities i_{corr} and decreasing polarization resistance Rp with increasing Co content typical for conventional hardmetals. Microstructural characteristics, in this case WC grain size, has shown to have a greater influence on the sintered samples corrosion resistance [6,7]. To obtain better insight into the electrochemical corrosion resistance of nanostructured hardmetals, previously researched samples designated as WC-5Co-1 and WC-10Co-1 were introduced in the analysis. WC-5Co-1 and WC-10Co-1 samples were consolidated using same production procedure and characterized by the same methods described in Section 2. The only alteration was the use of lower C and GGIs content added to the starting mixture, which caused different chemical nature of the Co binder and lower values of relative magnetic saturation. Characteristics of the additionally introduced samples and comparison with previously characterized nanostructured hardmetals with $d_{WC} < 200$ nm are presented in Tables 5 and 6.

Table 5. Comparison of WC-5Co samples with different characteristics.

Sample	GGI, wt.%	Added C, wt.%	Density, g/cm^3	ρ, %	Relative Magnetic Saturation, %	Coercive Force, kA/m	v_{corr} [mm/y]
WC-5Co-1	0.41%VC, 0.80% Cr$_3$C$_2$	0.150	14.96	100	48.0	44.9	0.1181
WC-5Co	0.30%VC, 0.50% Cr$_3$C$_2$	0.275	14.91	100	92.0	52.0	0.1748

Table 6. Comparison of WC-10Co samples with different microstructural characteristics and magnetic saturation.

Sample	GGI, wt.%	Added C, wt.%	Density, g/cm^3	ρ, %	Relative Magnetic Saturation, %	Coercive Force, kA/m	v_{corr} [mm/y]
WC-10Co-1	0.37%VC, 0.72% Cr$_3$C$_2$	0.225	14.35	100	74.7	35.1	0.3463
WC-10Co	0.5%VC, 0.75% Cr$_3$C$_2$	0.250	14.32	100	79.0	40.0	0.3888

As indicated in Table 5, the WC-5Co-1 sample has a significantly lower relative magnetic saturation of 48.0%, attributed to the lower added C content and the different chemical nature of the Co binder. As mentioned, typical relative magnetic saturation/percentage saturation ranges from 80–100%. Saturation percentage values lower than 70% indicate the presence of microstructural irregularity η-phase, confirmed by optical, FESEM, and XRD analysis [6,7]. Previous research found that η-phase most likely enhances the passive layer formation on the sample surfaces, thereby reducing the tendency of sample dissolution and increasing the stability of oxides forming in addition to the existing passive layer on the surface [2,6,7]. The slightly higher measured density of the WC-5Co-1 sample is also associated with η-phase presence in the structure since W_6Co_6C or W_3Co_3C has higher density when compared to a two-phase WC-Co hardmetal. The coercive force obtained for the WC-5Co-1 sample amounts to 44.9 kA/m and is lower than that of WC-5Co, suggesting a coarser grain size of the WC-5Co-1 sample. Thus, its microstructure can be classified as near nano, in the ultrafine range from 200 to 500 nm. Tafel extrapolation curves of WC-5Co samples are presented in Figure 12.

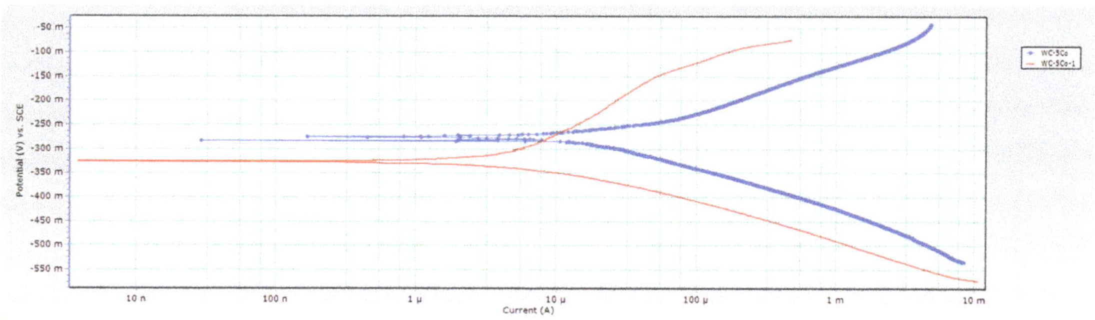

Figure 12. Tafel extrapolation curves of WC-5Co samples with different microstructural characteristics.

The WC-5Co-1 sample with lower magnetic saturation value and consequently higher W and C content in the Co binder showed approximately 30% lower corrosion rate, which is in line with previous research. Sutthiruangwong and Mori found that the magnetic saturation related to Co binder composition plays an essential role in the corrosion properties of hardmetals [15,23,24]. F.J.J. Kellner et al. found that electrochemical corrosion resistance of hardmetals is influenced by the W and C diffusion in the Co binder amount which is increased by decreasing the WC grain size [25]. They concluded that W and C dissolved in Co binder during the sintering process stabilize the thermodynamically unstable FCC Co crystal structure at room temperature, the amount of which is increased by an increase of W and C content in the binder. FCC Co is characterized by better corrosion resistance compared to the HCP crystal structure of Co, thermodynamically stable at room temperature [25], when a HCP+FCC Co layer around the HCP Co binder is formed [25]. Their research was performed in an alkaline medium. Still, since in this study the tests were performed in an acidic solution where the lower corrosion resistance of hardmetals is attributed to the dissolution of the HCP Co matrix, these claims could be expected to be more pronounced.

Besides lower magnetic saturation, the WC-5Co-1 sample has a higher content of GGIs, VC, Cr_3C_2 in the starting mixture, as presented in Table 5. It is well-known that GGIs are dissolved and distributed among the WC phase and binder during sintering and influence characteristics of hardmetals [26]. Sutthiruangwong and Mori have found that higher corrosion resistance can be assigned to binders that experience higher chromium dissolution rates during sintering [23,24]. Tomlinson and Ayerst found that small additions of Cr_3C_2 improve the electrochemical corrosion resistance of hardmetals due to the formation of Cr_2O_3 film on the Co binder surface [26]. On the other hand, a small addition of VC in combination with Cr_3C_2 decreases the positive Cr_3C_2 influence in acidic solution [26]. The WC-5Co-1 sample has a 0.3 wt.% higher content of Cr_3C_2, which dissolved in the binder and contributed to a 30% lower corrosion rate than the WC-5Co sample. In this research, it is hard to distinguish which factor has the most substantial influence on electrochemical corrosion resistance. To specify more clearly, WC-10Co samples with different microstructural characteristics and magnetic saturation were compared in Table 6.

The WC-10Co-1 sample has a slightly lower relative magnetic saturation of 74.7% than the WC-10Co sample, which is related to marginally lower C content in the amount of 0.025 wt.% added to the starting mixture. Lower C content resulted in more W dissolved in the Co binder, which, as in the case of the WC-5Co sample, most probably stabilized Co's thermodynamically unstable FCC crystal structure. Consequently, the WC-10Co-1 sample is characterized by a marginally higher measured density. Density values vary within the two-phase region of the WC-Co phase, and is increased with an increasing amount of W which remains in the Co binder [27]. Sample WC-10Co-1 is located at the lower end of the two-phase WC-Co region in the isothermal part of the WC-Co phase diagram. Therefore, the measured density value is slightly higher than the theoretical density, despite η-phase not detected. The coercive force of WC-10Co-1 amounts to 35.1 kA/m, i.e., lower when compared to WC-10Co, which indicates a coarser grain size of the WC-10Co-1 sample. The same was noted for WC-5Co samples. Tafel extrapolation curves of WC-10Co samples are presented in Figure 13.

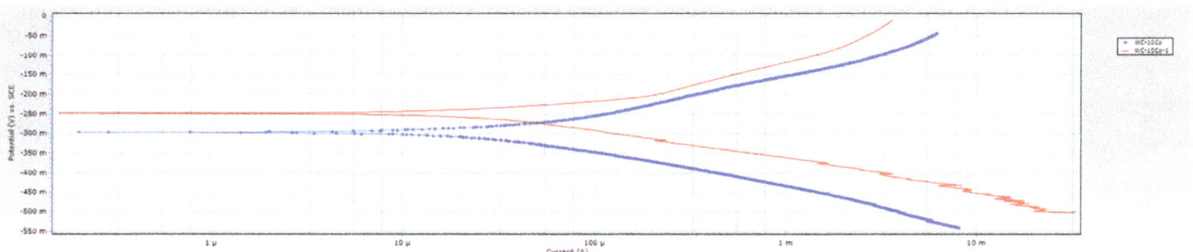

Figure 13. Tafel extrapolation curves of WC-10Co samples with different microstructural characteristics.

The WC-10Co-1 sample with a lower relative magnetic saturation value showed approximately 10% lower corrosion rate. Both microstructures consist of two phases, WC and Co, with no η-phase detected in the microstructure. The difference in GGIs content is relatively small and amounts to an extra 0.13 wt.%VC and 0.03 wt.% Cr_3C_2 added to the WC-10Co mixture. It can be concluded that the slight increase of VC wt.% did not contribute to corrosion resistance which corresponds to previously published research. Machio et al. found that small VC addition of 0.4 wt.% increase i_{corr} and make hardmetals more sensitive to pitting corrosion due to VC influence on the W dissolution in the Co matrix and formation of (V,W)C layer around the WC grains. VC decreases the dissolution of W atoms in the Co binder during the sintering process and increases the magnetic saturation compared to pure WC-Co hardmetal without GGIs [1,28]. D.S. Konadu et al. found that WC-Co hardmetal possesses nobler corrosion resistance compared to 0.4 wt.%VC containing hardmetal in both HCl and H_2SO_4 [1]. Accordingly, better corrosion resistance in this research may be related to W and C dissolution in the Co binder, magnetic saturation, or WC grain size in the sintered sample.

5. Conclusions

The following conclusions can be drawn from the conducted research:

(1) Fully dense nanostructured hardmetals with a WC grain size $d_{WC} \leq 200$ nm were developed utilizing the single-cycle sinter-HIP process. For different Co contents, a homogeneous microstructure of equal and uniform grain size without microstructural defects in the form of carbide agglomerates, abnormal grain growth, or Co lakes was successfully obtained.

(2) The importance of GGIs content adjustment was established as a key factor of obtaining a homogeneous microstructure with WC grain size retained at the same values as in starting mixtures of different Co binder content.

(3) The Co content in the starting mixture proved to have a significant influence on the electrochemical corrosion resistance of nanostructured hardmetals in acidic solution. A noticeable trend of polarization resistance R_p decrease, and current density i_{corr} and corrosion rate v_{corr} increase has been established with increasing Co content. Nanostructured hardmetals with the grain size $d_{WC} < 200$ nm showed the same corrosion behavior as coarser grain-size conventional WC hardmetals depending on the Co content.

(4) The chemical composition of the Co binder showed a significant influence. Samples with lower relative magnetic saturation related to lower added C content and more W dissolved in the Co binder showed better corrosion resistance. Significant differences in magnetic saturation for samples with the same Co content lead to more pronounced differences in the corrosion rates. A slight difference in magnetic saturation and WC grain size changed the Taffel curves.

(5) Co content was shown to be the dominant influential factor governing electrochemical corrosion resistance of nanostructured hardmetals when compared to the chemical composition of the Co binder and WC grain size. Samples with lower Co content exhibited lower corrosion rates.

(6)   The slight increase of GGIs content, Cr_3C_2, and VC did not improved the corrosion resistance significantly for the samples with the same Co content. Higher content of Cr_3C_2 dissolved in the binder contributed to a lower corrosion rate. Slight VC increase did not contribute to corrosion resistance. Superior corrosion resistance is attributed to W and C dissolved in the Co binder, lower magnetic saturation, or WC grain size of the sintered sample.

Author Contributions: Conceptualization, T.A.F.; formal analysis, T.A.F., M.K., M.S., and M.Š.M.; investigation, T.A.F., M.K., and M.S.; writing—original draft, T.A.F.; writing—review & editing, M.Š.M. All authors have read and agreed to the published version of the manuscript.

Funding: This work is supported in part by the Croatian Science Foundation under Project Number UIP-2017-05-6538 Nanostructured hardmetals–New challenges for Powder Metallurgy.

Acknowledgments: The authors acknowledge the infrastructure and research support of Fraunhofer IKTS, Dresden, Germany, group Hardmetal and Cermet.

Conflicts of Interest: The authors declare no conflict of interest.

References

1.   Konadu, D.S.; van der Merwe, J.; Potgieter, J.H.; Potgieter-Vermaak, S.; Machio, C.N. The corrosion behaviour of WC-VC-Co hardmetals in acidic media. *Corros. Sci.* **2010**, *52*, 3118–3125. [CrossRef]
2.   Alar, Ž.; Alar, V.; Fabijanić, T.A. Electrochemical Corrosion Behavior of Near-Nano and Nanostructured WC-Co Cemented Carbides. *Metals* **2017**, *7*, 69. [CrossRef]
3.   Hochstrasser-Kurz, S.; Mueller, Y.; Latkoczy, C.; Virtanen, S.; Schmutz, P. Analytical characterisation of the corrosion mechanism and inductively coupled plasma mass spectroscopy. *Corros. Sci.* **2007**, *49*, 2002–2020. [CrossRef]
4.   Zhang, L.; Wan, Q.; Huang, B.; Liu, Z.; Zhu, J. Effects of WC Grain Sizes and Aggressive Media on the Electrochemical Corrosion Behaviours of WC-Co Cemented Carbides. In Proceedings of the EURO PM2015-HM-Modelling and Characterisation, Reims, France, 4–7 September 2015.
5.   Tarrag', J.M.; Fargas, G.; Isern, L.; Dorvlo, S.; Tarres, E.; Muller, C.M.; Jim'enez-Piqu'e, E.; Llanes, L. Microstructural influence on tolerance to corrosion-induced damage in hardmetals. *Mater. Des.* **2016**, *111*, 36–43. [CrossRef]
6.   Fabijanić, T.A.; Pötschke, J.; Alar, V.; Alar, Ž. Influence of C Content on Electrochemical Corrosion Resistance of Nanostructured Hardmetals. In Proceedings of the EURO PM2017, Milano, Italy, 10 October 2017.
7.   Fabijanić, T.A.; Sakoman, M.; Kurtela, M.; Marciuš, M. Electrochemical Corrosion Resistance of Nanostructured Hardmetals in Acid Media. In Proceedings of the EURO PM2018, Bilbao, Spain, 14–18 October 2018.
8.   Human, A.M.; Exner, H.E. Electrochemical behavior of tungsten-carbide hardmetals. *Mater. Sci. Eng. A* **1996**, *209*, 180–191. [CrossRef]
9.   Human, A.M. The Corrosion of Tungsten Carbide-Based Cemented Carbides. Ph.D. Thesis, Technische Hochschule Darmstadt, Darmstadt, Germany, 1994.
10.   Tarrago, J.M.; Fargas, G.; Jimenez-Pique, E.; Felip, A.; Isern, L.; Coureaux, D.; Roal, J.J.; Al-Dawery, I.; Fair, J.; Llanes, L. Corrosion damage in WC–Co cemented carbides: Residual strength assessment and 3D FIB-FESEM tomography characterization. *Powder Metall.* **2014**, *57*, 5. [CrossRef]
11.   Sacks, N. The Wear and Corrosive-Wear Response of Tungsten Carbide-Cobalt Hardmetals under Woodcutting and Three Body Abrasion Conditions. Ph.D. Thesis, Faculty of Engineering of the University of Erlangen-Nürnberg, Erlangen, Germany, 2002.
12.   Tarrago, J.M.; Fargas, G.; Jimenez-Pique, E.; Dorvlo, S.; Llanes, L. Influence of Carbide Grain Size on the Corrosion damage in WC–Co cemented carbides: Electrochemical Measurement, residual strength and 3D FIB/FESEM characterisation of Induced Damage. In Proceedings of the World PM2016, Hamburg, Germany, 9–13 October 2016; pp. 509–514.
13.   Hashiva, M.; Kubo, Y. The Influence of Carbon Content and Additions of Growth Inhibitors (V, Cr) on the Formation of Melt in WC-Co and WC-Ni alloys. In Proceedings of the 17th Plansee Seminar 2009, Reutte, Austria, 29 May 2009; Plansee Group: Reutte, Austria, 2009. HM 55/1–HM 55/17.
14.   Toller, L. *Alternative Binder Hardmetals for Steel Turning*; Uppsala University: Uppsala, Sweden, 2017.
15.   Peterson, A. Cemented Carbide Sintering: Constitutive Relations and Microstructural Evolution. Ph.D. Thesis, Royal Institute of Technology, Stockholm, Sweden, 2004.
16.   Sutthiruangwong, S.; Mori, G. Corrosion properties of Co-based cemented carbides in acidic solutions. *Int. J. Refract. Met. Hard Mater.* **2003**, *21*, 135–145. [CrossRef]
17.   Imasato, S.; Sakaguchi, S.; Okada, T.; Hayashi, Y. Effect of WC Grain Size on Corrosion Resistance of WC-Co Cemented Carbide. *J. Jpn. Soc. Powder Powder Metall.* **2001**, *48*, 609–615. [CrossRef]
18.   Gille, G.; Szesny, B.; Dreyer, G.; van den Berg, H.; Schmidt, J.; Gestrich, T.; Leitner, G. Submicron and ultrafine grained hardmetals for microdrills and metal cutting inserts. *Int. J. Refract. Met. Hard Mater.* **2002**, *20*, 3–22. [CrossRef]

19. Roebuck, B.; Gee, M.; Bennett, E.G.; Morrell, R. *A National Measurement Good Practice Guide No. 20 Mechanical Tests for Hardmetals*; National Physical Laboratory: Teddington, UK, 1999; Revised February 2009.
20. Sakoman, M.; Ćorić, D.; Šnajdar Musa, M. Plasma-Assisted Chemical Vapor Deposition of TiBN Coatings on Nanostructured Cemented WC-Co. *Metals* **2020**, *10*, 1680. [CrossRef]
21. Sakoman, M. *Development of PACVD Coatings on Nanostructured Hardmetals*; Faculty of Mechanical Engineering and Naval Architecture: Zagreb, Republika Hrvatska, 2020.
22. Fabijanić, T.A.; Kurtela, M.; Škrinjarić, I.; Pötschke, J.; Mayer, M. Electrochemical corrosion resistance of Ni and Co bonded near-nano and nanostructured cemented carbides. *Metals* **2020**, *10*, 224. [CrossRef]
23. Sutthiruangwong, S.; Mori, G. Influence of refractory metal carbide addition on corrosion properties of cemented carbides. *Mater. Manuf. Process.* **2005**, *20*, 47–56. [CrossRef]
24. Mori, G.; Zitter, H.; Lackner, A.; Schretter, M. Influencing the Corrosion Resistance of Cemented Carbides by addition of Cr_2C_3, TiC and TaC. In Proceedings of the 15th International Plansee Seminar, Reutte, Austria, 15–17 May 2001; Volume 2.
25. Kellner, F.J.J.; Hildebrand, H.; Virtanen, S. Effect of WC grain size on the corrosion behavior of WC–Co based hardmetals in alkaline solutions. *Int. J. Refrac. Met. Hard Mater.* **2009**, *27*, 806–812. [CrossRef]
26. Tomlinson, W.J.; Ayerst, N.J. Anodic polarization and corrosion of WC-Co hardmetals containing small amounts of Cr3C2 and/or VC. *J. Mater. Sci.* **1989**, *24*, 2348–2352. [CrossRef]
27. Fabijanić, T.A.; Pötschke, J.; Alar, V.; Alar, Ž. Potentials of nanostructured WC–Co hardmetal as reference material for Vickers hardness. *Int. J. Refrac. Met. Hard Mater.* **2015**, *50*, 126–132. [CrossRef]
28. Machio, C.N.; Konadu, D.S.; Potgieter, J.H.; Potgieter-Vermaak, S.; van der Merwe, J. *Corrosion of WC-VC-Co Hardmetal in Neutral Chloride Containing Media*; Hindawi Publishing Corporation: London, UK, 2013.

materials

Review

Comparison of the Values of Solar Cell Contact Resistivity Measured with the Transmission Line Method (TLM) and the Potential Difference (PD)

Małgorzata Musztyfaga-Staszuk 

Welding Department, Silesian University of Technology, Konarskiego 18A, 44-100 Gliwice, Poland; malgorzata.musztyfaga@polsl.pl

Abstract: This work presents a comparison of values of the contact resistivity of silicon solar cells obtained using the following methods: the transmission line model method (TLM) and the potential difference method (PD). Investigations were performed with two independent scientific units. The samples were manufactured with silver front electrodes. The co-firing process was performed in an infrared belt furnace in a temperature range of 840 to 960 °C. The electrical properties of a batch of solar cells fabricated in two cycles were investigated. This work focuses on the different metallisation temperatures of co-firing solar cells and measurements were carried out using the methods mentioned. In the TLM and PD methods, the same calculation formulae were used. Moreover, solar cell parameters measured with these methods had the same, similar, or sometimes different but strongly correlated values. Based on an analysis of the selected databases, this article diagnoses the recent and current state of knowledge regarding the employment of the TLM and PD methods and the available hardware base. These methods are of interest to various research centres, groups of specialists dealing with the optimisation of the electrical properties of silicon photovoltaic cells, and designers of measuring instruments.

Keywords: transmission line model (TLM) method; potential difference (PD) method; contact resistance; resistivity; silicon solar cells; I–V characteristics

check for
updates

Citation: Musztyfaga-Staszuk, M. Comparison of the Values of Solar Cell Contact Resistivity Measured with the Transmission Line Method (TLM) and the Potential Difference (PD). *Materials* **2021**, *14*, 5590. https://doi.org/10.3390/ma14195590

Academic Editors: Claudio Mele and Artur Czupryński

Received: 23 August 2021
Accepted: 23 September 2021
Published: 26 September 2021

Publisher's Note: MDPI stays neutral with regard to jurisdictional claims in published maps and institutional affiliations.

1. TLM and PD Methods—Presentation of Statistical Data Based on Electronic Databases

Information on the use of TLM transmission lines in electronics is collected in the IEEE Explore database by IEEE (Institute of Electrical and Electronics Engineers) and IET (Institution of Engineering and Technology). The method was originally proposed by Shockley in 1964 [1] and later modified by Berger [2]. Based on the data contained in the database, it can be concluded that in the period from 1990 to 2020, 17 related articles were published in journals from the Philadelphia list, including the *Journal of Photovoltaics, Electron Device Letters, Transactions on Electron Devices*, and the *Journal of Display Technology*.

Information on the use of the transmission line method was also analysed based on data contained in the Scopus database using the SciaVal tool. Based on studies from 100 countries from 2015 to 2020, a ranking of the scientific achievements of 35 countries was prepared in which Poland placed 14th. In the case of Poland, only the author of this publication published three works in the period of 2015–2020, so this is an interesting issue and a subject that requires further investigation.

Articles covering the application of the transmission line method in various fields of science from 2015 to 2020 were also analysed from the Scopus database, following the ASJC (All Science Journal Classification) classification of thematic areas. Based on this analysis, it was found that the most significant applicability of the TLM method was observed in the field of science—materials engineering (32%). In contrast, the value did not exceed 5% in energy, chemistry, or other fields. In the field of photovoltaics, from 1989 to 2020 the most significant number of publications on the application of the transmission line method in the Scopus database was recorded in 2019.

According to the Web of Science database, the topic of TLM covers the literature from 1973 to 2021, for a total of 12,996 publications. After limiting the search to the electronics category, we obtained 7722 scientific papers.

From 1983 to 2020, the IEEE Explore database in the field of electronics and electrics collected 34 publications; we searched these using the term "contact resistance and PD method". However, from 1978 to 2020, 30 publications were recorded, of which four were in the field of photovoltaics. These were found using the keyword "contact resistance scanning,". According to the search "contact resistance scanning" in the Scopus database, there were eight publications from 1979 to 2021, while in the Web of Science database, there were seven publications from 2005 to 2020.

Certain discrepancies in the scientific circulation of published research results in the form of scientific journals within the aforementioned bibliographic databases may result from the emergence of an increasing number of scientific journals which themselves fulfil the role of a sieve, separating the valuable scientific publications while omitting those of poor quality or that are non-scientific [3,4].

This work aims to test the value of contact resistivity of the front metallisation of solar cells. These cells were made only for measurement purposes. This work uses two methods, TLM and PD, to determine the same parameters in different ways and with different measuring stands, depending on the needs and expectations of the person performing the measurement.

The correct performance of this technological process requires the precise selection of parameters. The production of metallisation requires the correct execution of the front electrode of the correct size and shape and proper cavity in the semiconductor material. In the area where the front electrode connects to the semiconductor, junctions are formed, introducing additional resistances to the electrical circuit that limit the photocurrent flow. The use of electrode pastes minimises these losses and the level of emitter doping can be determined. There are also leakages of separated charge carriers in the junction area, causing a photovoltage drop in the solar cell. The power loss of a photovoltaic cell is influenced, among other things, by resistivity, which is a feature of cells that has a fundamental impact on their quality.

The determined parameters of the technological process strongly depend on the composition of the conductive paste or the metallic powder from which the paste is made. This is because the conductivity of the tested paste depends on the granulation, particle shape, and powder content of the paste. Ceramic glaze (e.g., SiO_2) binds the particles of the base material with the silicon substrate; the rest is an organic carrier mixture that makes the paste sticky. The thickness of the applied layer on the front electrode and the morphology of the substrate affect the resistivity value obtained as a result of the measurements.

2. Methods Applied to Measuring Selected Parameters of the Electrical Properties of Photovoltaic Cells

Currently, research and development in electronics and photovoltaics are focused on developing and producing electrical contacts using various techniques and determining the dependence of the resistivity of these contacts between multicomponent metallic components and conductive layers. The goal is to optimise the technology of contact metallisation. The metallic element layer should meet various requirements to ensure the low resistivity in the vicinity of the metal–semiconductor junction. Of particular importance in a correctly performed technological process are the proper selection of the material (conductive coating and substrate); its thickness; its coating geometry (shape and size); the conditions of its production; the adhesion of the metallic element to the substrate; and the substrate morphology (e.g., structure and roughness).

The external operating parameters that characterise a silicon solar cell include the cell open-circuit voltage (V_{oc}), short circuit current (I_{sc}), fill factor (FF), maximum obtainable power (Pm), and photovoltaic conversion efficiency (E_{ff}). These values depend, for example, on technological factors (the reflectance factor; cover factor; incomplete absorption factor due to limited material thickness; and the Q_i collection factor, taking into account that not all generated charge carriers reach the pn junction, where they can be separated) and are determined by the influence of the manufacturing process on the parameters, the material and physical properties of the individual layers, and the structural elements of a solar cell. The most important of these physical parameters include: the electromagnetic radiation reflection coefficient (R_{ref}); the thickness of the base material of the cell (D_c); the concentration of charge carriers (ni); the mobility (μ) and lifetime of charge carriers (τ); the length of the carrier diffusion path charge (L); the speed of surface recombination (SRV); the resistance (R) and resistivity (ρ) of areas and structural elements of the cell; and the types and concentration of defects and the resulting density of recombination centres (Nit) [5,6].

Basic measurements of photovoltaic cells include the current–voltage characteristics (I–V), which determine the physical parameters of the manufactured solar cell [7]. A one- or two-diode model can be matched numerically to the measured I–V characteristics. In the measuring system of the stand used for measuring the I–V characteristics of light and dark cells, four basic elements that determine the quality of the measurement can be distinguished: a light source, measuring system, table, and contact probes.

The I–V characteristics must be measured under the strictly defined conditions of a specific radiation spectrum and temperature. The standard used is the AM1.5 spectrum, which has an intensity of 1000 W/m^2 at a cell temperature of 25 °C. The measured cell is placed on a brass table that acts as a current electrode for the back contact, while gold-plated probes provide the contact to the front electrode with telescopic pressure. The lighting elements are four independently positioned halogen lamps powered by a highly stabilised power supply with a "ramp" voltage increase (the so-called soft start). A reverse-biased reference photodetector controls the stability of the light intensity during the measurement. The measurement of the I–V characteristics of the cell consists of the simultaneous measurement of the voltage biasing the reverse link in the range of ± 0.75 V and the measurement of the current, the value of which is calculated for the selected load resistance and the measured voltage drop.

One of the operations used for producing photovoltaic cells is the application of electrical contacts. As numerous studies show [8–11], the electrode layer should meet various requirements to ensure the low resistance of the electrode connection zone with the substrate. Of particular importance is the appropriate selection of the material, including the electrode and substrate; the conditions of its production; the shape and size of the electrode and its adhesion to the substrate; and the substrate morphology [12–14]. The resistivity is understood to be the quantity that characterises the metal–semiconductor junction, considering the area above and below the junction. The value of contact resistance, which depends on the type of paste used, the substrate resistance, and the temperature of the metallisation process, can be determined experimentally—for example, by using the TLM (transmission line model) method [15–25] or the PD (potential differences) method [26–32]. In the case of the TLM method, the measurement consists of forcing an electric current signal between the selected pair of adjacent front conductive lines on the tested sample through the supply soda and the spontaneous generation of a potential difference in them through the measuring probes (Figure 1a). In the PD method, local lighting produces a current and the voltage is measured with a metal probe placed on the front metallisation of the test sample (Figure 1b). During the measurement, the sample is short-circuited with an external receiver.

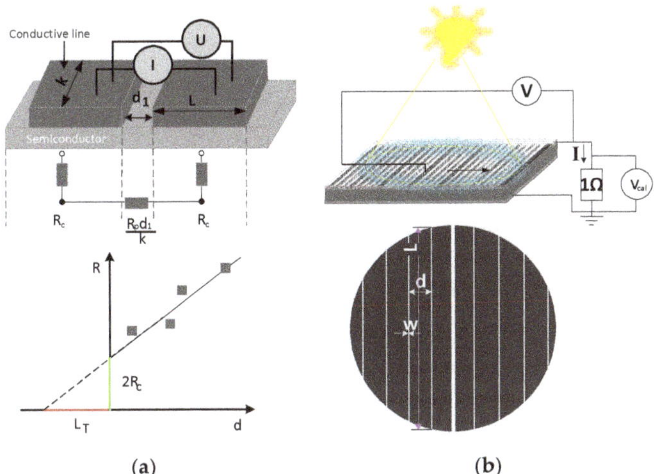

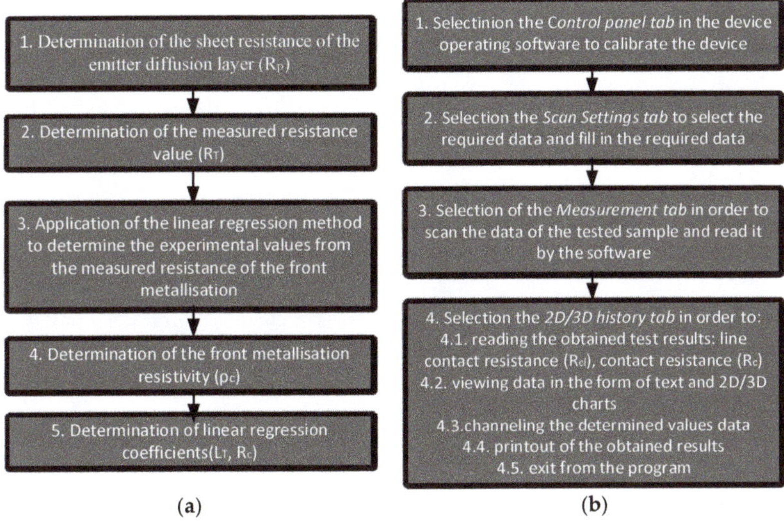

(a)   (b)

Figure 1. Graphical illustration of the methods of (**a**) TLM (where R—resistance; R_p—surface resistance; I—current; U—voltage; k—electrode line length; d—distance between electrode path lines; L_T—electrode line width to the effect of current; R_c—contact resistance) and (**b**) PD (where w—path width; d—distance between the electrode path lines; L—path length; V—voltage on the front electrode, measured through a metal probe in direct contact with its surface) [33,34].

Figure 2 shows the method for measuring using two measurement techniques [33,34].

(a)   (b)

Figure 2. Sequence of actions when carrying out measurements using the (**a**) TLM and (**b**) PD methods.

Table 1 shows the procedure for determining the selected electrical parameters (including resistivity) of semiconductor structure contacts using the aforementioned measurement methods. In order to standardise the results obtained from the electrical properties tests, this article adopts one determination of contact resistance (R) and resistivity (ρ) for two methods.

Table 1. Formulae and quantities characterising the metal–semiconductor junction according to the method applied.

No	TLM Method	Literature	No	PD Method	Literature
A1.	$R_p = \frac{U}{I} \cdot K,\ \Omega/\square$			$R_{cl} = C \cdot \frac{V}{I'} = C \cdot \frac{V}{d \cdot J},\ \Omega\cdot cm$	
A2.	$R_T = \frac{U}{I},\ \Omega$			$I' = \frac{I}{L} = J \cdot d,\ A/cm$	
A3.	$R_T = 2R_c + \frac{d \cdot R_p}{k},\ \Omega$	[17,20–25]	B1	$R_c = C \cdot \frac{V}{I''} = C \cdot \frac{V \cdot w}{J \cdot d},\ \Omega\cdot cm^2$	[26–34]
A4.	$L_T = \left(\rho_c / R_p\right)^{0.5},\ cm$ $R_c = \frac{\sqrt{\rho_c \cdot R_p}}{k} \coth\left(L \cdot L_T\right),\ \Omega$			$I'' = \frac{I}{(L \cdot w)} = \frac{J \cdot d}{w},\ A/cm^2$	
A5.	If, $L_T \geq 2\,L$, to $\rho_c = R_c \cdot k \cdot L_T\ \Omega\cdot cm^2$ If, $L_T < 2\,L$, to $\rho_c = R_c \cdot k \cdot L$ $\Omega\cdot cm^2$				

Where C—correction factor ($\cdot1.8$) for the current leakage out of the spot and shading by the probe.

Measurements of the resistivity of the front contacts of the semiconductor structures were carried out in two research centres and designated as the research centre A_TLM method and the research centre B_PD method. Table 2 presents the differences between the methods used and the research equipment used for this purpose.

Table 2. Primary specifications of the available research equipment [26,33].

No	Feature	Type of Measuring Stand	
		Laboratory	Industrial
1.	Test sample size (thickness, length × width)	200–1000 µm, 50 mm × 50 mm	200–1000 µm, 25–215 mm × 40–215 mm
2.	Pattern of the produced front metallisation	A series of parallel track lines with varying distances between them	Busbar with collecting tracks
3.	Measurement method used	TLM	PD
4.	Measurement mode	Manual	Automatic
5.	Time consumption	Short measurement time	The optimal one depending on the operator settings in the software
6.	Measurement data output	Graphical and textual	Graphical 2D/3D and textual
7.	Printout of measurement data	No	Yes
8.	Method type	Destructive/Nondestructive *	Destructive
9.	Test device cost	Low	High
10.	Dimensions L × W × H (length, width, height)	TLM stand: 1000 mm × 1000 mm × 300 mm	Corescan system: 515 mm × 809 mm × 350 mm
11.	Measurement accuracy	(1) Digital multimetr: ±(0.02% of the set value + 2 digits) (2) Calibrator: ±(0.1% of the set value + 6 digits)	-

* depends on the application of the probes.

In the case of the TLM method, the measurement is performed manually but is repeatable. The mere collection and processing of the measurements were time-consuming. The user has to develop a form for the graphical presentation of the received data.

In the PD method, the measurement is performed automatically, and the user decides how long it is to last. This stand enables the immediate recording of the measurement results in 2D and 3D formats with the calculated data and their direct printout. However, it is impossible to perform the measurement again because the tested sample is very damaged. The patterns of the applied front metallisation, sample size, cost and size of the test stand, number of patterns used to determine the parameters sought, and the symbols of these parameters (Table 1) are the main differences between these methods. The symbols of the same parameters may differ because they often depends on the manufacturers bringing something new to the market. In the case of Mechatronics, this was a stand equipped with a Corescan device.

3. Methodology

The test apparatus with the applied method was used to test the selected electrical parameters (i.e., resistance and resistivity): (1) TLM [11] (Figure 3a) and (2) PD (Figure 3b) [26,33].

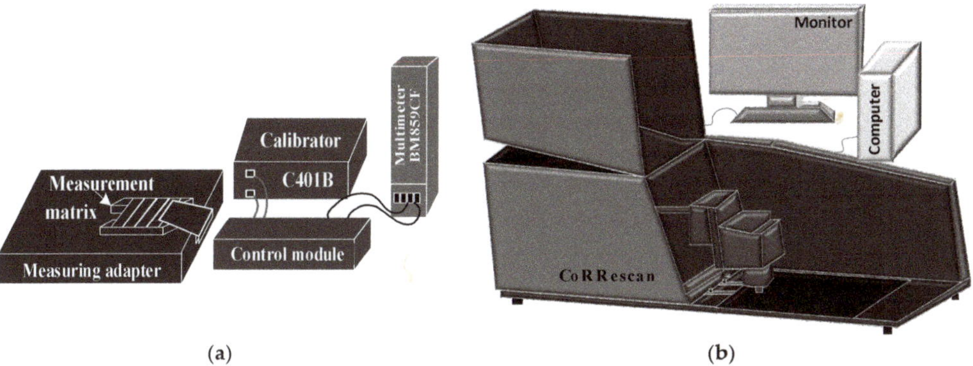

Figure 3. Determination of electrical parameters by the method of (**a**) TLM with the use of laboratory equipment and (**b**) PD with the use of industrial equipment.

The measuring stand used for the TLM method is a measuring system developed as part of a research project [16]. The proposed solution used for measuring the selected electrical parameters with the TLM method was granted a patent by the Patent Office of the Republic of Poland in 2014—P.398223. On the other hand, the measuring stand used for the PD method was introduced to the market in 2000. It was designed by SunLab (an ECN spin-off) and manufactured and delivered to the market by Mechatronics (under license from SunLab) [26].

As part of the experiment, the selected electrical parameters (i.e., resistance and resistivity of photovoltaic cells) were tested, and then, using the formulae contained in Table 1, their values were determined [34]. In the TLM method, a row of five front electrodes (strip size—0.1 mm × 8 mm) with a variable distance between them (2.5; 5; 10; 30 mm) was used. The electrode height was 15 μm and the surface resistance of the emitter diffusion layer was 50 $\Omega/\square$. A current value of 30 mA was used for the TLM method, while for the PD method a current density value of 30 mA/cm^2 was used. The front electrode was made of DuPont standard PV19b silver paste, which is commonly used in the industry. The front electrode was applied using a template. The tests were carried out on two series of samples, which differed in terms of the firing temperature used in the front metallisation in the belt furnace, selected to obtain the objectively most advantageous product feature—i.e., the solar cell and the measurements performed.

A total of 32 samples were used for this investigation. This paper presents the selected results of samples produced with the use of the methods described. The research was carried out in a narrow scope. For this article, samples were selected with layer parameter distributions similar to a Gaussian distribution. The limit values from the interval were given for each series, which meant the minimum value of the resistivity of the front electrode connection with the substrate.

4. List of Test Results Performed with the Use of the TLM and PD Methods

Figure 4 shows the results of measurements of the electrical properties of the first series of photovoltaic cells [34]. Based on the results obtained by testing the electrical properties of this series of photovoltaic cells, it was found that the lowest values for resistance and resistivity were obtained from samples with a firing temperature of 940 °C for both methods (TLM and PD). These values were R = 0.4 Ω, ρ = 18 mΩ cm^2 for the TLM method and R = 2.5 Ω cm, ρ = 20 mΩ cm^2 for the PD method (Figure 5). In the case of cells where the electrodes were fired at a low temperature—e.g., 840 °C—it can be concluded that the discrepancies in the measurements of the resistance and resistivity values from both methods resulted from the poor connection of the electrode with the substrate.

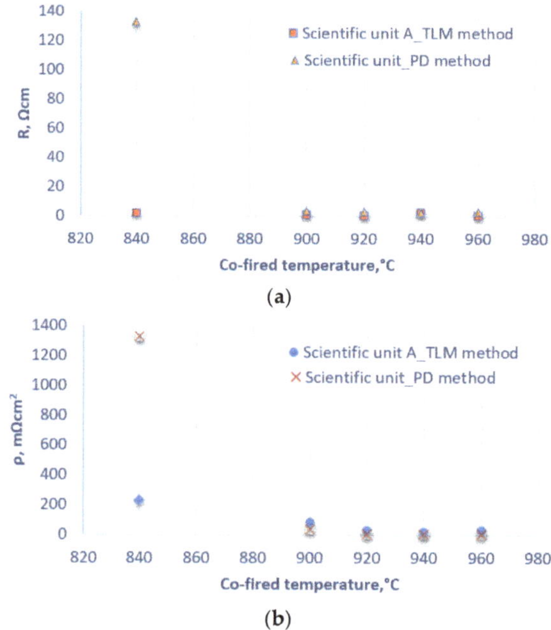

Figure 4. List of electrical parameters—(**a**) resistance and (**b**) resistivity—of photovoltaic cells with a front electrode applied using a template and made of commercial PV19B paste [34].

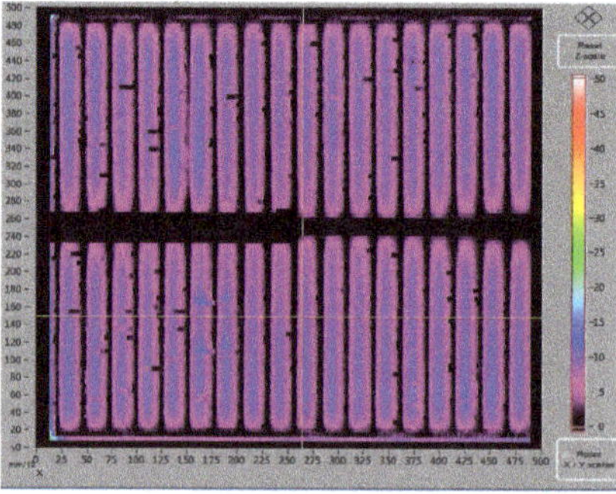

Figure 5. An example of the graphical distribution of the resistance of a photovoltaic cell with front metallisation fired in an infrared IR furnace at a temperature of 940 °C (original 2D printout from the Corescan device software).

Figure 6 shows the results of measurements of the electrical properties of the second series of photovoltaic cells. Based on the results of this series of tests, it was found that the lowest values of resistance and resistivity were obtained for samples with a firing temperature of 930 °C for both methods (TLM and PD) (Figure 6b). These values were R = 0.5 Ω, ρ = 30 mΩ cm^2 for the TLM method and R = 4 Ωcm, ρ = 35 mΩ cm^2 for the PD method. Figure 7 presents an original measurement printout from the Corescan device of the resistance distribution of the cells.

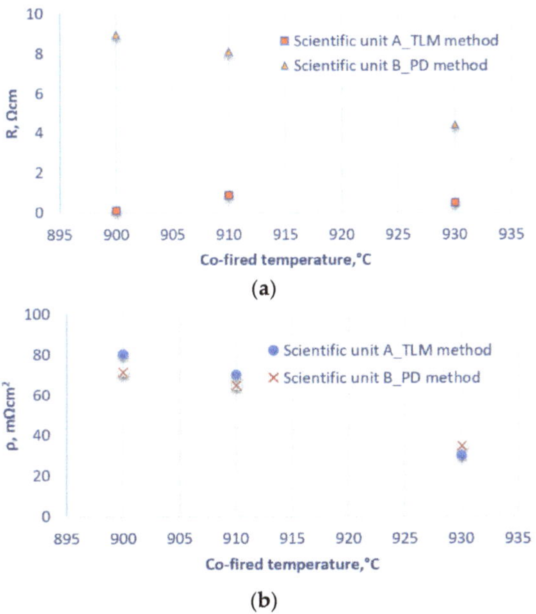

Figure 6. List of electrical parameters—(**a**) resistance and (**b**) resistivity—of photovoltaic cells with a front electrode applied using a template and made of commercial PV19B paste.

To summarise, based on the comparative analysis of the first series of photovoltaic cells using the TLM and PD methods, it was found that the measurements of resistivity in a temperature range of 900 to 960 °C are of the same or similar order. It can also be stated that in a temperature range of 840 to 900 °C, the cells showed the highest and most non-uniform contact resistance, which resulted in the dispersion of the obtained resistivity values. At 940 °C, the cell with the lowest contact resistance without damage was obtained.

In the second series, the results of the resistance values at the metal–semiconductor contact were scattered. Discrepancies in the obtained data may have resulted from, among other things, the manual application of the front metallisation—e.g., using a template—or another stage in the technological process of producing a finished photovoltaic cell. The results of the resistivity measurements in the accepted temperature range are of a similar order. One can also notice some defects in all samples. This is confirmed by the results of the risk at the metal–semiconductor interface. The lowest resistance value was obtained at 930 °C.

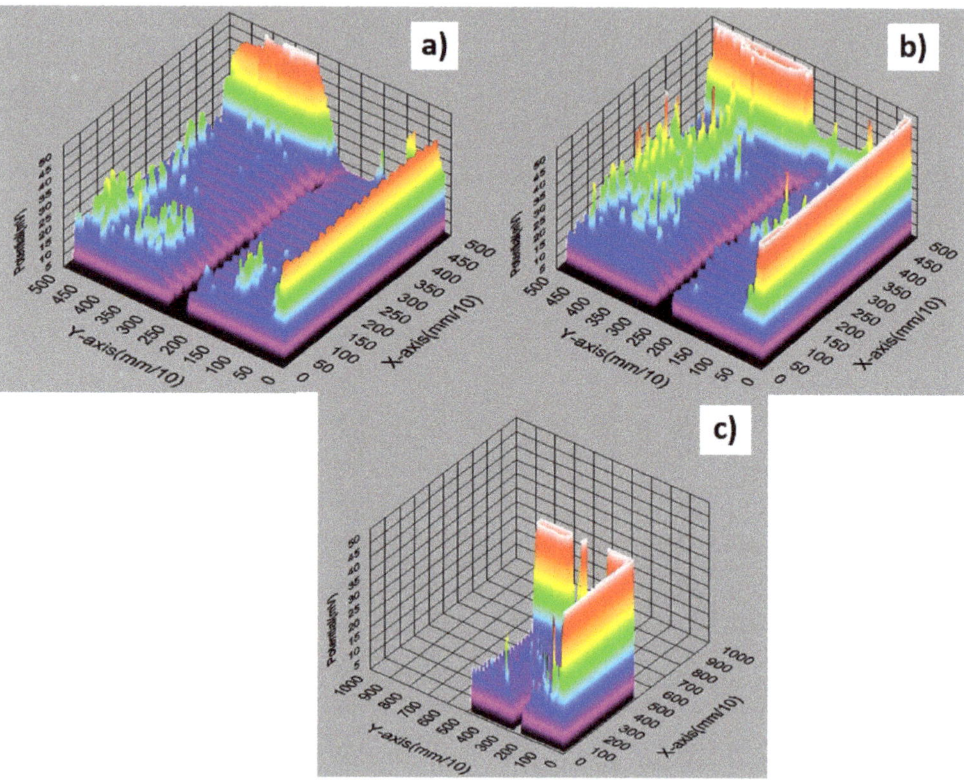

Figure 7. An example of the graphical distribution of the resistance of photovoltaic cells with front metallisation fired in an infrared IR furnace at different temperatures: (**a**) 900 °C, (**b**) 910 °C, (**c**) 920 °C (original 3D print from the Corescan device software).

After the analysis, it can be unequivocally stated that each of the described methods can be used to collect data and information that can be used in the cell manufacturing process. The dispersion of their values may result from, among other things, the technology used in manufacturing the finished product; production automation (which reduces costs and significantly increases productivity); the robotisation of control stations; the methods and indicators used in the product quality assessment at every stage of production; and the mathematical formulae available in the literature for their calculation and analysis.

5. Summary

1. Both series of samples were made using the same technology. The measurement methods used in two independent research centres made it possible to compare their advantages and disadvantages. The potential difference method allows the modelling of the contact resistance of the front electrode's frontal contact mesh and the optimisation of the burnout process. In addition, it is possible to graphically present the contact resistance measurements of the front electrode of the photovoltaic cell in 2D/3D, which is very useful for detecting possible defects at this stage of the technological process. This is not possible in the transmission line method. The PD method is destructive, while in the TLM method, depending on the measuring probes used, the measurement can be repeated. Automatic measurement and adjustment of measurement parameters is another advantage of the PD method, but the dimensions and the cost of purchasing the entire stand can be classified as disadvantages. Performing the measurements manually is associated with an extended time of im-

    plementation, so this can be considered a disadvantage of the TLM method, while the cost of purchasing the station itself is low, which is clearly an advantage.

2. The researcher/research team, depending on their requirements or the needs of the industrial market, decide on the choice of method and a suitable measuring station.

3. Based on the research analysis (Figures 4b and 6b), it can be concluded that the measured values of resistivity may differ slightly from each other. This is mainly due to the formulae available in the literature, the number of measurements performed, and the technological processes used, often the lack of automatic line technology in the cell manufacturing process. The analysis mentioned above proves the credibility of the obtained results of electrical measurements using the TLM method in research centre A and the PD method in research centre B, as well as the legitimacy of their use in testing photovoltaic cells.

4. The TLM method has a few problems. Firstly, the fit of the straight line to the measurement points, expressed by the value of its slope, is subject to uncertainty. A more significant problem, however, is the correct determination of the value of the layered resistance used to calculate the resistance between the two contact lines. This is because, in the process of forming the contact between the paste and the substrate, there is a change in the layer resistance directly below the contact line. Secondly, the method does not take into account the resistance of the metal on the contact line. The advantage of the TLM method is a simple sample preparation procedure (no need to use a ready-made sample of solar cell) and quick and direct measurement of the resistance values. It is also a non-destructive method. The required components of the stand are calibrated according to the guidelines of the manufacturers of these components. The measuring station enables:

- Setting the current value on the C401B calibrator in a range from 0 to 110 mA with a resolution of 0.01 mA, with an accuracy of $\pm$ (0.1% of the set value + 6 digits);
- Voltage measurement with the BM859CF voltmeter in a range from 0 to 500 mV with resolution of 0.01 mV, and in a range from 0.5 to 50 V with a resolution of 0.1 mV and an accuracy of $\pm$ (0.02% of the set value +2 digits).

    In the case of the PD method, the problem is the precise determination of the current density value necessary in the measurement procedure. This can be performed using a high-class monochromator, measuring the spectral efficiency of a cell—e.g., silicon—in a range of 300–1100 nm; multiplying it by the photon flux; and then integrating it over the entire range and calculating the J_{sc} of the cell. The second method is to measure the bright I–V characteristic and calculate J_{sc}, but this is also subject to uncertainty. The second disadvantage of the method is its destructive procedure that does not allow repeated measurements of the sample in the same area. This necessitates the use of samples in the form of a ready cell. The manufacturer is responsible for the calibration of the device, and the measurement accuracy refers to a constant value in the formulae (i.e., "correction factor" (* 1.8) for current leakage out of the spot and shading by the probe), which is beyond the control of the user. On the other hand, the advantage of the PD method is the ability to determine the quality of the sample (e.g., its defects) based on the graphical observation of the result and, more importantly, the determination of contact homogeneity (i.e., the low or high uniformity of the contact resistance value).

Funding: This research received no external funding.

Institutional Review Board Statement: Not applicable.

Informed Consent Statement: Not applicable.

Conflicts of Interest: The authors declare no conflict of interest.

References

1. Shockley, W. Research and Investigation of Inverse Epitaxial UHF Power Transistors. Air Force At. Lab. Wright-Patterson Air Force Base Ohio, 1964. Available online: https://www.readcube.com/articles/10.21236%2Fad0605376 (accessed on 29 July 2021).
2. Berger, H.H. Contact resistance on diffused resistors. In Proceedings of the IEEE Solid-State Circuits Conference, Philadelphia, PA, USA, 19–21 February 1969; pp. 160–161.
3. Available online: https://depot.ceon.pl/bitstream/handle/123456789/15614/Aneta_Drabek_Indeksowanie_czasopism_w_referencyjnych_bazach_danych.pdf?sequence=1&isAllow (accessed on 29 July 2021).
4. Available online: https://publications.jrc.ec.europa.eu/repository/bitstream/JRC123157/jrc123157_online_2.pdf (accessed on 29 July 2021).
5. Panek, P. *Polish Photovoltaics 2011*; Publikatech: Warszawa, Poland, 2011. (In Polish)
6. Bayod-Rújula, A.A. *Solar Photovoltaics (PV)*; ScienceDirect Topics in Solar Hydrogen Production; Elsevier: Amsterdam, The Netherlands, 2019.
7. Jordan, D.C.; Kurtz, S.R. Photovoltaic Degradation Rates—An Analytical Review. *Prog. Photovolt. Res. Appl.* **2012**, *21*, 12–29. [CrossRef]
8. Enebish, N.; Agchbayar, D.; Dorjkhand, S.; Baatar, D.; Ylemj, I. Numerical Analysis of Solar Cell Current-Voltage Characteristics. *Sol. Energy Mater. Sol. Cells* **1993**, *29*, 201–208. [CrossRef]
9. Rodacki, T.; Kandyba, A. *The Energy Processing in Solar Station*; Silesian University of Technology Press: Gliwice, Poland, 2000. (In Polish)
10. Green, M.A. *Technology and System Applications: Solar Cells. Operating Principles*; The University of New South Wales, Kensington: San Mateo, CA, USA, 1986; p. 269.
11. Pysch, D.; Mette, A.; Filipovic, A.; Glunz, S.W.A. Comprehensive Analysis of advanced solar cell contacts consisting of printed fine-line seed layers thickened by silver plating. *Prog. Photovolt. Res. Appl.* **2009**, *17*, 101–114. [CrossRef]
12. Schroder, D.K.; Meier, D.L. Solar cell contact resistance—A review. *IEEE Trans. Electron Devices* **1984**, *31*, 637–646. [CrossRef]
13. Urban, T.; Heitmann, J.; Müller, M. Numerical Simulations for In-Depth Analysis of Transmission Line Method Measurements for Photovoltaic Applications—The Influence of the p–n Junction. *Phys. Status Solidi A* **2020**, *217*, 1900600. [CrossRef]
14. Scharlack, R.S. The Optimal Design of Solar Cell Grid Lines. *Sol. Energy* **1979**, *23*, 199–201. [CrossRef]
15. Szlufcik, J. Study of the Conditions of Using Thick-Film Technology for the Production of Solar Cells from Monocrystalline Silicon. Ph.D. Thesis, Library of the Wroclaw University of Science and Technology, Wroclaw, Poland, 1989. Unpublished work. (In Polish).
16. Musztyfaga-Staszuk, M. Laser Micromachining of Silicon Photovoltaic Cells. Ph.D. Thesis, The Silesian University of Technology, Gliwice, Poland, 8 July 2011.
17. Dobrzański, L.A.; Musztyfaga, M.; Drygała, A.; Panek, P. Investigation of the screen printed contacts of silicon solar cells from Transmissions Line Model. *J. Achiev. Mater. Manuf. Eng.* **2010**, *41*, 57–65.
18. Woelk, E.G.; Kräutle, H.; Beneking, H. Measurement of low resistive ohmic contacts on semiconductors. *IEEE Trans. Electron Devices* **1986**, *33*, 19–22. [CrossRef]
19. Available online: http://tuttle.merc.iastate.edu/ee432/topics/metals/tlm_measurements.pdf (accessed on 29 July 2021).
20. Schroder, D.K. *Semiconductor Material and Device Characterisation*, 3rd ed.; Wiley-IEEE Press: Piscataway, NJ, USA, 2006; ISBN 9780471749080.
21. Grover, S.; Sahu, S.; Zhang, P.; Davis, K.O.; Kurinec, S.K. Standardisation of Specific Contact Resistivity Measurements using Transmission Line Model (TLM). In Proceedings of the 33rd IEEE International Conference on Microelectronic Test Structures, Edinburgh, UK, 4–18 May 2020.
22. Gregory, G.; Li, M.; Gabor, A.; Anselmo, A.; Yang, Z.; Ali, H.; Iqbal, N.; Davis, K. Nondestructive Contact Resistivity Measurements on Solar Cells Using the Circular Transmission Line Method. *J. Photovolt.* **2019**, *9*, 1800–1805. [CrossRef]
23. Hocine, R.; Belkacemi, K.; Kheris, D. 3D-analytical method analysis of thermal effect in space shaded solar panel. In Proceedings of the 9th International Conference on Recent Advances in Space Technologies, RAST, Istanbul, Turkey, 11–14 June 2019.
24. Mir, H.; Arya, V.; Höffler, H.; Brand, A. A Novel TLM Analysis for Solar Cells. *J. Photovolt.* **2019**, *9*, 1336–1342. [CrossRef]
25. Takaloo, A.V.; Joo, S.K.; Es, F.; Turan, R.; Lee, D.W. A Study on Characterisation of Light-Induced Electroless Plated Ni Seed Layer and Silicide Formation for Solar Cell Application. *J. Korean Phys. Soc.* **2018**, *72*, 615–621. [CrossRef]
26. Available online: https://www.energy-xprt.com/downloads/sunlab-model-corescan-mapping-of-metal-grid-contact-resistance-datasheet-736296 (accessed on 29 July 2021).
27. Musztyfaga, M.; Dobrzański, L.A.; Rusz, S.; Staszuk, M. Application examples for the different measurement modes of electrical properties of the solar cells. *Arch. Metall. Mater.* **2014**, *59*, 247–252. [CrossRef]
28. Van der Heide, A.S.H.; Bultman, J.H.; Hoornstra, J.; Schonecker, A.; Wyers, G.P.; Sinke, W.C. Optimising the front side metallisation process using the Corescan. In Proceedings of the 29th IEEE Photovoltaic Specialists Conference, New Orleans, LA, USA, 19–24 May 2002.
29. Van der Heide, A.S.H.; Goris, M.J.A.A. Contact optimisation on lowly doped emitters using the corescan on non-uniform emitter cells, Nineteenth European Photovoltaic Solar Energy Conference. In Proceedings of the International Conference, Paris, France, 7–11 June 2004.

30. Van der Heide, A.S.H. Dutch Patent NL1013204, Applied 4 October 1999, Granted 5 April 2001, Worldwide Patent Pending. Available online: https://patentimages.storage.googleapis.com/bf/9e/60/5bd068bd0580b6/NL1013204C2.pdf (accessed on 29 July 2021).
31. Van der Heide, A.S.H.; Bultman, J.H.; Hoornstra, J.; Schönecker, A.; Wyers, G.P.; Sinke, W.C. Locating losses due to contact resistance, shunts and recombination by potential mapping with the Corescan. In Proceedings of the 12th Workshop on Crystalline Silicon Solar Cell Materials and Proceses, Breckenridge, CO, USA, 11–14 August 2002.
32. Van der Heide, A.S.H.; Schönecker, A.; Wyers, G.P.; Sinke, W.C. Mapping of Contact Resistance and Locating Shunts on Solar Cells Using Resistance Analysis by Mapping of Potential (RAMP) Techniques. In Proceedings of the 16th European Photovoltaic Solar Energy Conference, Glasgow, UK, 1–5 May 2000.
33. Musztyfaga-Staszuk, M. *New Copper-Based Composited in Use for Fabrication of the Silicon Photovoltaic Cells*; Silesian University of Technology Press: Gliwice, Poland, 2019; p. 161. (In Polish)
34. Musztyfaga-Staszuk, M.; Janicki, D.; Panek, P. Correlation of different electrical parameters of solar cells with silver front electrodes. *Materials* **2019**, *12*, 366. [CrossRef] [PubMed]

MDPI

St. Alban-Anlage 66

4052 Basel

Switzerland

www.mdpi.com

Materials Editorial Office

E-mail: materials@mdpi.com

www.mdpi.com/journal/materials